Random vibrations of elastic systems

MONOGRAPHS AND TEXTBOOKS ON MECHANICS OF SOLIDS AND FLUIDS
Editor-in-Chief: G. Æ. Oravas

MECHANICS OF ELASTIC STABILITY
Editor: H.H.E. Leipholz

Also in this series:

H.H.E. Leipholz, Theory of elasticity. 1974.
ISBN 90-286-0193-7

L. Librescu, Elastostatics and kinetics of anisotropic and heterogeneous shell-type structures. 1975.
ISBN 90-286-0035-3

C.L. Dym, Stability theory and its application to structural mechanics. 1974.
ISBN 90-286-0094-9

K. Huseyin, Nonlinear theory of elastic stability. 1975.
ISBN 90-286-0344-1

H.H.E. Leipholz, Direct variational methods and eigenvalue problems in engineering. 1977.
ISBN 90-286-0106-6

K. Huseyin, Vibrations and stability of multiple parameter systems. 1978.
ISBN 90-286-0136-8

H.H.E. Leipholz, Stability of elastic systems. 1980.
ISBN 90-286-0050-7

Random vibrations of elastic systems

by

V.V. Bolotin

Moscow Energetics Institute
Moscow, USSR

Translated from the Russian edition by

I. Shenkman

English translation edited by

H.H.E. Leipholz
University of Waterloo
Waterloo, Ontario, Canada

1984 **MARTINUS NIJHOFF PUBLISHERS**
a member of the KLUWER ACADEMIC PUBLISHERS GROUP
THE HAGUE / BOSTON / LANCASTER

Distributors

for the United States and Canada: Kluwer Academic Publishers, 190 Old Derby Street, Hingham, MA 02043, USA
for the UK and Ireland: Kluwer Academic Publishers, MTP Press Limited, Falcon House, Queen Square, Lancaster LA1 1RN, England
for all other countries: Kluwer Academic Publishers Group, Distribution Center, P.O. Box 322, 3300 AH Dordrecht, The Netherlands

Library of Congress Cataloging in Publication Data

Bolotin, V. V. (Vladimir Vasil'evich)
Random vibrations of elastic systems.

(Mechanics of elastic stability ; 8)
Translation of: Sluchaĭnye kolebaniîa uprugikh sistem.
Bibliography: p.
Includes index.
1. Random vibration. 2. Elasticity. 3. Elastic analysis (Theory of structures) I. Leipholz, H. H. E. (Horst H. E.), 1919- . II. Title. III. Series.
QA935.B64213 1984 531'.3823 84-8191
ISBN 90-247-2981-5

ISBN 90-247-2981-5 (this volume)
ISBN 90-247-2743-X (series)

PRINTED IN THE NETHERLANDS

Editor's Preface

The subject of random vibrations of elastic systems has gained, over the past decades, great importance, specifically due to its relevance to technical problems in hydro- and aero-mechanics. Such problems involve aircraft, rockets and oil-drilling platforms; elastic vibrations of structures caused by acoustic radiation of a jet stream and by seismic disturbances must also be included. Applications of the theory of random vibrations are indeed numerous and the development of this theory poses a challenge to mathematicians, mechanicists and engineers. Therefore, a book on random vibrations by a leading authority such as Dr. V.V. Bolotin must be very welcome to anybody working in this field. It is not surprising that efforts were soon made to have the book translated into English.

With pleasure I acknowledge the co-operation of the very competent translater, I Shenkman; of Mrs. C. Jones, who typed the first draft; and of Th. Brunsting, P. Keskikiikonen and R. Piché, who read it and suggested where required, corrections and changes. I express my gratitude to Martinus Nijhoff Publishers BV for entrusting me with the task of editing the English translation, and to F.J. van Drunen, publishers of N. Nijhoff Publishers BV, who so kindly supported my endeavours. Special acknowledgement is due to Mrs. L. Strouth, Solid Mechanics Division, University of Waterloo, for her competent and efficient preparation of the final manuscript.

Last, but not least, thanks to Dr. V.V. Bolotin for his advice and the updating of the contents of the Russian version of this book.

H.H.E. Leipholz
University of Waterloo
January 1984

Author's Preface

The theory of random vibrations is finding more and more applications in engineering. It is fundamental in the analysis of aircraft subjected to atmospheric turbulence, pressure pulsations in a turbulent boundary layer, acoustic radiation of a jet stream, etc.; in the analysis of road vehicles subjected to vibrations when moving on a rough path, etc. The methods of the theory of random vibrations are also extensively used in the analysis of tall buildings and structures subjected to wind pressure, in the analysis of ships and other (ocean) structures subjected to the effect of waves, and in the analysis of buildings and structures under seismic effects. Publications on random vibrations and vibration reliability in scientific journals have been growing in number. The theory of random vibrations is one of the rapidly developing branches of contemporary applied mechanics.

This book considers systematically the problems of the theory of random vibrations and the methods of solving them, with particular attention devoted to continuous systems. There are eight chapters in this book. The first chapter serves as an introduction by providing a general idea of random loadings of mechanical systems and of methods for describing these loadings probabilistically. The second chapter deals with methods to be used for solving random vibration problems (mostly, those applied to continuous systems).

The next two chapters are devoted to random vibrations of linear continuous systems. The third chapter deals with the mutual coupling of generalized coordinates in elastic and viscoelastic systems, with the problems of transmission of random vibrations through spatially distributed elastic systems, and with vibrations of elastic-acoustic systems. The fourth chapter describes the asymptotic method for the analysis of broad-band random vibrations of linear continuous systems. The specific feature of such vibrations is that a large number of natural modes are excited simultaneously, which makes it possible to discover certain new regularities (laws) and asymptotic estimates for the characteristics of a vibration field.

The fifth chapter deals with parametrically excited random vibrations. The methods of investigating the stability of stochastic systems and methods effective for constructing stochastic instability regions are described here. Special attention is devoted to parametric resonances excited by narrow-band stationary random processes. The sixth chapter treats random vibrations in nonlinear systems. These chapters describe not only systems with a finite number of degrees of freedom but also continuous systems.

The last two chapters are devoted to problems associated with the application of the theory of random vibrations. The seventh chapter describes the foundations of reliability theory of mechanical systems as applied to systems subjected to random vibrations. The methods of this theory are used for the analysis of optimal random vibration isolation of systems. In the last chapter, the principles of planning the measurements of random vibration fields are discussed. The selection of the number of sensors and their optimal location in vibrating structures, and also the making of corrections to account for changes in the vibration field due to the introduction of sensors, are considered.

The list of references at the end of the book is not at all comprehensive, but it gives a general idea of all new schools of thought and of major developments both in our country and abroad. Also, literature used in each section of the book is referred to in footnotes.

This book sums up the author's work on the theory of random vibrations and its application since 1959. Part of the material is taken from the author's lectures for senior level undergraduate and graduate students at the Moscow Energetics Institute Some of the problems were developed by the author in co-operation with his co-workers from the Moscow Energetics Institute, which is reflected in the references. Parts of the manuscript were read by V. Chirkov, V. Radin, N. Ginger, V. Volohovsky, V. Moskvin, V. Chromatov, V. Vasenin, V. Semenov and A. Scherbatov, to whom the author expresses his sincere thanks.

V.V. Bolotin
June 1978

Contents

Chapter 1
Random Loadings Acting on Mechanical Systems

1.1 LOADINGS AS RANDOM FUNCTIONS OF TIME

Introductory Remarks

The basic methods for an analytical description of stochastic loadings acting on mechanical systems as well as experimental data on certain classes of loadings will be considered in this chapter. Loadings will be interpreted not only as external forces, but also as external kinematic effects (i.e., the prescribed displacements of a system or its points), and as the effects of heat, radiation, etc. If the intensity of a source of external influences is sufficiently high as compared to the intensity of an excited vibration process, so that the influence of the system's behaviour on the process of loading can be ignored, the loading can be considered as stochastically prescribed. Otherwise it would be necessary to take into account the interaction with the environment, or to consider the vibrations of the system and processes taking place in the environment simultaneously. Problems of this kind often arise in acoustics and hydro-aero-elasticity. Henceforth, loadings, if it is not specially mentioned, are assumed to be stochastically given.

We shall distinguish two basic classes of stochastic loadings: those given as stochastic functions of time, and those given as stochastic functions of both time and spatial coordinates. To the first class belong, for example, kinematic impacts on a system resulting from the displacement of a rigid body foundation or from the effect of an application of concentrated forces to fixed points, etc. Examples of random loadings of the second class are wind pressure or wave effects on structures and pressure on the surface of flying vehicles caused by pulsations in a turbulent boundary layer.

Loadings of the first class are described by methods in the theory of random processes and for the description of loadings of the second class, the random field theory is used.

Two Basic Methods of Description of Multidimensional Random Processes

Let the loading within a certain interval T of time t be represented as a set of m functions of time $q_1(t), q_2(t), \ldots, q_m(t)$. This set forms an m-dimensional random process $\underset{\sim}{q}(t)$. The random process $\underset{\sim}{q}(t)$ can be described in two ways. The first is by prescribing a full system of joint distributions (joint probability densities) for the components of the process at any instants of time within the considered time interval T. The second is by prescribing a full system of moment functions of the components of the process within the interval T. Moment functions are obtained by multiplying the process components $\underset{\sim}{q}(t)$ at different instants of time and averaging over the set of realizations:

$$\langle \underset{\sim}{q}(t) \rangle , \quad \langle \underset{\sim}{q}(t_1) \otimes \underset{\sim}{q}(t_2) \rangle , \quad \langle \underset{\sim}{q}(t_1) \otimes \underset{\sim}{q}(t_2) \otimes \underset{\sim}{q}(t_3) \rangle , \ \ldots \qquad (1.1)$$

Here the angular brackets denote the averaging over the set of realizations; the symbol $\otimes$ stands for tensor multiplication. That is, expression $\underset{\sim}{q}(t_1) \otimes \underset{\sim}{q}(t_2) \otimes \cdots \otimes \underset{\sim}{q}(t_r)$ is an ordered set

m^r of all possible products of the components $q_1(t), q_2(t), \ldots, q_m(t)$ of the process $\mathbf{q}(t)$, at the instants of time $t_1, t_2, \ldots, t_r$. In (1.1) the following quantities are listed: the mathematical expectation of the process $\mathbf{q}(t)$, the set m^2 of moment functions of the second order, the set m^3 of moment functions of the third order, etc. The connection between the two methods of description is given by the formulae

$$\langle \mathbf{q}(t) \rangle = \int \mathbf{q}p(\mathbf{q};t)d\mathbf{q} \; ,$$

$$\langle \mathbf{q}(t_1) \otimes \mathbf{q}(t_2) \rangle = \iint (\mathbf{q}_1 \otimes \mathbf{q}_2)p(\mathbf{q}_1,\mathbf{q}_2;t_1,t_2)d\mathbf{q}_1 d\mathbf{q}_2 \; , \tag{1.2}$$

$$\langle \mathbf{q}(t_1) \otimes \mathbf{q}(t_2) \otimes \mathbf{q}(t_3) \rangle =$$

$$\iiint (\mathbf{q}_1 \otimes \mathbf{q}_2 \otimes \mathbf{q}_3)p(\mathbf{q}_1,\mathbf{q}_2,\mathbf{q}_3;t_1,t_2,t_3) \times d\mathbf{q}_1 d\mathbf{q}_2 d\mathbf{q}_3 \; ,$$

where $p(\mathbf{q};t)$ is the joint probability density of the components of the process $\mathbf{q}(t)$ at the instant of time t; $p(\mathbf{q}_1,\mathbf{q}_2;t_1,t_2)$ is the joint probability density of the components of the process $\mathbf{q}(t)$ at the instants of time t_1 and t_2, etc,; $d\mathbf{q} \equiv dq_1\, dq_2 \ldots dq_m$.

Correlation Functions of a Multidimensional Random Process

We do not make use of a full description of random functions when solving applied problems because of two reasons. First, the amount of statistical information required for a sufficiently detailed description of random functions is so large that it is practically impossible to obtain it. Secondly, a full description of external influences for a general case makes the solution of problems of statistical dynamics too cumbersome. Therefore, in applications, first approximations are sufficient when dealing with both external influences and system responses. Within the scope of correlation theory, random functions are described by their mathematical expectations and correlation functions. The latter are introduced as the second order moments of the centralized process components $\tilde{q}(t) = q(t) - \langle q(t) \rangle$, i.e.,

$$K_{jk}(t_1,t_2) = \langle \tilde{q}_j(t_1)\tilde{q}_k(t_2) \rangle \ . \tag{1.3}$$

The matrix $K(t_1,t_2)$ composed of correlation functions is called a correlation matrix. Note the terms "covariant function" and "covariant matrix" are often used especially in foreign literature.

It is known that normal processes are fully determined by their mathematical expectations and correlation functions and that a normal process remains normal after passing through a linear system. This makes a correlation description a convenient technique for the analysis of random vibrations of linear systems. When considering nonlinear systems one usually has to use different techniques.

Description of Multidimensional Stationary Processes

Consider a process $q(t)$ specified on $T = (-\infty,\infty)$ with the property that all its probability characteristics (joint probability densities, moment and correlation functions) are independent of the choice of the initial time. Such processes are called stationary and stationarily constrained. For brevity, we shall call them stationary. The elements of the correlation matrix $K(\tau)$ treated as functions of the time shift $\tau = t_2-t_1$ are defined by the relation

$$K_{jk}(\tau) = \langle \tilde{q}_j(t)\tilde{q}_k(t+\tau) \rangle \ . \tag{1.4}$$

The stationary random process is called ergodic if the result of the averaging of any functions of the process over time coincides with the result of the averaging of the corresponding functions over a set of realizations. The ergodic stationary process manifests itselt in time so that one sufficiently long realization of the process contains the information about the entire process. In applications, the ergodicity hypothesis is often used in statistical data processing. Note that for ergodic processes $||K(\tau)|| \to 0$ as $|\tau| \to \infty$.

Let the process q(t) admit of a spectral representation in the form of the generalized stochastic integral of Fourier-Stieltjes

$$\underset{\sim}{q}(t) = \langle \underset{\sim}{q} \rangle + \int_{-\infty}^{\infty} e^{i\omega t} \underset{\sim}{Z}(d\omega) \; . \tag{1.5}$$

Here $\underset{\sim}{Z}(\omega)$ denotes an m-dimensional distribution function (a generalized random vector-function of frequency ω), $\langle \underset{\sim}{Z}(\omega) \rangle = \underset{\sim}{0}$, $\underset{\sim}{Z}(\Delta\omega) = \underset{\sim}{Z}(\omega+\Delta\omega) - \underset{\sim}{Z}(\omega)$. Since we assume $\underset{\sim}{q}(t)$ to be a real process, $\underset{\sim}{Z}^*(\omega) = \underset{\sim}{Z}(-\omega)$, (asterisk denotes a complex-conjugate value). The distribution function satisfies the relation

$$\langle \underset{\sim}{Z}^*(d\omega) \otimes \underset{\sim}{Z}(d\omega') \rangle = \begin{cases} \underset{\sim}{0} \text{ when } d\omega \cap d\omega' = \emptyset \; , \\ S(\omega)d\omega \text{ when } d\omega = d\omega' \; , \end{cases} \tag{1.6}$$

where $S(\omega)$ is a square matrix of dimension $m \times m$, designated as the spectral matrix of the $\underset{\sim}{q}(t)$ process. In the literature dealing with applied problems, the spectral representation of (1.5) is often written in a form analogous to the ordinary Fourier integral

$$\underset{\sim}{q}(t) = \langle \underset{\sim}{q} \rangle + \int_{-\infty}^{\infty} Q(\omega) e^{i\omega t} d\omega \; , \tag{1.7}$$

the Fourier spectrum $\underset{\sim}{Q}(\omega)$ being interpreted as a generalized random vector-function of the frequency ω, and satisfying the relations

$$\langle \underset{\sim}{Q}(\omega) \rangle = \underset{\sim}{0} \; , \qquad \langle \underset{\sim}{Q}^*(\omega) \otimes \underset{\sim}{Q}(\omega') \rangle = S(\omega)\delta(\omega-\omega') \; , \tag{1.8}$$

where $\delta(\omega)$ is the delta-function.

The elements $S_{jk}(\omega)$ of the matrix $S(\omega)$ are the joint spectral densities of the process components $\underset{\sim}{q}(t)$. They are related to the elements of the correlation matrix by the Wiener-Khinchin relations,

$$K_{jk}(\tau) = \int_{-\infty}^{\infty} S_{jk}(\omega) e^{i\omega\tau} d\omega \ ,$$
$$S_{jk}(\omega) = \frac{1}{2\pi} \int_{-\infty}^{\infty} K_{jk}(\omega) e^{-i\omega\tau} d\tau \ . \qquad (1.9)$$

Let us list the basic characteristics of the matrix $S(\omega)$ for the real process $\underset{\sim}{q}(t)$. All the diagonal elements of this matrix are real and non-negative, and the non-diagonal elements are, generally, complex. The non-diagonal elements satisfy the relations:

$$S_{jk}(\omega) = S^*_{jk}(-\omega) = S^*_{kj}(\omega) \ , \qquad (1.10)$$

i.e., the matrix $S(\omega)$ is Hermitian. It is expedient to separate the real and the imaginary parts, representing the matrix elements $S(\omega)$ as

$$S_{jk}(\omega) = S^{(R)}_{jk}(\omega) + iS^{(I)}_{jk}(\omega) \ . \qquad (1.11)$$

If we introduce the functions

$$K^{(C)}_{jk}(\tau) = \frac{1}{2} [K_{jk}(\tau) + K_{kj}(\tau)] \ ,$$
$$K^{(S)}_{jk}(\tau) = \frac{1}{2} [K_{kj}(\tau) - K_{jk}(\tau)] \ , \qquad (1.12)$$

the first being even and the second odd functions of τ, we obtain relations analogous to those in formulae (1.9):

$$S_{jk}^{(R)}(\omega) = \frac{1}{\pi}\int_0^\infty K_{jk}^{(C)}(\tau)\cos\omega\tau d\tau ,$$
$$S_{jk}^{(I)}(\omega) = \frac{1}{\pi}\int_0^\infty K_{jk}^{(S)}(\tau)\sin\omega\tau d\tau . \tag{1.13}$$

The inverse relations are

$$K_{jk}^{(C)}(\tau) = 2\int_0^\infty S_{jk}^{(R)}(\omega)\cos\omega\tau d\omega ,$$
$$K_{jk}^{(S)}(\tau) = 2\int_0^\infty S_{jk}^{(I)}(\omega)\sin\omega\tau d\omega . \tag{1.14}$$

Note that

$$S_{jk}^{(R)}(\omega) = S_{jk}^{(R)}(-\omega) = S_{kj}^{(R)}(\omega) ,$$
$$S_{jk}^{(I)}(\omega) = -S_{jk}^{(I)}(-\omega) = -S_{kj}^{(I)}(\omega) . \tag{1.15}$$

Some Models of Stationary Random Processes

Let the one-dimensional process $q(t)$ consist of a stationary sequence of uncorrelated impulses whose duration tends to zero, and whose magnitude remains finite. The correlation function of such a process is expressed in terms of the delta function $\delta(t)$, and the spectral density is constant over the entire frequency axis:

$$K(\tau) = s\delta(\tau) , \qquad S(\omega) = s/(2\pi) . \tag{1.16}$$

Such a process is called delta-correlated or "white noise", and the constant s in the equations (1.16) is the intensity of white noise. The correlation function of white noise becomes infinite at zero, i.e., variance does not exist. A more realistic model is truncated white noise, whose spectral density is constant in the interval $[-\omega_c,\omega_c]$, and equal to zero outside this interval

(ω_c is the cut-off frequency). The graphs of the correlation function and spectral density of truncated white noise are shown in Figure 1(a).

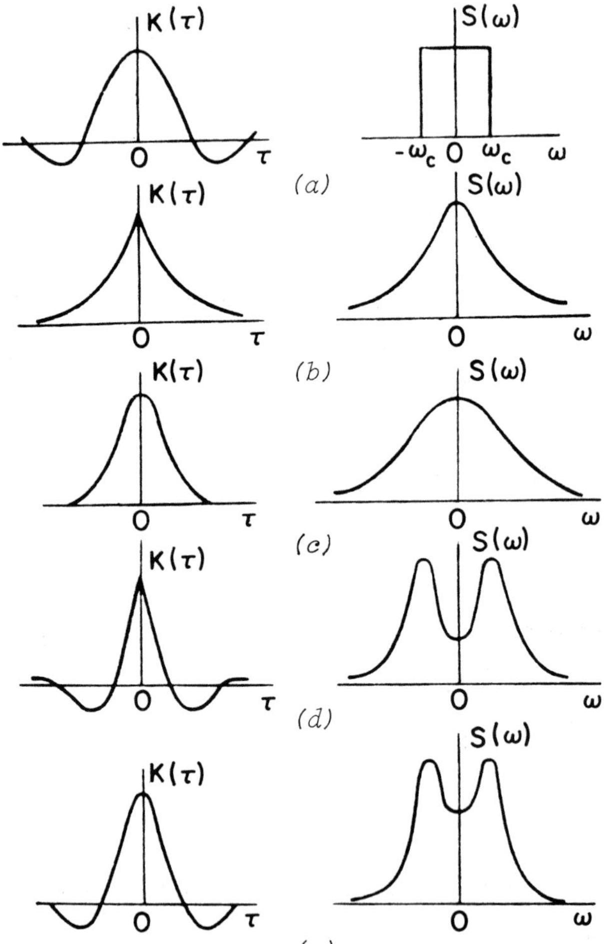

Figure 1

An example of a process with finite variance is an exponentially-correlated process, whose correlation function and spectral density are expressed by

$$K(\tau) = \sigma_0^2 e^{-\alpha|\tau|}, \qquad S(\omega) = \frac{\sigma_0^2}{\pi} \frac{\alpha}{\alpha^2+\omega^2}. \tag{1.17}$$

Here σ_0^2 is the process variance and $\alpha > 0$ is a parameter characterizing the stochastic interdependence of process values at different moments of time. The constant $\tau_0 = \alpha^{-1}$ is the process correlation time. Another example is the normal process, characterized by

$$K(\tau) = \sigma_0^2 e^{-\tau^2/\tau_0^2}, \qquad S(\omega) = \frac{\sigma_0^2 \tau_0}{2\sqrt{\pi}} e^{-\frac{1}{4}\omega^2\tau_0^2}. \tag{1.18}$$

The graphs of the correlation functions and spectral densities of these processes are given in Figures 1(b) and (c).

For applications, models of stationary random processes with latent periodicity are of interest. One such model is given by the formulae

$$K(\tau) = \sigma_0^2 e^{-\alpha|\tau|}\cos\theta_\alpha\tau,$$

$$S(\omega) = \frac{\sigma_0^2}{\pi}\,\frac{\alpha(\omega^2+\theta^2)}{(\omega^2-\theta^2)^2+4\alpha^2\omega^2} \tag{1.19}$$

and another one by

$$K(\tau) = \sigma_0^2 e^{-\alpha|\tau|}\left(\cos\theta_\alpha\tau + \frac{\alpha}{\theta_\alpha}\sin\theta_\alpha|\tau|\right),$$

$$S(\omega) = \frac{2\sigma_0^2}{\pi}\,\frac{\alpha\theta^2}{(\omega^2-\theta^2)^2+4\alpha^2\omega^2}. \tag{1.20}$$

Here σ_0^2 is the process variance as before, $\theta > 0$ is the characteristic frequency of the process, $\alpha > 0$ is the correlation parameter and $\theta^2 = \theta_\alpha^2+\alpha^2$. The graphs of functions (1.19) and (1.20) are given in Figures 1(d) and (e). For $\alpha \sim \theta$ the correlation time is comparable with the period of latent periodicity. If $\alpha \ll \theta$, the realizations of the processes are similar to periodic functions of time or, to be more exact, they are oscillating functions with slowly varying random amplitudes and phases. Processes of this type are called narrow-band processes.

There are differentiable and non-differentiable random processes. The condition for the existence of the variance of the first derivative for a stationary random process with given spectral density $S(\omega)$ is

$$\int_{-\infty}^{\infty} \omega^2 S(\omega) d\omega < \infty \ . \tag{1.21}$$

The processes whose spectral densities are given by the formulae (1.18) and (1.20) satisfy the condition (1.21). The processes with spectral densities (1.17) and (1.19) are non-differentiable ones.

Normal stationary random processes with rational spectral densities can be interpreted as the result of the passage of normal white noise through a certain linear system (filter) with constant parameters. For instance, an exponentially-correlated process (1.17) can be considered as the output of a first-order linear system with white noise input. Analogously, if white noise is transmitted through a second order linear system with constant parameters, we have the process with correlation function and spectral density (1.20). These considerations help in selecting models of multidimensional stationary random processes with appropriate spectral characteristics. To obtain an m-dimensional normal process with the matrix $S(\omega)$ whose elements are rational functions of the frequency ω, it is sufficient to apply an m-dimensional normal stationary white noise to a linear system with m inputs and m outputs. This method is also convenient for obtaining realizations of random processes if, for an investigation, modelling by means of digital or analog computers is used.

1.2 LOADINGS AS RANDOM FUNCTIONS OF SPATIAL COORDINATES AND TIME

Methods of Describing Random Fields

Loadings acting on continuous systems are as a rule, functions of spatial coordinates and time. Let a loading be characterized by the vector function $\underset{\sim}{q}(\underset{\sim}{x},t)$, which may represent either three-dimensional, or surface, or linearly distributed external forces. In the first case, the loading is a function of time $t \in T$ and of the radius vector $\underset{\sim}{x} = (x_1, x_2, x_3)$ prescribed in a certain region V of the space $\underline{R}^3$. In the second case, the loading is prescribed on the surface Ω of a certain region $V \subseteq \underline{R}^3$, with $\underset{\sim}{x} = (x_1, x_2)$, where x_1 and x_2 are generally curvilinear coordinates on Ω. In the third case, the loading is a function $\underset{\sim}{q}(x,t)$ of coordinate x and time t. In the following, unless specifically mentioned, we shall consider all the three cases jointly, treating $x \in V \subseteq \underline{R}^n$ as a set of coordinates $\underset{\sim}{x} = (x_1, \ldots, x_n)$ of a point at which the loading is specified.

Let $\underset{\sim}{q}(\underset{\sim}{x},t)$ be a random function of the spatial coordinates $\underset{\sim}{x}$ and of time t (a space-time random field, a non-stationary random field, or just a random field). For a full description of this field it is necessary to prescribe a full system of joint probability distributions of the values of the vector $\underset{\sim}{q}$ for any set of points of the field $\underset{\sim}{x}_1, \underset{\sim}{x}_2, \ldots, \in V$, and any instants of time $t_1, t_2, \ldots, \in T$. This system of distributions starts from one point, two-point, etc., distributions, i.e., $p(\underset{\sim}{q}; \underset{\sim}{x}, t)$, $p(\underset{\sim}{q}_1, \underset{\sim}{q}_2; \underset{\sim}{x}_1, t_1; \underset{\sim}{x}_2, t_2) \ldots$.

Another method is based on prescribing mathematical expectations and a full system of moment functions of field values at different points and at different instants of time:

$$\langle \underset{\sim}{q}(\underset{\sim}{x},t) \rangle \; , \quad \langle \underset{\sim}{q}(\underset{\sim}{x}_1,t_1) \otimes \underset{\sim}{q}(\underset{\sim}{x}_2,t_2) \rangle \; ,$$
$$\langle \underset{\sim}{q}(\underset{\sim}{x}_1,t_1) \otimes \underset{\sim}{q}(\underset{\sim}{x}_2,t_2) \otimes \underset{\sim}{q}(\underset{\sim}{x}_3,t_3) \rangle \; , \; \ldots \; . \tag{1.22}$$

Here, as in (1.1), angular brackets stand for averaging over a set of field realizations, and symbol $\otimes$ for tensor multiplication.

Within the scope of a correlation approximation it is sufficient to prescribe the mathematical expectation of the field $\langle \underset{\sim}{q}(\underset{\sim}{x},t) \rangle$ and the correlation field tensor $K_q(\underset{\sim}{x}_1,t_1;\underset{\sim}{x}_2,t_2)$ to be equal to the mathematical expectation of the tensor product of values of the centred field $\tilde{q}(x,t) = q(x,t) - \langle q(x,t) \rangle$ at two different points of the field and at two different moments of time:

$$K_q(x_1,t_1;x_2,t_2) = \langle \tilde{q}(x_1,t_1) \otimes \tilde{q}(x_2,t_2) \rangle \ . \tag{1.23}$$

Since $\tilde{q}(x,t)$ is a vector field in $V \subseteq R^n$, $K_q(x_1,t_1;x_2,t_2)$ is a two-point tensor of the second rank, prescribed in the region V.

Let a system of basis (coordinate) non-random vector functions $\phi_1(x)$, $\phi_2(x)$, ... be constructed in the region V where the field $q(x,t)$ is prescribed, such that for any realization of this field the following expansion is valid:

$$q(x,t) = \langle q(x,t) \rangle + \sum_\alpha Q_\alpha(t) \phi_\alpha(x) \ . \tag{1.24}$$

Here $Q_\alpha(t)$ are random functions of time. Then, instead of the description of the field $q(x,t)$, the description of the set of random time functions $Q_\alpha(t)$ can be used. For example, the correlation tensor (1.23) is expressed in terms of the correlation matrix of the multidimensional random process $Q_1(t), Q_2(t)$, ... in this way:

$$K_q(x_1,t_1;x_2,t_2) = \sum_\alpha \sum_\beta \langle Q_\alpha(t_1) Q_\beta(t_2) \rangle \phi_\alpha(x_1) \otimes \phi_\beta(x_2) \ . \tag{1.25}$$

Stationary and Homogeneous Random Fields

A certain specialization in the description of space-time fields is possible if a field is stationary in time and/or homogeneous in the coordinates. If the field $q(x,t)$ is prescribed on $T = (-\infty,\infty)$ and is stationary in t, it is possible to introduce the result of the corresponding Fourier transformation (with $\tau = t_2-t$ instead of the correlation tensor $K_q(x_1,t_1;x_2,t_2)$:

$$S_q(\underset{\sim}{x}_1,\underset{\sim}{x}_2;\omega) = \frac{1}{2\pi}\int_{-\infty}^{\infty} \langle\tilde{q}(\underset{\sim}{x}_1,t) \otimes \tilde{q}(\underset{\sim}{x}_2,t+\tau)\rangle e^{-i\omega\tau}d\tau \ . \quad (1.26)$$

The newly introduced characteristic $S_q(\underset{\sim}{x}_1,\underset{\sim}{x}_2;\omega)$ has the properties of the correlation tensor (correlation matrix) with respect to coordinates $\underset{\sim}{x}_1$ and $\underset{\sim}{x}_2$, and the properties of the spectral density with respect to the variable ω. The inverse relation is

$$K_q(\underset{\sim}{x}_1,\underset{\sim}{x}_2;\tau) = \int_{-\infty}^{\infty} S_q(\underset{\sim}{x}_1,\underset{\sim}{x}_2,\omega)e^{i\omega\tau}d\omega \ . \quad (1.27)$$

Let us briefly consider fields homogeneous in $\underline{R}^n$ and stationary in $T = (-\infty,\infty)$. These fields admit a spectral representation of the type (1.5),

$$\underset{\sim}{q}(\underset{\sim}{x},t) = \langle\underset{\sim}{q}(\underset{\sim}{x},t)\rangle + \int_{\underline{R}^{n+1}} e^{i(kx+\omega t)}\underset{\sim}{Z}(d\underset{\sim}{k},d\omega) \ . \quad (1.28)$$

Here $\underset{\sim}{k}$ is a wave vector, $\underset{\sim}{k}\underset{\sim}{x} = k_1x_1 + \cdots + k_nx_n$, $d\underset{\sim}{k} = dk_1\cdots dk_n$, and ω is a frequency parameter. The distribution function $\underset{\sim}{Z}(\underset{\sim}{k},\omega)$ is related to the space-time spectral matrix $S(\underset{\sim}{k},\omega)$ by relations of the type (1.6).

Formulae of the type (1.9), in which the elements $S_{jk}(\underset{\sim}{k},\omega)$ of the spectral matrix are related to the elements $K_{jk}(\underset{\sim}{\xi},\tau)$ of the correlation matrix, have the form

$$K_{jk}(\underset{\sim}{\xi},\tau) = \int_{\underline{R}^{n+1}} S_{jk}(\underset{\sim}{k},\omega)e^{i(\underset{\sim}{k}\underset{\sim}{\xi}+\omega\tau)}d\underset{\sim}{k}d\omega \ ,$$

$$S_{jk}(\underset{\sim}{k},\omega) = \frac{1}{(2\pi)^{n+1}}\int_{\underline{R}^{n+1}} K_{jk}(\underset{\sim}{\xi},\tau)e^{-i(\underset{\sim}{k}\underset{\sim}{\xi}+\omega\tau)}d\underset{\sim}{\xi}d\tau \ , \quad (1.29)$$

where $\underset{\sim}{\xi} = \underset{\sim}{x}_2-\underset{\sim}{x}_1$, $d\underset{\sim}{\xi} = d\xi_1\cdots d\xi_n$.

Further specialization of the properties of random fields is based on the concept of an isotropic random field. The probabilistic properties of a homogeneous and isotropic field will be prescribed if the properties of its section are known, i.e., properties of a one-dimensional random coordinate function, evaluated along a certain straight line in space. Detailed descriptions of random fields can be found in [62,65].

Simplest Models of Random Loadings

For definiteness let us assume that the intensity of a loading is characterized by the scalar field $q(\underset{\sim}{x},t)$. It may be, say, pressure on a surface exposed to a gas stream; temperature of the environment, etc. It may be the intensity of one of the components of a surface loading, one of the components of the velocity in a turbulent stream, etc. Let us consider some very simple situations with random loadings which are stationary in time and described correlationally.

Let the field of loadings $q(\underset{\sim}{x},t)$ be prescribed in the form

$$q(\underset{\sim}{x},t) = Q(t)\phi(\underset{\sim}{x}) , \tag{1.30}$$

where $Q(t)$ is a stationary random function of time and $\phi(\underset{\sim}{x})$ is a deterministic function of the spatial coordinates. From formula (1.25) we find that

$$K_q(\underset{\sim}{x}_1,\underset{\sim}{x}_2;\tau) = K_Q(\tau)\phi(\underset{\sim}{x}_1)\phi(\underset{\sim}{x}_2) , \tag{1.31}$$

where $K_Q(\tau)$ is the correlation function for $Q(t)$.

Now let us assume that the field $q(\underset{\sim}{x},t)$ is delta-correlated with respect to the spatial coordinates. Then,

$$K_q(\underset{\sim}{x}_1,\underset{\sim}{x}_2;\tau) = K_0(\tau)\delta(\underset{\sim}{x}_1-\underset{\sim}{x}_2) , \tag{1.32}$$

where $K_0(\tau)$ is the time component of the correlation function, and where $\delta(\underset{\sim}{x})$ is the n-dimensional delta-function.

For reasons of dimensionality, a multiplier having the dimension of volume in the space $\underline{R}^n$ can be introduced into the right side of the formula (1.32). The time spectral density (1.26) is determined in this case as

$$S_q(\underset{\sim}{x}_1,\underset{\sim}{x}_2;\omega) = S_0(\omega)\delta(\underset{\sim}{x}_1-\underset{\sim}{x}_2) \;, \tag{1.33}$$

where $S_0(\omega)$ denotes the spectral density corresponding to the correlation function $K_0(\tau)$:

$$S_0(\omega) = \frac{1}{2\pi}\int_{-\infty}^{\infty} K_0(\tau)e^{-i\omega\tau}d\tau \;. \tag{1.34}$$

The above considered schemes of random loadings are two extreme cases in the sense that the first scheme corresponds to full correlation with respect to spatial coordinates, while the second corresponds to the absence of correlation between loading values at any points of the field, however close to each other. For non-extreme cases, schemes constructed, say, by means of correlation functions and spectral densities considered in Section 1.1 may be used.

Moving Random Loads

Many real loads (see further below) move with respect to a structure at a certain speed. The simplest modelling of such loads is based on the assumption that there exists a system of coordinates translating rectilinearly at a constant speed in which the loading field is independent of time. Let us consider the characteristics of these "frozen" loads in detail. Let the loading $r(\underset{\sim}{y})$ be prescribed by means of the correlation function $K_r(\underset{\sim}{y}_1,\underset{\sim}{y}_2)$ in the system of coordinates Y, moving with respect to the fixed system of coordinates X at the speed $\underset{\sim}{v}$ = const. Taking the fixed system of coordinates, we find that $q(\underset{\sim}{x},t) = r(\underset{\sim}{x}-\underset{\sim}{v}t)$. Hence

$$K_q(\underset{\sim}{x}_1,t_1;\underset{\sim}{x}_2,t_2) = K_r(\underset{\sim}{x}_1-\underset{\sim}{v}t_1,\underset{\sim}{x}_2-\underset{\sim}{v}t_2) \ . \tag{1.35}$$

Now let us consider some special cases of a "frozen" loading. Let $r(\underset{\sim}{y})$ be a homogeneous field with the correlation function $K_r(\underset{\sim}{\eta})$. Here $\underset{\sim}{\eta} = \underset{\sim}{y}_2-\underset{\sim}{y}_1$. Then the field $q(\underset{\sim}{x},t)$ will be homogeneous in spatial coordinates $\underset{\sim}{x}$ and in time t. Instead of the formula (1.35) we have

$$K_q(\underset{\sim}{\xi},\tau) = K_r(\underset{\sim}{\xi}-\underset{\sim}{v}\tau) \ . \tag{1.36}$$

Let $S_r(\underset{\sim}{k})$ denote the spatial coordinate spectral density corresponding to the correlation function $K_r(\underset{\sim}{\eta})$, i.e.,

$$S_r(\underset{\sim}{k}) = \frac{1}{(2\pi)^n}\int_{\underline{R}^n} K_r(\underset{\sim}{\eta})e^{-i\underset{\sim}{k}\underset{\sim}{\eta}}d\underset{\sim}{\eta} \ .$$

The space-time spectral density of the field $q(\underset{\sim}{x},t)$ is expressed by $S_r(\underset{\sim}{k})$ in this way:

$$S_q(\underset{\sim}{k},\omega) = S_r(\underset{\sim}{k})\delta(\omega+\underset{\sim}{k}\underset{\sim}{v}) \ . \tag{1.37}$$

Validity of this relation is not difficult to verify by considering $K_q(\underset{\sim}{\xi},\tau)$ instead of $S_q(\underset{\sim}{k},\omega)$

$$K_q(\underset{\sim}{\xi},\tau) = \int_{\underline{R}^{n+1}} S_q(\underset{\sim}{k},\omega)e^{i(\underset{\sim}{k}\underset{\sim}{\xi}+\omega\tau)}d\underset{\sim}{k}d\omega \ .$$

Substituting (1.37) here yields

$$K_q(\underset{\sim}{\xi},\tau) = \int_{\underline{R}^n}\left[S_r(\underset{\sim}{k})e^{i\underset{\sim}{k}\underset{\sim}{\xi}}\int_{-\infty}^{\infty}\delta(\omega+\underset{\sim}{k}\underset{\sim}{v})e^{i\omega\tau}d\omega\right]d\underset{\sim}{k} \ ,$$

and from this expression we come to (1.36).

Note some characteristics of the time spectral density

$$S_q(\underset{\sim}{\xi},\omega) = \frac{1}{2\pi}\int_{-\infty}^{\infty} K_r(\underset{\sim}{\xi}-\underset{\sim}{v}\tau)e^{-i\omega\tau}d\tau \ . \tag{1.38}$$

Let vector $\underset{\sim}{v}$ be directed along the x_1 axis. Then $\xi_1 - v\tau = \eta_1$, $\xi_2 = \eta_2$ ... where $v = |\underset{\sim}{v}|$. Substituting the integration variable in formula (1.38) by $\tau = (\xi_1 - \eta_1)/v$ leads to the expression

$$S_q(\underset{\sim}{\xi},\omega) = \Psi(\omega,\xi_2,\ldots)\exp(-i\omega\xi_1/v) , \tag{1.39}$$

where

$$\Psi(\omega,\eta_2,\ldots) = \frac{1}{2\pi v}\int_{-\infty}^{\infty} K_r(\eta)\exp\left(\frac{i\omega\eta_1}{v}\right) d\eta_1 . \tag{1.40}$$

The characteristics of Ψ depend both on frequency and on spatial coordinates evaluated along the normal to the vector $\underset{\sim}{v}$. This function is always real. As far as the spectral density $S_q(\underset{\sim}{\xi},\omega)$ is concerned, it is a complex-valued function, and we have

$$\begin{aligned} \operatorname{Re} S_q(\underset{\sim}{\xi},\omega) &= \Psi(\omega,\xi_2,\ldots)\cos(\omega\xi_1/v) , \\ \operatorname{Im} S_q(\underset{\sim}{\xi},\omega) &= -\Psi(\omega,\xi_2,\ldots)\sin(\omega\xi_1/v) . \end{aligned} \tag{1.41}$$

The imaginary part of the function $S_q(\underset{\sim}{\xi},\omega)$ is different from zero; in fact it is a joint spectral density at two different points of the field. Therefore, it should have all the characteristics of the joint time spectral density (see Section 1.1).

We have a more special particular case of a "frozen" load when we consider a wave-type loading whose values are completely correlated in the direction of the normal to the vector $\underset{\sim}{v}$. Let the vector $\underset{\sim}{v}$ again be directed along the axis x_1. Then, the formula (1.39) takes the form

$$S_q(\underset{\sim}{\xi},\omega) = \Psi(\omega)\exp(-i\omega\xi_1/v) , \tag{1.42}$$

i.e., the function Ψ depends only on the frequency. It is not difficult to extend this formula to the case when the vector $\underset{\sim}{v}$ is arbitrarily oriented with respect to the coordinate axes of the system X.

Denoting the unit velocity vector in the system X by $\underset{\sim}{n}$, and the velocity modulus by v, we obtain

$$S_q(\underset{\sim}{\xi},\omega) = \Psi(\omega)\exp(-i\omega\underset{\sim}{\xi}\underset{\sim}{n}/v) . \tag{1.43}$$

Real moving loads, as a rule, evolve with time, i.e., in a moving system of coordinates their intensity r depends stochastically on time. If the dependence on time is stationary and ergodic, the correlation in the fixed system of spatial coordinates will be damped out with increasing distance for all frequencies ω. The natural extension of formula (1.42) and (1.43) to this case takes the form

$$S_q(\underset{\sim}{\xi},\omega) = \Psi(\omega)f(\omega\xi_1/v,\omega\xi_2/v,\ldots) . \tag{1.44}$$

To this class, for example, belongs the formula which is often used for processing experimental data

$$S_q(\underset{\sim}{\xi},\omega) = \Psi(\omega)\exp\left[-\frac{\omega}{v}(\alpha_1|\xi_1| + \alpha_2|\xi_2| + i\beta\xi_1)\right] , \tag{1.45}$$

where α_1, α_2 and β are positive real constants. The parameters α_1^{-1} and α_2^{-2} characterize the field correlation scale in the direction of the coordinates x_1 and x_2, respectively. We obtain the formula (1.42) from (1.45) if we assume that $\alpha_1 = \alpha_2 = 0$, $\beta = 1$.

1.3 EXPERIMENTAL DATA ON CERTAIN STATIONARY RANDOM LOADINGS

Approximating Loadings by Means of Stationary and Homogeneous Random Functions

By definition, stationary random processes are prescribed on the entire time axis, and homogeneous random functions are prescribed on the whole space. For this reason alone, the description of loadings by stationary and/or homogeneous random functions can only be an idealization. Quasi-stationary and/or quasi-homogeneous random functions are a more realistic approximation.

The random time function $\underset{\sim}{q}(t)$ is called quasi-stationary if its realizations can be divided into intervals of duration T, sufficiently long as compared to the typical correlation time, but sufficiently small to allow the realization to be treated within every interval as stationary. Thus, the quasi-stationary random function is a function with sufficiently slowly varying probabilistic characteristics. Let us represent it as $\underset{\sim}{q} = \underset{\sim}{q}(t,\mu t)$, where μ is a small parameter. Within intervals of duration T, the "slow" time μt can be assumed as equal to a certain mean value. If the interval T is sufficiently long as compared to the characteristic response time of the system, the response of the system to $\underset{\sim}{q}(t,\mu t)$ can be determined by treating $\underset{\sim}{q}(t,\mu t)$ as a stationary random function of t, and treating the "slow" time μt as a constant parameter.

Analogous considerations apply to quasi-homogeneous functions of the spatial coordinates, i.e., to functions whose typical scale of variation and correlation is small as compared to distances at which the probabilistic characteristics vary perceptibly. Henceforth, when prescribing real loadings in the form of stationary and/or homogeneous random functions, we shall assume that the parameters of these functions, in their turn, may be deterministic or stochastic functions of "slow" arguments.

Loadings Due to Atmospheric Turbulence

A wind loading acting on structures consists of a steady component and a pulsation component. For sufficiently rigid structures, the design loading is mainly associated with the steady component, the pressure on the surface of the structure being, as a first approximation, proportional to the square of the mean wind velocity. If the frequencies of natural vibrations of the structure are close to typical frequencies for turbulent pulsations in the atmosphere, the design should be made with due regard for both steady and pulsation components. This situation is typical for high towers, radio masts, etc. For modern aircraft, the most dangerous components of wind loadings are pulsations caused by large-scale turbulence. It is these pulsations (gusts) that are the cause of "air pockets".

In regions sufficiently distant from the earth's surface and from other sources of perturbation, wind pulsations are described by means of a model of homogeneous (sometimes homogeneous and isotropic) turbulence. An additional simplification becomes possible due to the hypothesis of "frozen" turbulence, which permits the use of similarity of time and spatial spectral densities. The results of measuring a one-dimensional spatial spectral density of the vertical velocity component [65] are shown in Figure 2. The light circles show the measurements taken from a meterological tower 70 meters in height. These measurements yielded the time spectral density, whereas the coordinate spectral density was found by recalculation based on the hypothesis of "frozen" turbulence. The dark dots denote the results of measurements taken from an aircraft flying at great speed by the top of the tower. These measurements give the coordinate spectral density with sufficient precision. The experimental data show that the hypothesis is valid up to wave numbers of order $k = 10^{-5}\ \text{cm}^{-1}$. The straight line corresponds to the law of "five thirds" resulting from the theory of homogeneous isotropic turbulence.

The mean wind velocity and the parameters of turbulent pulsations are determined by meteorological conditions and depend on the geographical location, the elevation level, the season, the time of the day, etc. Statistical regularities of these parameters are very complicated and not sufficiently investigated. Figure 3 shows the dependence of the normalized time spectral density s for the vertical velocity component [65]. $\underset{\sim}{\text{Sh}}$ denotes the dimensionless frequency (Struhal's number), $\underset{\sim}{\text{Ri}}$ a dimensionless parameter characterizing the atmosphere stratification (Richardson's number):

$$\mathrm{Sh} = \frac{\omega z}{v}\,, \qquad \mathrm{Ri} = -\frac{g}{\rho}\frac{d\rho}{dz}\left(\frac{dv}{dz}\right)^{-2} .$$

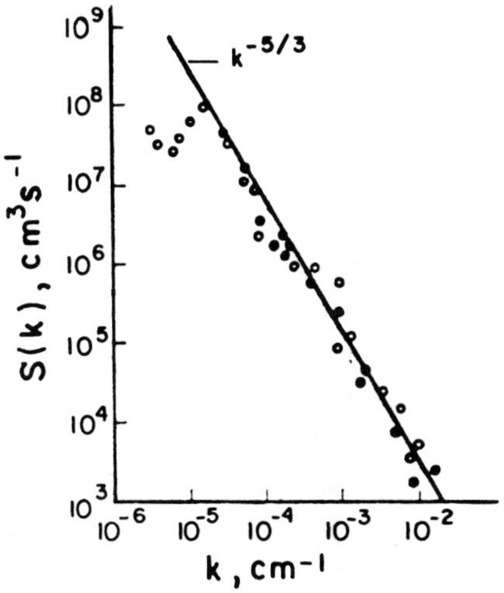

Figure 2

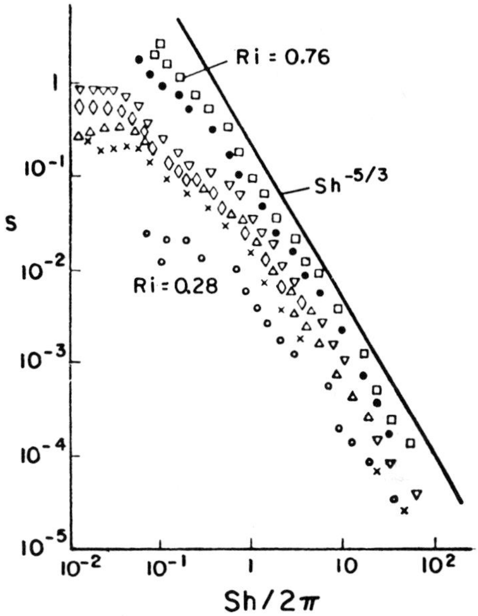

Figure 3

Here v is a mean horizontal velocity, z the elevation, $\rho(z)$ the mean atmosphere density, g the gravitational acceleration. Positive values of Richardson's number correspond to statically stable stratification, at which air density decreases with height. If Richardson's number is negative, it implies that stratification is statically unstable. Test points correspond to various values of Richardson's number from $\mathrm{Ri} = 0.28$ to $\mathrm{Ri} = -0.76$, i.e., to a transition from a very stable to a very unstable stratification, the spectral density values varying by one or two orders.

Due to the above-mentioned reasons, certain standard representations of a spectrum are used in the analysis of structures subjected to atmospheric turbulence. For instance, in the analysis of tall structures subjected to wind pressure, the time spectral density of the longitudinal component of velocity is often taken as [88]

$$S_x(\omega) = \frac{2\sigma_x^2\lambda_z}{3v_0}\,\frac{\bar{\omega}^2}{(1+\bar{\omega}^2)^{4/3}}\,, \qquad \bar{\omega} = \frac{\omega\lambda_z}{v_0} .$$

Here σ_x^2 is the variance of the longitudinal component, v_0 the mean wind velocity, λ_z a typical vertical scale for nonhomogeneity. In the analysis of aircraft construction subjected to atmospheric turbulence, various approximations of the spectral density of the vertical component of gusty wind are used. Among them, for instance, is the expression [63]

$$S_z(\omega) = \frac{\sigma_z^2 \lambda_x}{\pi v_0} \frac{1+3\bar{\omega}^2}{(1+\bar{\omega}^2)^2}, \qquad \bar{\omega} = \frac{\omega\lambda_x}{v_0},$$

where σ_z^2 is the variance of the vertical component, and λ_x a typical longitudinal scale for turbulence. Numerical values of the parameters in these equations are given by established standards.

In analysis of land structures it is necessary to take into consideration the interaction of air masses with structures and the land surface. If the structure is not a well-streamlined one, the interaction may be followed by a boundary layer separation, or by the formation of a vortex wake with a relatively regular location of vortices. This is accompanied by strong pulsations originating in the wake, their frequency depending on both the wind velocity v and a characteristic dimension of the structure b across the flow. The typical pulsation frequency in the vortex wake is determined by the condition $\underset{\sim\sim}{Sh} = \omega b/v = \text{const}$. The constant on the right side is close to 0.2. If there is another structure in the vortex wake, it is necessary to take into account the change of spectral components of pulsations in its analysis. The well-known phenomenon of buffeting in the aircraft tail unit is caused by the vortices sliding off the parts of the aircraft situated in the flow ahead.

Loadings Due to Pulsations in a Turbulent Boundary Layer

Pulsations of pressure in a boundary layer on supersonic flying vehicles have significantly large values [63,153]. For the root-mean-square value σ_p of these pulsations a rough estimate yields $\sigma_p \sim 0.02\rho_0 v_0^2/2$, where v_0 is the velocity, ρ_0 the density of a non-turbulent flow. The spectrum of pulsations in a turbulent boundary layer lies in the range of sonic frequencies and partly extends into the supersonic region. It is believed that pulsations in a turbulent boundary layer are the main source of noise in the cabins of supersonic passenger airplanes. These pulsations also cause vibrations in the panel covering, which may lead to fatigue damage.

Extensive literature is devoted to the measurements of pulsations in a turbulent boundary layer. Figure 4 [122] gives an idea of the spectral composition of pulsations. Here the dimensionless time spectral density $s = S_p(\omega)v_0/(q_0^2\delta^*)$ is represented as a function of dimensionless frequency $\bar{\omega} = \omega\delta^*/v_0$, where δ^* denotes a typical thickness of the boundary layer, $q_0 = 1/2\rho_0 v_0^2$ the velocity head.

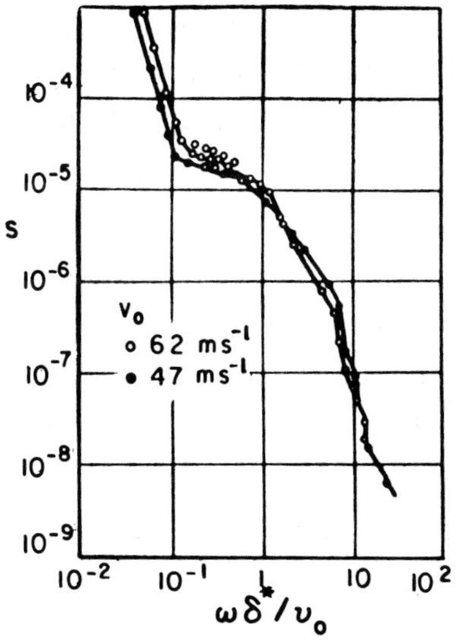

Figure 4

Consideration of the mechanism of the vortices' displacement on the surface leads to the model of a moving random loading. Expressions of the type (1.44) and (1.45) are extensively used for the approximation of experimental data, the velocity v of the moving load being identified with the mean velocity of the vortices' convection, which, in its turn, is associated with the velocity of the unperturbed flow v_0. As a rule, it turns out that $v = (0.6 \text{ to } 0.8)v_0$. A more detailed analysis reveals that high frequency components move faster than low frequency ones. Typical values for the coefficients in equation (1.45) are $\alpha_1 \approx 0.1$, $\alpha_2 \approx 0.7$, $\beta \approx 1$. Another representation for the correlation function of pulsations in a turbulent boundary layer has the form

$$K_p(\xi_1, \xi_2; \tau) = \sigma_p^2 \frac{\sin[(\xi_1 - v\tau)/\lambda_1]}{(\xi_1 - v\tau)/\lambda_1} \exp\left(-\frac{\xi_2}{\lambda_2} - \frac{\tau}{\tau_0}\right),$$

where λ_1 and λ_2 are turbulence scales along the flow and across it, τ_0 being the time correlation scale. The thickness of the boundary layer and, hence, the correlation function parameters, depend on the coordinates (say, on the distance to the point of the transition of the laminar layer to the turbulent one). Therefore, a field of pulsations can be considered uniform only within a relatively small region; for instance, within one panel covering.

Note that pulsations in a turbulent boundary layer are not a sound in the sense of physical acoustics. The velocity of their propagation is determined mainly by the velocity of an unperturbed flow, and not by the acoustic velocity in the medium. The fraction of pure acoustic energy is relatively small here. Phenomena of this type are called "pseudosound" [48].

Loadings Due to Acoustic Radiation of a Jet Stream

The jet stream of a jet engine or rocket engine is the source of intensive acoustic radiation. The noise level in the vicinity of a modern transport airplane is 160 decibels and higher. It will be remembered that the noise level L (in decibels) is related to the pressure variance σ_p^2 by the expression

$$L = 10 \log \sigma_p^2/p_*^2 \quad . \tag{1.46}$$

Here p_* is the pressure corresponding to the "zero" noise level. It is usually assumed that $p_* = 2 \cdot 10^{-5}\ N \cdot m^{-2}$.

The acoustic radiation is one percent of the engines' total power. Thus an increase of power implies an increase of the noise level. Besides the negative effect on man, this noise can cause strong vibrations in aircraft panels and in instruments, which result in fatigue damage [63,122]. For instance, at a noise level of 174 decibels, the root-mean-square value of the acoustic pressure pulsations is about $10^4\ N \cdot m^{-2} \approx 0.1\ kg \cdot cm^{-2}$.

The field of jet stream pressures is essentially non-uniform [48]. This is shown schematically in Figure 5, where the lines of equal noise level is decibels are plotted. The pulsations' spectrum is very wide; in fact it covers the entire acoustic range.

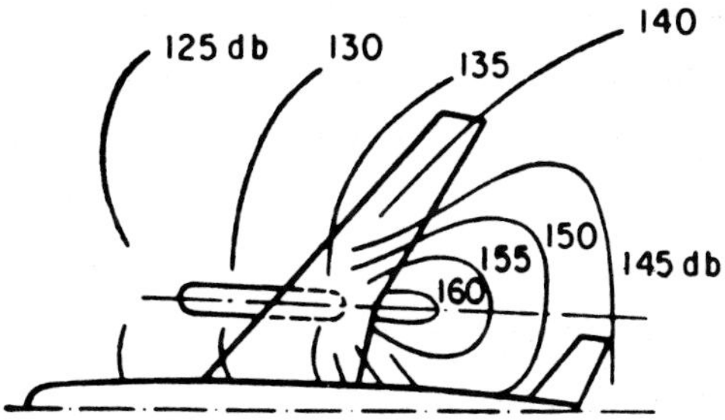

Figure 5

Let us consider the propagation of perturbations caused by some external sources in an unbounded stationary acoustic medium. The perturbed pressure $p(\underset{\sim}{x},t)$ is determined from the non-homogeneous wave equation

$$\frac{\partial^2 p}{\partial t^2} - c_0^2 \Delta p = \sigma \ , \tag{1.47}$$

where c_0 is the acoustic velocity, and $\sigma(\underset{\sim}{x},t)$ is the density of the sources. Assuming that the sources are located in some region G, we shall represent the solution of the equation (1.47) in the external region by means of the potential with delay,

$$p(\underset{\sim}{x},t) = \frac{1}{4\pi c_0^2} \int_G \sigma\left(\underset{\sim}{y}, t - \frac{|\underset{\sim}{x}-\underset{\sim}{y}|}{c_0}\right) \frac{d\underset{\sim}{y}}{|\underset{\sim}{x}-\underset{\sim}{y}|} . \qquad (1.48)$$

For the correlation function of the pressure $p(\underset{\sim}{x},t)$ we obtain the formula

$$K_p(\underset{\sim}{x}_1,t_1;\underset{\sim}{x}_2,t_2) = \frac{1}{16\pi^2 c_0^4} \int_G \int_G \left\langle \sigma\left(\underset{\sim}{y}_1, t_1 - \frac{|\underset{\sim}{x}_1-\underset{\sim}{y}_1|}{c_0}\right) \times \right.$$

$$\left. \times\ \sigma\left(\underset{\sim}{y}_2, t_2 - \frac{|\underset{\sim}{x}_2-\underset{\sim}{y}_2|}{c_0}\right)\right\rangle \frac{d\underset{\sim}{y}_1 d\underset{\sim}{y}_2}{|\underset{\sim}{x}_1-\underset{\sim}{y}_1||\underset{\sim}{x}_2-\underset{\sim}{y}_2|} . \qquad (1.49)$$

The formula (1.49) is much simplified in the case of a distant acoustic field, when the dimensions of the radiation region can be neglected relative to the distance to the points where the pressure is measured. Assuming $|\underset{\sim}{x}_1-\underset{\sim}{y}_1| \approx |\underset{\sim}{x}_1|$, $|\underset{\sim}{x}_2-\underset{\sim}{y}_2| \approx |\underset{\sim}{x}_2|$, we obtain the formula for time spectral density:

$$S_p(\underset{\sim}{x}_1,\underset{\sim}{x}_2;\omega) \approx \frac{\Psi(\omega)}{|\underset{\sim}{x}_1||\underset{\sim}{x}_2|} \exp\left[\frac{i\omega}{c_0}(|\underset{\sim}{x}_1|-|\underset{\sim}{x}_2|)\right] . \qquad (1.50)$$

Here $\Psi(\omega)$ is a certain function giving spectral characteristics of the radiation region. In the approximation (1.50), the region G is treated as a point source. Further simplification is achieved by substituting spherical waves by a system of plane waves. As a result, we obtain equations of the type (1.42) and (1.43).

The generation of a sound by hydrodynamic sources was considered by Lighthill [148]. According to Lighthill, the density of the sources is approximately determined as

$$\sigma = c_0^2 \frac{\partial^2 \tau_{jk}}{\partial x_j \partial x_k} , \qquad (1.51)$$

where $\tau_{jk} = \rho_0 v_j v_k$, v_j being the vector components of the velocity in the region G. The radiation characterized by the density (1.51) is of quadrupole character. The consequences of this and further generalizations are considered in [48,148].

The theory permits the predicting of the order of the values under measurement and gives some similarity relations. However, the analysis of the vibrations in airplane panels and of the noise level in the cabin is based on statistical data obtained by the direct measurement of the pressure on the surface of the airplane's skin. To analyze the data, expressions like (1.42) - (1.45) can be used. The typical velocity v is in this case proportional to the acoustic velocity c_0. Pressure fluctuations, except for the region of the jet stream and its immediate vicinity, are actual sound. Sensors placed on the surface of the skin measure the total pressure of direct and reflected waves. Hence, it becomes possible to treat the pressure from acoustic radiation in applied analyses as a stochastically prescribed external loading.

Loadings Due to Pressure of Sea Waves

These loadings are considered in the analyses of the strength and seaworthiness of ships and other hydraulic-engineering structures. Over a long period of time, these loadings have been investigated with the aid of special automatic wavemeters, stereoscopic aerophotography, etc. A survey of these investigations can be found in [59].

High sea is characterized by the height $h(\underset{\sim}{x},t)$ above the calm sea level. The function $h(\underset{\sim}{x},t)$ depends stochastically on coordinates $x \in \underline{R}^2$ and time t. This function will be considered as stationary and homogeneous. Its representation in terms of a Fourier integral has the form

$$h(\underset{\sim}{x},t) = \int_{\underline{R}^3} H(\underset{\sim}{k},\omega) e^{i(\underset{\sim}{k}\underset{\sim}{x}+\omega t)} d\underset{\sim}{k} d\omega, \tag{1.52}$$

where $H(\underset{\sim}{k},\omega)$ is a stochastic spectrum. Instead of (1.52), we can take as a simple scheme of linearly polarized waves the expression

$$h(x,t) = \int_{-\infty}^{\infty}\int_{-\infty}^{\infty} H(k,\omega)e^{i(kx+\omega t)}dkd\omega \ . \tag{1.53}$$

The wave number k is associated with the frequency ω by the relation $|k| = \omega^2/g$, resulting from the theory of waves of small amplitude on the surface of a deep fluid (g is the acceleration of gravity). Hence, rough sea can be characterized by both the time spectral density $S_h(\omega)$ and the spatial spectral density.

For the function $S_h(\omega)$, various analytical expressions are used, which are based partly on statistical data and partly on the consequences of hydrodynamic wave theory. As an example we can take

$$S_h(\omega) = C\omega^{-6}\exp(-2g^2/(\omega^2v^2)) \ , \tag{1.54}$$

where v is the wind velocity and C a normalizing constant. The latter is determined from the condition that the dispersion of wave heights (or their mean energy) should be equal to the prescribed value that corresponds to a certain high sea, sufficiently rare in the given region. For the functions h(x,t), the normal distribution law is generally assumed to hold; with some additional assumptions, this leads to a Rayleigh distribution of amplitudes.

In general, the spectral composition of rough sea and its intensity depend on the wind velocity, its duration, on the dimensions of the water area, etc. Therefore, the parameters in an expression of the type (1.54) must, in their turn, be considered as stochastic values. Statistical investigations of these values are only in their initial stage.

Random Loadings Acting on Mechanical Systems

Loadings to Which Road Vehicles are Subjected

Vehicles moving on rough roads undergo vibrations. Since the profile of a road is a random function of the spatial coordinates, these vibrations are also random. As an example, let us consider the motion of a two-axle vehicle (Figure 6). For simplification, we assume that the microprofile of the road h(x) depends only on the longitudinal coordinate x, that the motion of the vehicle in the direction of the x axis is uniform (with speed v), and that the contact between the wheels and the road is a point contact, the contact points moving vertically with displacements $w_1(t)$ and $w_2(t)$, which are related to the microprofile h(x) by the expression

$$w_1(t) = h(vt) , \qquad w_2(t) = h(\ell+vt) . \tag{1.55}$$

Here ℓ is the distance between the axles. If h(x) is a homogeneous random function of the coordinate, the displacements $w_1(t)$ and $w_2(t)$ will be stationary random functions of time. Taking into consideration the formulae (1.55), we find that the correlation functions of these displacements are expressed by the correlation function $K_h(\xi)$ of the microprofile of the road in the following:

$$K_{w_1}(\tau) = K_{w_2}(\tau) = K_h(v\tau) ,$$

$$K_{w_1w_2}(\tau) = K_{w_2w_1}(-\tau) = K_h(\ell+v\tau) .$$

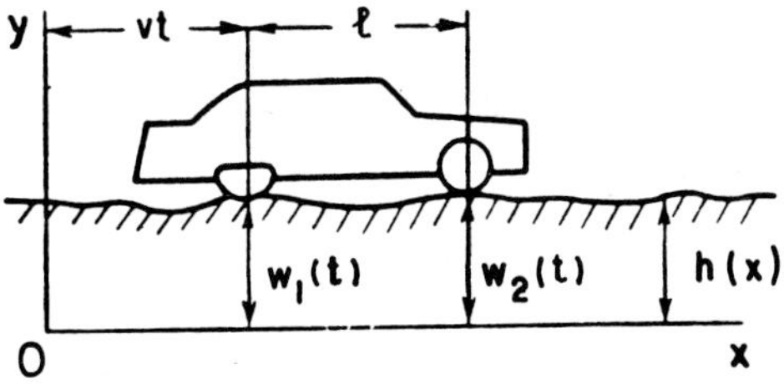

Figure 6

Using the Fourier transformation, we find the relation between the corresponding spectral densities as

$$S_{w_1}(\omega) = S_{w_2}(\omega) = \frac{1}{v} S_h\left(\frac{\omega}{v}\right),$$
$$S_{w_1 w_2}(\omega) = S^*_{w_2 w_1}(\omega) = \frac{1}{v} S_h\left(\frac{\omega}{v}\right)\exp\left(\frac{i\ell\omega}{v}\right). \tag{1.56}$$

The statistics of road microprofiles has been investigated by various authors (see, for instance [92,109]). One of the possible representations of the correlation function $K_h(\xi)$ is

$$K_h(\xi) = \sigma_h^2 e^{-\alpha|\xi|}\left(\cos \beta\xi + \frac{\alpha}{\beta}\sin \beta|\xi|\right). \tag{1.57}$$

The parameters σ_h, α and β depend on the type of road surface and its condition.[1]

1.4 EXPERIMENTAL DATA ON SOME NON-STATIONARY RANDOM LOADINGS

Almost Stationary Processes

In Section 1.3 the notion of a quasi-stationary stochastic process was introduced. This type of process is often encountered in practice. For instance, dynamic loadings acting on road vehicles depend on the speed of motion and the conditions of the road. The periods of time during which these factors vary significantly are generally longer than typical periods of loading, so that in the longevity analysis of machines as well as in the design of suspension systems the parameters σ_h, α and β in the formula (1.57) should be considered as stochastic functions of time, varying slowly as compared to $w_1(t)$ and $w_2(t)$. Thus a "slow process" $\{\sigma_h(t), \alpha(t), \beta(t), v(t)\}$ should be considered in one's calculations.

---

[1] In reference [92], one finds the following data: $\alpha = 0.01$ to $0.1\ \mathrm{m}^{-1}$, $\beta = 0.025$ to $0.140\ \mathrm{m}^{-1}$.

Random Loadings Acting on Mechanical Systems

The simplest representation of a non-stationary stochastic process which is close to a stationary one is

$$q(t) = A(t)\phi(t) , \tag{1.58}$$

where $\phi(t)$ is a stationary stochastic function with the spectral density $S_\phi(\omega)$, and where $A(t)$ is a deterministic function whose typical variation is sufficiently long as compared to the characteristic time for the process $\phi(t)$. The function $A(t)$ may be interpreted as a quasienvelope of the stochastic process. Treating the argument of $A(t)$ as a parameter, we introduce an instantaneous spectral density $S_q(\omega;t)$ of the process $q(t)$ by the relation:

$$S_q(\omega;t) = A^2(t)S_\phi(\omega) . \tag{1.59}$$

Thus, the expression (1.58) describes a quasistationary stochastic process with a slowly varying variance and a constant spectrum. A more general case can be obtained by taking the sum of processes of the type (1.58). Hence, it becomes possible to describe the variation of the spectrum composition with time.

Another possible approach is to make use of a special class of spectrum representations

$$q(t) = \langle q(t)\rangle + \int_{-\infty}^{\infty} A(t;\omega)\Phi(\omega)e^{i\omega t}d\omega , \tag{1.60}$$

where $\Phi(\omega)$ is the spectrum of a certain stationary stochastic process $\phi(t)$ with spectral density $S_\phi(\omega)$ and $A(t;\omega)$ is a deterministic complex function of frequency and time. Taking relation (1.8) into consideration, we find that the variance of the process $q(t)$ is

$$\sigma_q^2(t) = \int_{-\infty}^{\infty} |A(t;\omega)|^2 S_\phi(\omega)d\omega .$$

Hence, the function

$$S_q(\omega; t) = |A(t;\omega)|^2 S_\phi(\omega) \tag{1.61}$$

can be interpreted as the instantaneous spectral density of the process $q(t)$.

The concept of an instantaneous spectral density is useful for the interpretation of the results for processes which are almost stationary. Suppose that the realization of a quasistationary stochastic process can be divided into intervals, each interval being treated as a sufficiently accurate realization of some stationary stochastic process. Let us estimate the spectral density for each of the intervals using the methods of statistics of stationary stochastic processes. Let this spectral density be related to the moments of time $t_1, t_2, \ldots,$ of the corresponding intervals. Interpolation over all t_k gives an estimate for the instantaneous spectral density $S_q(\omega;t)$. Then, selecting the appropriate stationary process $\phi(t)$, we can estimate the functions $A(t)$ or $A(t;\omega)$ from the formulae (1.59) and (1.61).

Seismic Loadings

The motion of the ground caused by violent tectonic phenomena or underground explosions is the result of the superposition of waves generated from some source and propagating in a nonhomogeneous stratified medium. Due to multiple interference and diffraction of waves, this motion is of stochastic nature. In considering earthquakes, we distinguish single tremors, relatively long vibrations of small amplitude, and strong irregular vibrations lasting from several seconds to several scores of seconds. The first case is typical for weak earthquakes and in the regions close to the epicentre; the second for the effects of remote strong earthquakes; the third for the region close to the epicentre of a strong earthquake. The latter case is of the greatest interest, since it is these earthquakes that have the most disastrous effects. The El Centro earthquake of 1940 belongs to this type. In Figure 7 are recorded the acceleration of the ground a(t), of the velocity v(t) and the displacement u(t) during this earthquake [121]. The most informative characteristic of an earthquake is its accelerogram. As seen in Figure 7, the earthquake starts with weak oscillations which rapidly develop into strong vibrations of relatively low frequency. Then the intensity of vibrations gradually decreases and high-frequency components become predominant in the spectrum. Some data on strong earthquakes can be found, for example, in [157].

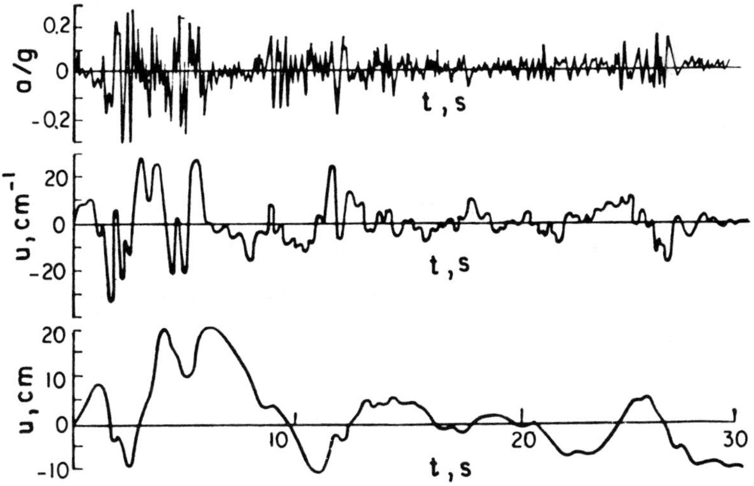

Figure 7

Various methods of approximation have been suggested for the representation of seismic motions; one of them is the assumption that the ground acceleration a(t) is stationary white noise. At the present time the representation most often used is [7,87,163]:

$$a(t) = \sum_j A_j(t)\phi_j(t)\eta(t) , \qquad (1.62)$$

where $\phi_j(t)$ are realizations of certain stationary stochastic processes with prescribed spectral densities, $A_j(t)$ are deterministic functions of time, $\eta(t)$ is the Heaviside step function. One of the simplest special cases of the formula (1.62) is

$$a(t) = A_0 e^{-ct}\phi(t)\eta(t) , \tag{1.63}$$

where A_0 and c are some positive constants and $\phi(t)$ is a stationary stochastic function of time.

The relations (1.62) and (1.63) imply that acceleration of the ground is represented as a realization of stationary stochastic processes modulated by means of certain functions of time, the quasienvelopes of the process. The correlation function and spectral density of the process $\phi(t)$ can be taken, for instance, as (1.19) or (1.20), where Θ is a typical frequency, α^{-1} is typical correlation time. For strong earthquakes, as a rule, $\Theta = (5 \text{ to } 30)s^{-1}$ $\alpha = (2 \text{ to } 10)s^{-1}$ so that the relative width of the spectrum is $\alpha/\Theta = 0.3$ to 0.5. If the analytical expression in (1.63) is taken for the quasienvelope, then $c = 0.1$ to $1.0s^{-1}$. Functions of the type $te^{-ct}\eta(t)$, $(e^{-c_1 t} - \alpha e^{-c_2 t})\eta(t)$, etc., can also be taken as quasienvelopes.

Each earthquake is a random phenomenon in the sense that the location of its epicentre, the depth of its focus, the amount of energy released, geological conditions, etc., are stochastic parameters. Therefore, a general statistical analysis of seismic phenomena is made along with the analysis of statistical properties of each strong earthquake. So far this analysis has resulted in the establishment of seismic zones. The next problem is to find for each zone the intensity distributions and spectral components of the earthquakes expected within a given period of time.

Chapter 2
Methods in the Theory of Random Vibrations

2.1 LINEAR SYSTEMS WITH A FINITE NUMBER OF DEGREES OF FREEDOM

Introductory Remarks

A review of the formulation of the problem and methods in the theory of random vibrations are given in this chapter. Methods developed for the analysis of continuous systems as well as methods for systems with a finite number of degrees of freedom are considered. Many of these methods are used in other fields of applied mathematics, such as, for instance, the theory of automatic control, statistical radiophysics, radioengineering and others. Here, the mentioned methods are considered in terms of the theory of vibrations of mechanical systems. Some methods developed especially for nonlinear systems (the method of statistical linearization, for instance) will be dealt with in Chapter 6.

Consider a certain system which interacts with the environment. Assume, at first, that the properties of this system and its interaction with the environment are deterministic. The influence of the environment (input) is characterized by the elements $\underset{\sim}{q}$ of the space $\underline{Q}$, and the state of the system (output) by the elements

$\underset{\sim}{u}$ of the space $\underline{U}$. The system is characterized by the operator H, by which the realization of the state $\underset{\sim}{u}$ is related to each of the realizations of the influence $\underset{\sim}{q}$:

$$\underset{\sim}{u} = H\underset{\sim}{q} \ . \tag{2.1}$$

Scalars, vectors and tensors, functions of one or several variables can be elements of the spaces $\underline{Q}$ and $\underline{U}$. Henceforth, the elements $\underset{\sim}{q}$ and $\underset{\sim}{u}$ will be treated as functions of time t, and the spaces $\underline{Q}$ and $\underline{U}$ as phase spaces. They will be Euclidean, finite-dimensional spaces for the systems with a finite number of degrees of freedom, whereas for continuous systems they will be either functional spaces or their coordinate realizations. The operator H is specified by the equations of the theory of vibrations, theory of elasticity, theory of shells, etc., together with corresponding initial and boundary conditions.

Assume that there exists an inverse operator $L = H^{-1}$. The equation (2.1) can be solved with respect to $\underset{\sim}{q}$ to yield

$$L\underset{\sim}{u} = \underset{\sim}{q} \ . \tag{2.2}$$

Many of the problems of statistical dynamics, including the problems of the theory of random vibrations, are formulated as differential equations for the states of the system, i.e., in the form (2.2).

If the input process $\underset{\sim}{q}(t)$ is a stochastic one, then a certain probabilistic measure is brought in to correspond to each element of $\underset{\sim}{q}(t)$, i.e., $\underline{Q}$ becomes a probabilistic phase space. For instance, the stochastic function of time is specified by means of the full system of joint probability densities of its values at arbitrarily chosen moments of time. Methods of specification by means of integrals with respect to the probabilistic measure, i.e., by means of mathematical expectations, moment and correlation functions, etc. are also used. If $\underset{\sim}{q}(t)$ is stochastic, the equations

(2.1) and (2.2) become stochastic, and the space $\underline{U}$ becomes a probabilistic phase space. The basic problem is to determine the probabilistic characteristics of the output $\underset{\sim}{u}(t)$ based on the probabilistic characteristics of the input $\underset{\sim}{q}(t)$. This problem follows the usual formulation of a problem in the theory of vibrations, where the external forces are assumed to be prescribed, and where displacements, strains and stresses are to be determined. An analogous problem arises if the system itself is stochastic, i.e., if the operator H is prescribed in the probabilistic sense. A general case when both the external influence and the properties of the system are stochastic is also possible.

The choice of the method depends to a great extent on the properties of the system. The system, described by the operator equation (2.1), will be linear if the operator H satisfies the condition

$$H(\alpha_1\underset{\sim}{q}_1+\alpha_2\underset{\sim}{q}_2) = \alpha_1 H\underset{\sim}{q}_1+\alpha_2 H\underset{\sim}{q}_2 \ , \tag{2.3}$$

where α_1 and α_2 are arbitrary nonstochastic numbers, and $\underset{\sim}{q}_1(t)$ and $\underset{\sim}{q}_2(t)$ are external actions. Note that in general, linearity of the differential equations and of the corresponding additional conditions on $\underset{\sim}{u}(t)$ does not imply that the operators H and L are linear. For example, the parametric system described by linear differential equations with variable coefficients (i.e., function of time) turns out to be nonlinear if these functions represent external action.

First, consider methods of solution to the problems of stochastic vibrations as applied to linear systems with a finite number of degrees of freedom. (Later, these methods will be extended to linear continuous systems and then to nonlinear systems; application of these methods to the latter causes some mathematical difficulties). As an illustration, consider the system of equations

$$A \frac{d^2\underset{\sim}{u}}{dt^2} + B \frac{d\underset{\sim}{u}}{dt} + C\underset{\sim}{u} = \underset{\sim}{q}(t) \ , \tag{2.4}$$

where $\underset{\sim}{u}(t)$ is the vector (column-matrix) of generalized displacements, and $\underset{\sim}{q}(t)$ is the vector (column-matrix) of generalized forces. The square matrices A, B and C can be interpreted respectively as inertial, dissipative and quasielastic matrices. The number of degrees of freedom which equals the dimension of the vectors $\underset{\sim}{u}(t)$ and $\underset{\sim}{q}(t)$ will be denoted by n. The vectors $\underset{\sim}{u}(t)$ and $\underset{\sim}{q}(t)$ will be treated as random functions of time; the matrices A, B and C, with elements a_{jk}, b_{jk} and c_{jk}, as deterministic. Some methods will be illustrated by the special case of a system with one degree of freedom:

$$\frac{d^2u}{dt^2} + 2\varepsilon \frac{du}{dt} + \omega_0^2 u = q(t) \ , \tag{2.5}$$

where ω_0 is the natural frequency and ε the coefficient of damping.

The systems (2.4) and (2.5) are stationary in the sense that their properties do not change in time. This does not restrict the applicability of most methods considered below. Their applicability to nonstationary systems will in each case depend on the context. The methods are applicable to deterministic systems, although they can be applied to certain classes of stochastic systems as well. For instance, let a system form a statistical ensemble, i.e., its properties vary stochastically in passing from one realization to another. Then, coefficients of the equations describing the system will be stochastic variables. The joint probability density of these variables will be assumed known. The solution of the problem consists of two stages. At the first stage, assuming coefficients of the equations as specified, we solve the problem as if it were a deterministic system. This solution, in which the coefficients appear as parameters, will give us conditional distributions of probabilities, conditional mathematical expectations, etc. At the second stage, a formula of total probability for the entire ensemble will be used to determine the probabilistic characteristics.

Method of Differential Equations for Moment Functions

The probabilistic description of input and output processes is usually based on the specification of the moment functions of first order, or functions which can be easily expressed in terms of these (correlation functions, spectral densities, etc.). Most extensively used methods for the solution of problems of statistical dynamics make use of this description.

Let us consider a method based on the solution of differential equations for moment (correlation) functions. For the sake of brevity, we shall call it henceforth, the method of moment functions. Let the relation between input and output processes be given in the form (2.2), where L is a differential operator. Then, equation (2.2) relating the stochastic functions $\underset{\sim}{q}(t)$ and $\underset{\sim}{u}(t)$ will be a stochastic differential equation. Applying the operator of mathematical expectation to equation (2.2), and to the equation which can be obtained from (2.2), multiplying these by $\underset{\sim}{u}(t_1)$, $\underset{\sim}{u}(t_2)$, etc., we shall obtain differential equations for the moment functions of the output process.

The relation between mathematical expectations of the processes $\underset{\sim}{q}(t)$ and $\underset{\sim}{u}(t)$ will be found by averaging (2.2) over a set of realizations. If L is a linear deterministic operator, it is commutable with the operator of averaging. As a result, we obtain the equation

$$L\langle \underset{\sim}{u} \rangle = \langle \underset{\sim}{q} \rangle , \tag{2.6}$$

which shows that mathematical expectations are related by the same differential equation as the corresponding realizations. Initial conditions for the function $\langle \underset{\sim}{u}(t) \rangle$ are to be obtained by averaging the initial conditions for realizations. For instance, if initial conditions for realizations are zero, mathematical expectations must also satisfy zero initial conditions.

Differential equations for the moment functions of the second and higher orders can be obtained analogously. Denote by L_{t_k} the operator transforming the process $\underset{\sim}{u}(t_k)$ into the process $\underset{\sim}{q}(t_k)$. This operator transforms only the functions of the variable t_k. Writing out the equations (2.2) for the two instants of time $t_1 \neq t_2$, multiplying them and applying the operator of mathematical expectation to the result, we obtain the equation for the moment functions of the second order:

$$L_{t_1} L_{t_2} \langle \underset{\sim}{u}(t_1) \otimes \underset{\sim}{u}(t_2) \rangle = \langle \underset{\sim}{q}(t_1) \otimes \underset{\sim}{q}(t_2) \rangle \ . \tag{2.7}$$

In general, if $t_1, t_2, \ldots, t_k$ are not coincident instants of time, the moment function of the k-th order is determined from the equation

$$L_{t_1} L_{t_2} \cdots L_{t_k} \langle \underset{\sim}{u}(t_1) \otimes \underset{\sim}{u}(t_2) \otimes \cdots \otimes \underset{\sim}{u}(t_k) \rangle$$

$$= \langle \underset{\sim}{q}(t_1) \otimes \underset{\sim}{q}(t_2) \otimes \cdots \otimes \underset{\sim}{q}(t_k) \rangle \ . \tag{2.8}$$

If $\underset{\sim}{u}(t)$ and $\underset{\sim}{q}(t)$ are one-dimensional stochastic processes, the products in (2.7), (2.8) and below are to be understood in their general sense. If $\underset{\sim}{u}(t)$ and $\underset{\sim}{q}(t)$ are multi-dimensional stochastic processes, the products in the formulae (2.7) and (2.8), as in formulae (1.1), must be treated as tensor products. If, for example, equation (2.2) in a detailed form is

$$\sum_{k=1}^{n} L_{jk} u_k = q_j \ , \qquad (j = 1, 2, \ldots, n) \ , \tag{2.9}$$

then, the detailed form of equation (2.7) will be

$$\sum_{\alpha=1}^{n} \sum_{\beta=1}^{n} \underset{t_1}{L_{j\alpha}} \underset{t_2}{L_{k\beta}} \langle u_\alpha(t_1) u_\beta(t_2) \rangle = \langle q_j(t_1) q_k(t_2) \rangle \,, \qquad (j,k = 1,2,\ldots,n) \,, \tag{2.10}$$

where $\underset{t_k}{L_{j\alpha}}$ is operator $L_{j\alpha}$ acting on functions of the variable t_k. For the linear system described by equation (2.4) we have

$$\underset{t_k}{L_{j\alpha}} = a_{j\alpha} \frac{d^2}{dt_k^2} + b_{j\alpha} \frac{d}{dt_k} + c_{j\alpha} \,. \tag{2.11}$$

For the correlation function $K_u(t_1,t_2)$ of the one-dimensional processes u(t) we have the equation

$$\underset{t_1}{L} \underset{t_2}{L} K_u(t_1,t_2) = K_q(t_1,t_2) \,, \tag{2.12}$$

which is analogous to equation (2.7). Here $K_q(t_1,t_2)$ is a correlation function for the one-dimensional process q(t). From now on, letters without the tilde underlining will be used in the formulas specialized for one-dimensional processes.

The equations (2.7), (2.8), etc., are partial differential equations; the equations (2.7), (2.10) and (2.12) are the exceptions for the case when $\underset{\sim}{q}(t)$ and $\underset{\sim}{u}(t)$ are stationary processes. In the latter case, moment functions of the second order depend only on the time difference $\tau = t_2 - t_1$, and we obtain ordinary differential equations. The equations are to be supplemented by initial conditions for the variables $t_1, t_2, \ldots, t_k$, derived from the initial conditions by averaging. If the process $\underset{\sim}{u}(t)$ is stationary, the conditions of boundedness at infinity and symmetry relations should be used instead of initial conditions.

Let us illustrate the above-described method by an example of a system with one degree of freedom (2.5), initial conditions being zero, and the function $K_q(t_1,t_2)$ analytical everywhere for $t_1 \geq 0$, $t_2 \geq 0$. The equation (2.12) in this case is

$$\left(\frac{\partial^2}{\partial t_1^2} + 2\varepsilon \frac{\partial}{\partial t_1} + \omega_0^2\right)\left(\frac{\partial}{\partial t_2^2} + 2\varepsilon \frac{\partial}{\partial t_2} + \omega_0^2\right) K_u(t_1,t_2) = K_q(t_1,t_2) .$$

The initial conditions for the correlation function $K_u(t_1,t_2)$ can be formulated as follows:

$$\begin{aligned} K_u &= \partial K_u/\partial t_1 = 0 , \qquad (t_1 = 0) , \\ K_u &= \partial K_u/\partial t_2 = 0 , \qquad (t_2 = 0) . \end{aligned} \tag{2.14}$$

It is expedient to replace some of the conditions (2.14) by the conditions obtained by differentiating the conditions (2.14) along the corresponding axis. The equation (2.13) supplemented by additional conditions can be solved, for example, by the method of separation of variables. If $q(t)$ and $u(t)$ are stationary processes, then, introducing a new variable $\tau = t_2 - t_1$, we can rewrite equation (2.13) in the form

$$\left(\frac{d^2}{d\tau^2} - 2\varepsilon \frac{d}{d\tau} + \omega_0^2\right)\left(\frac{d^2}{d\tau^2} + 2\varepsilon \frac{d}{d\tau} + \omega_0^2\right) K_u(\tau) = K_q(\tau) . \tag{2.15}$$

The correlation function $K_u(\tau)$ and its derivative must be bounded at infinity; also, the correlation function must be even: $K_u(-\tau) = K_u(\tau)$.

The solution of the equation (2.15) will be constructed on the assumption that the external action is a stationary delta-correlated process (white noise). The correlation function of the latter has the form (1.16),

$$K_q(\tau) = s\delta(\tau) , \tag{2.16}$$

where s is the intensity of the white noise. As can be seen from equation (2.15), the fourth derivative of $K_u(\tau)$ contains a singular component $s\delta(\tau)$. Hence, the third derivative has a discontinuity s at zero. Let us consider, instead of the equation (2.15), the corresponding homogeneous equation with nonhomogeneous initial conditions

$$dK_u/d\tau = 0\ , \qquad d^3K_u/d\tau^3 = \frac{1}{2}\, s\ , \qquad (\tau = 0^+)\ .$$

The solution of the equation yields

$$K_u(\tau) = \frac{s}{4\omega_0^2\varepsilon}\, e^{-\varepsilon|\tau|}\left(\cos\,\omega_\varepsilon\tau + \frac{\varepsilon}{\omega_\varepsilon}\sin\,\omega_\varepsilon|\tau|\right)\ , \qquad (2.17)$$

where $\omega_\varepsilon = (\omega_0^2-\varepsilon^2)^{1/2}$ is the natural frequency calculated with the correction for damping ($\varepsilon < \omega_0$). Thus, at the output of the system we have obtained the process with the correlation function (1.20). The process is characterized by the carrying frequency ω_ε and the correlation time $\tau_0 = \varepsilon^{-1}$.

Method of Green Functions (Pulse Response Functions)

We shall proceed from an equation for realizations in the form (2.1). If (2.2) is a linear differential equation, the operator H in (2.1) is the Volterra linear integral operator. For instance, the relation of type (2.1) for equation (2.5) will have the form

$$u(t) = \int_{-\infty}^{t} h(t,\tau)q(\tau)d\tau\ , \qquad (2.18)$$

where the notation

$$h(t,\tau) = \begin{cases} 0 & (t \leq \tau)\ , \\ \dfrac{1}{\omega_\varepsilon}\, e^{-\varepsilon(t-\tau)}\sin\,\omega_\varepsilon(t-\tau) & (t > \tau)\ , \end{cases} \qquad (2.19)$$

is introduced. The function $h(t,\tau)$ describes the response of the the system for $t > \tau$. This function is usually called the impulse response function. It is an analog of the Green function in boundary-value problems. In problems in the theory of vibrations for continuous systems we shall deal with integral operators whose kernels have the properties of an impulse response function when considered as functions of time, and properties of the Green function when considered as functions of the spatial coordinates. For uniformity, the term "Green function" will be used henceforth for both cases. The Green function (2.19) depends only on the difference $t - \tau$. This is a consequence of the stationarity of the system (2.5). For non-stationary systems, the Green function $h(t,\tau)$ depends on each of the arguments separately.

Let H be a linear deterministic operator. Applying the procedure of averaging over a set of realizations to relation (2.1), we obtain the equation

$$\langle \underset{\sim}{u} \rangle = H \langle \underset{\sim}{q} \rangle \,, \tag{2.20}$$

relating the input and output mathematical expectations. Formula (2.20) represents an inversion of equation (2.6). In order to calculate the moment function of the second order, we multiply the relations (2.1) written for two different instants of time t_1 and t_2 and perform averaging. As a result, we obtain

$$\langle \underset{\sim}{u}(t_1) \otimes \underset{\sim}{u}(t_2) \rangle = \overset{t_1}{\underset{\tau_1}{H}} \; \overset{t_2}{\underset{\tau_2}{H}} \langle \underset{\sim}{q}(\tau_1) \otimes \underset{\sim}{q}(\tau_2) \rangle \,, \tag{2.21}$$

where $\overset{t_k}{\underset{\tau_k}{H}}$ is an operator transforming the function $\underset{\sim}{q}(\tau_k)$ into the function $\underset{\sim}{u}(t_k)$. The formulae for moment functions of higher orders are of analogous structure.

Consider the realization of these formulae for the case when $n = 1$ and the relation (2.1) has the form (2.18). We obtain, instead of (2.20),

$$\langle u(t) \rangle = \int_{-\infty}^{t} h(t,\tau) \langle q(\tau) \rangle d\tau \,. \tag{2.22}$$

Instead of (2.21), we obtain correspondingly the formula which relates the correlation functions of the input and the output:

$$K_u(t_1,t_2) = \int_{-\infty}^{t_2}\int_{-\infty}^{t_1} h(t_1,\tau_1)h(t_2,\tau_2)K_q(\tau_1,\tau_2)d\tau_1 d\tau_2, \quad (2.23)$$

etc. If the system is stationary, $h(t,\tau) = h(t-\tau)$, and formula (2.23) can be rewritten as

$$K_u(t_1,t_2) = \int_{-\infty}^{t_2}\int_{-\infty}^{t_1} h(t_1-\tau_1)h(t_2-\tau_2)K_q(\tau_1,\tau_2)d\tau_1 d\tau_2. \quad (2.24)$$

Finally, if both the input and the output are stationary stochastic functions, it is expedient to change the variables to $t_2-t_1 = \tau$, $t_1-\tau_1 = \Theta_1$, $t_2-\tau_2 = \Theta_2$; as a result, formula (2.24) becomes

$$K_u(\tau) = \int_0^{\infty}\int_0^{\infty} h(\Theta_1)h(\Theta_2)K_q(\tau+\Theta_1-\Theta_2)d\Theta_1 d\Theta_2 \ . \quad (2.25)$$

Consider again, as an example, equation (2.5). Substituting (2.19) into (2.25), we obtain:

$$K_u(\tau) = \frac{1}{\omega_\varepsilon^2}\int_0^{\infty}\int_0^{\infty} e^{-\varepsilon(\Theta_1+\Theta_2)}\sin\omega_\varepsilon\Theta_1 \sin\omega_\varepsilon\Theta_2 \, K_q(\tau+\Theta_1-\Theta_2)d\Theta_1 d\Theta_2 \ .$$

The integral on the right side can easily be calculated if $q(t)$ is stationary white noise. Substituting $K_q(\tau)$ on the right side according to (2.16), we obtain the formula (2.17) again.

Formulae (2.23) and (2.25) can be extended to the case when $\tilde{q}(t)$ and $\tilde{u}(t)$ are multi-dimensional random processes. The products in these formulae are to be treated as tensor products. Let the differential equation relating the components of the vectors $\tilde{q}(t)$ and $\tilde{u}(t)$ have the form (2.4), and let us construct $H(t,\tau)$, the Green matrix, whose elements $h_{jk}(t,\tau)$ are the response of the j-th generalized coordinate to a unit impulse corresponding to the k-th generalized force. The realization of the operator relation (2.1) can be written as

$$u_j(t) = \sum_{k=1}^{n} \int_{-\infty}^{t} h_{jk}(t,\tau)q_k(\tau)d\tau \;, \quad (j = 1,2,\ldots,n). \quad (2.26)$$

Now the multi-dimensional analogs of formulae (2.23) - (2.25) can easily be written down. For instance, instead of (2.23), we obtain

$$K_{u_j u_k}(t_1,t_2) =$$
$$\sum_{\alpha=1}^{n} \sum_{\beta=1}^{n} \int_{-\infty}^{t_2} \int_{-\infty}^{t_1} h_{j\alpha}(t_1,\tau_1)h_{k\beta}(t_2,\tau_2)K_{q_\alpha q_\beta}(\tau_1,\tau_2)d\tau_1 d\tau_2 \;, \qquad (2.27)$$
$$(j,k = 1,2,\ldots,n)$$

where $K_{q_\alpha q_\beta}(t_1,t_2)$ are elements of a correlation matrix of generalized forces, and $K_{u_\alpha u_\beta}(t_1,t_2)$ elements of a correlation matrix of generalized displacements.

2.2 SPECTRAL METHODS IN THE THEORY OF RANDOM VIBRATIONS

Methods of Spectral Representation

Some elements in the theory of spectral representation of random processes were considered in Section 1.1 in connection with the description of stationary random processes. Techniques of spectral representation are very convenient for the solution of dynamics problems, since, to some extent, it permits substitution of operations on random functions by operations of some deterministic basis functions. The choice of a suitable system of basis functions gives additional advantages. The representation (1.5) or (1.7) of the stationary random process $\underset{\sim}{q}(t)$ as a stochastic Fourier integral is an example of this.

Let the random vector-function q(t) be represented as

$$\underset{\sim}{q}(t) = \sum_k Q_k \underset{\sim}{\phi}_k(t) \;, \qquad (2.28)$$

where $\underset{\sim}{\phi}_1(t)$, $\underset{\sim}{\phi}_2(t)$, are elements of a certain system of non-random vector functions, and Q_1, Q_2 are random variables whose joint probability density or the entire system of moments are assumed as known. Let the relation between the input and the output be specified by the linear operator equation (2.2). Substituting (2.28) into (2.2) and taking into account the linearity of the system, we obtain the following representation of the output process:

$$\underset{\sim}{u}(t) = \sum_k Q_k \underset{\sim}{\psi}_k(t) . \tag{2.29}$$

The basis functions $\underset{\sim}{\psi}_k(t)$ are determined from the deterministic equation

$$L\underset{\sim}{\psi}_k = \underset{\sim}{\phi}_k . \tag{2.30}$$

Taking into consideration (2.29), it is not difficult to find the mathematical expectation, the moment and correlation functions of the output process, for example

$$\langle \underset{\sim}{u}(t) \rangle = \sum_k \langle Q_k \rangle \, \underset{\sim}{\psi}_k(t) ,$$

$$\langle \underset{\sim}{u}(t_1) \otimes \underset{\sim}{u}(t_2) \rangle = \sum_j \sum_k \langle Q_j Q_k \rangle \, \underset{\sim}{\psi}_j(t_1) \otimes \underset{\sim}{\psi}_k(t_2) ,$$

etc. The distribution functions for $\underset{\sim}{u}(t)$ are calculated by the well-known formulae for non-random functions of random variables (time t being considered as a parameter). Thus, the problem of finding the probabilistic characteristics of the output process can be reduced to the solution of the auxiliary deterministic problem (2.30) and standard operations on random functions and random variables. Additional simplifications are introduced by the use of stochastically orthogonal canonical representations, whose coefficients satisfy the condition

$$\langle Q_j Q_k \rangle = 0 , \qquad (j \neq k) . \tag{2.31}$$

This method has been systematically developed by V. Pugachev [81].

Now, let us consider the continuous analog of the spectral representation (2.28). Let the input process $\tilde{q}(t)$ take the form

$$\tilde{q}(t) = \langle \tilde{q}(t) \rangle + \int_{-\infty}^{\infty} Q(\omega) \tilde{\phi}(t,\omega) d\omega , \tag{2.32}$$

where $\tilde{\phi}(t,\omega)$ is a real deterministic basis vector function, dependent on the parameter ω, and $Q(\omega)$ is a centred random function of this parameter. The representation (2.32) is assumed to be stochastically orthogonal, i.e., a condition of type (1.8) is assumed to be satisfied:

$$\langle Q(\omega) Q(\omega') \rangle = S_q(\omega) \delta(\omega - \omega') . \tag{2.33}$$

Here, $S_q(\omega)$ is the spectral density of the input process. At the output of the system, we obtain the spectral representation for the process $\tilde{u}(t)$ as

$$\tilde{u}(t) = \langle \tilde{u}(t) \rangle + \int_{-\infty}^{\infty} Q(\omega) \tilde{\psi}(t,\omega) d\omega , \tag{2.34}$$

where the mathematical expectation $\langle \tilde{u}(t) \rangle$ is determined from equation (2.6), and the basis function $\tilde{\psi}(t,\omega)$ is determined from the auxiliary deterministic problem

$$\tilde{L} \psi(t,\omega) = \tilde{\phi}(t,\omega) . \tag{2.35}$$

Using the representation (2.34), we shall calculate moment functions of the output process. Specifically, the correlation function of the one-dimensional process u(t) is determined as

$$K_u(t_1,t_2) = \int_{-\infty}^{\infty} S_q(\omega) \psi(t_1,\omega) \psi(t_2,\omega) d\omega . \tag{2.36}$$

Application of the Method of Spectral Representation to Stationary Systems

Consider a stationary deterministic system with a specified stationary random process at the input. The output will also be a stationary random process. The theory of random variables often deals with problems of this type. An effective solution in this case is given by the method of spectral representation. First, let us consider the case when u(t) and q(t) are one-dimensional processes.

Represent the stationary random process q(t) as a Fourier stochastic integral in the formal representation of the type (1.7), that is to say,

$$q(t) = \langle q(t)\rangle + \int_{-\infty}^{\infty} Q(\omega)e^{i\omega t}d\omega \ . \tag{2.37}$$

The spectrum of the input process Q(ω) satisfies the condition of stochastic orthogonality (2.33) extended to the case of complex-valued functions:

$$\langle Q^*(\omega)Q(\omega')\rangle = S_q(\omega)\delta(\omega-\omega') \ . \tag{2.38}$$

Here, as before, * denotes transition to a conjugate-complex value. Equation (2.35) takes the form

$$L\psi(t,\omega) = e^{i\omega t} \ . \tag{2.39}$$

Since no additional conditions are imposed on the function ψ(t,ω), except for boundedness at ±∞, the solution of equation (2.39) is

$$\psi(t,\omega) = e^{i\omega t}/L(i\omega) \ ,$$

where L(iω) is a Fourier transformation of the operator L. Let us now calculate the correlation function of the output process $K_u(\tau)$:

$$K_u(\tau) = \langle \tilde{u}^*(t)\tilde{u}(t+\tau) \rangle ,$$

where, as before $\tilde{u}(t) = u(t) - \langle u(t) \rangle$. Substituting here the expressions obtained, and using (2.38), we obtain

$$K_u(\tau) = \int_{-\infty}^{\infty} \frac{S_q(\omega)e^{i\omega\tau}d\omega}{|L(i\omega)|^2} .$$

Comparing the result with the first relation of the Wiener-Khinchin type (1.9),

$$K_u(\tau) = \int_{-\infty}^{\infty} S_u(\omega)e^{i\omega\tau}d\omega , \tag{2.40}$$

we find the relation between the input spectral density $S_q(\omega)$ and the output spectral density $S_u(\omega)$ to be given as

$$S_u(\omega) = \frac{S_q(\omega)}{|L(i\omega)|^2} . \tag{2.41}$$

Formula (2.41) is the basic relation for the solution to the problem of the passage of a stationary process through a linear, stationary, deterministic system, function $L(i\omega)$ being the dynamic stiffness of the system in relation to a sinusoidal action (the relation of the amplitudes of a generalized force and a generalized displacement). If L is a differential operator with constant coefficients, the dynamic stiffness $L(i\omega)$ is obtained by substituting the operator d/dt by $i\omega$ in the polynomial $L(d/dt)$. For example, in equation (2.5) we have

$$L(i\omega) = \omega_0^2 - \omega^2 + 2i\omega\varepsilon . \tag{2.42}$$

But, if the operator of the system is specified by the impulse response function $h(t)$, then,

$$L(i\omega) = H^{-1}(i\omega) , \tag{2.43}$$

where $H(i\omega)$ is the transfer function of the system, which has the meaning of dynamic compliance. The latter is related to the pulse response function by the expression

$$H(i\omega) = \int_0^\infty h(t)e^{-i\omega t}dt . \tag{2.44}$$

Thus, the basic formula (2.41) may also have the form

$$S_u(\omega) = |H(i\omega)|^2 S_q(\omega) . \tag{2.45}$$

Example

As a simple example, let us calculate the spectral density and displacement variance of $u(t)$ for a system with one degree of freedom whose motion is described by equation (2.5). The dynamic stiffness of the system is given by the expression (2.42). Hence, using the formula (2.41), we find the output spectral density

$$S_u(\omega) = \frac{S_q(\omega)}{(\omega_0^2-\omega^2)^2+(2\varepsilon\omega)^2} . \tag{2.46}$$

Using formula (2.46), we can find some other probabilistic characteristics of the process $u(t)$. The correlation function is determined by means of (2.40), and the displacement variance as the correlation function value for $\tau = 0$ is:

$$D[u] \equiv \sigma_u^2 = \int_{-\infty}^{\infty} \frac{S_q(\omega)\,d\omega}{(\omega_0^2-\omega^2)^2+(2\varepsilon\omega)^2} . \tag{2.47}$$

For white noise with intensity s, the spectral density $S_q(\omega)$ is determined according to (1.16) as $s/(2\pi)$. Substituting this expression in (2.47) and calculating the integral on the right side we obtain

$$\sigma_u^2 = s/(4\varepsilon\omega_0^2) . \tag{2.48}$$

If the dissipation in the system is sufficiently small, the output variance can be expressed by an approximate formula, valid for the arbitrary sufficiently slowly varying functions $S_q(\omega)$. Indeed, for $\varepsilon \ll \omega_0$, the spectral density (2.46) is high only in the small vicinity of the frequencies $\pm\omega_0$; the integral in the formula (2.47) can be approximately substituted by the sum of two integrals over small intervals $\Delta\omega$, covering the frequencies $\pm\omega_0$. If in these intervals the function $S_q(\omega)$ varies sufficiently slowly, it is possible to use the mean-value theorem, putting the slowly varying factor outside the sign of the integral:

$$\sigma_u^2 \approx 2S_q(\omega_0) \int_{\omega_0-\Delta\omega/2}^{\omega_0+\Delta\omega/2} \frac{d\omega}{(\omega_0^2-\omega^2)^2+(2\varepsilon\omega)^2} .$$

But the integrand takes on small values outside the mentioned intervals, so it is possible to extend again the region of integration to $[0,\infty]$. Finally we obtain

$$\sigma_u^2 \approx \pi S_q(\omega_0)/(2\varepsilon\omega_0^2) . \tag{2.49}$$

The formula (2.49) differs from (2.48) in that it includes the spectral density value $S_q(\omega)$ for $\omega = \omega_0$. This result has a clear mechanical meaning. Due to its high quality, the system only responds to those parts of the spectrum of external action the frequency of which is close to the natural frequency ω_0. Comparing the formulae (2.48) and (2.49), we can draw another conclusion: if $\varepsilon \ll \omega_0$, and $S_q(\omega)$ is a sufficiently slowly varying frequency function, then, for an approximate consideration, the external action can be substituted by white noise whose spectral density is equal to the spectral density of the input process at the frequency coinciding with the natural frequency of the system.

Extension to Multidimensional Random Processes

The above-described method can be extended to systems with an arbitrary finite number of degrees of freedom whose equations have the form (2.4). The generalized forces $q_j(t)$ are given as Fourier stochastic integrals

$$q_j(t) = \langle q_j(t)\rangle + \int_{-\infty}^{\infty} Q_j(\omega)e^{i\omega t}d\omega, \quad (j = 1,2,\ldots,n), \tag{2.50}$$

with the spectra $Q_j(\omega)$ satisfying the orthogonality conditions

$$\langle Q_j^*(\omega)Q_k(\omega')\rangle = S_{jk}^{(q)}(\omega)\delta(\omega-\omega') , \tag{2.51}$$

where the $S_{jk}^{(q)}(\omega)$ are the joint spectral densities of the n-dimensional process $\underset{\sim}{q}(t) = \{q_1(t), q_2(t),\ldots,q_n(t)\}$. The solutions of the equations (2.4) are

$$u_j(t) = \langle u_j(t)\rangle + \int_{-\infty}^{\infty} U_j(\omega)e^{i\omega t}d\omega, \quad (j = 1,2,\ldots,n), \tag{2.52}$$

The spectra of the output process $U_j(\omega)$ satisfy the system of linear algebraic equations

$$\sum_{k=1}^{n} L_{jk}(i\omega)U_k(\omega) = Q_j(\omega) , \qquad (j = 1,2,\ldots,n) , \tag{2.53}$$

where the $L_{jk}(i\omega)$ are the images in the Fourier space of the operators L_{jk}. The matrix consisting of the elements $L_{jk}(i\omega)$ will be denoted by $L(i\omega)$. The inverse matrix $H(i\omega)$ will be called a transfer matrix. Its elements $H_{jk}(i\omega)$ are up to a constant factor equal to Fourier transformations of the type (2.44) of the corresponding elements $h_{jk}(t)$ of the Green matrix $H(t)$. Solving the system (2.53) with respect to $U_j(\omega)$, we obtain

$$U_j(\omega) = \sum_{k=1}^{n} H_{jk}(i\omega)Q_k(\omega) , \qquad (j = 1,2,\ldots,n) .$$

Hence, taking into account the relations

$$\langle U_j^*(\omega)U_k(\omega')\rangle = S_{jk}^{(u)}(\omega)\delta(\omega-\omega') ,$$

where the $S_{jk}^{(u)}(\omega)$ are joint spectral densities of the output process, we obtain the final formula

$$S_{jk}^{(u)}(\omega) = \sum_{\alpha=1}^{n} \sum_{\beta=1}^{n} H_{j\alpha}^*(i\omega)H_{k\beta}(i\omega)S_{\alpha\beta}^{(q)}(\omega) . \tag{2.54}$$

This formula, analogous to (2.45), relates the spectral matrices of generalized forces and generalized displacements.

Response of Stationary Systems to Non-Stationary Actions

The method of spectral representation is convenient for solving some non-stationary problems. For example, let a linear stationary system with one degree of freedom be at rest for $t < 0$, and for $t \geq 0$ be subjected to the action of a realization of a stationary random process with spectral density $S_q(\omega)$. Then, in the representation (2.32) it is to be assumed that

$$\phi(t,\omega) = \begin{cases} 0 & (t < 0) , \\ e^{i\omega t} & (t \geq 0) , \end{cases} \tag{2.55}$$

holds. Equation (2.35), with consideration of (2.5), takes the form:

$$\frac{d^2\psi}{dt^2} + 2\varepsilon \frac{d\psi}{dt} + \omega_0^2\psi = \phi(t,\omega) .$$

Its solution, with the right side of (2.55) at zero initial condition, is

$$\psi(t,\omega) = \frac{e^{i\omega t} - e^{-\varepsilon t}\left(\cos \omega_\varepsilon t + \frac{i\omega+\varepsilon}{\omega_\varepsilon} \sin \omega_\varepsilon t\right)}{\omega_0^2-\omega^2+2i\omega\varepsilon} ,$$

where $\omega_\varepsilon^2 = \omega_0^2 - \varepsilon^2$. Then, according to formula (2.36), extended to the case of the complex functions $\psi(t,\omega)$, we find the correlation function of the generalized displacements as

$$K_u(t_1, t_2) = \int_{-\infty}^{\infty} S_q(\omega)\psi^*(t_1,\omega)\psi(t_2,\omega)\,d\omega \,. \tag{2.56}$$

2.3 LINEAR CONTINUOUS SYSTEMS

Preliminary Remarks

In this section, we shall consider some methods of solving problems for stochastic equations of the kind

$$L\underset{\sim}{u} = \underset{\sim}{q} \,, \tag{2.57}$$

where L is a linear deterministic operator, while $\underset{\sim}{u}(\underset{\sim}{x},t)$ and $\underset{\sim}{q}(\underset{\sim}{x},t)$ are scalar, vector or tensor fields, their components being functions of spatial coordinates $\underset{\sim}{x} \in V \subseteq \underline{R}^n$ and of time t.

As an illustrative example, we shall take the equation of flexural vibrations of an elastic rod of uniform cross-section

$$EI\,\frac{\partial^4 w}{\partial x^4} + m\,\frac{\partial^2 w}{\partial t^2} + 2m\varepsilon\,\frac{\partial w}{\partial t} = q \,, \tag{2.58}$$

where w(x,t) is the rod deflection, q(x,t) the magnitude of the external transverse loading, EI the stiffness of the rod in bending, m the mass per unit length, and ε the damping coefficient. We shall assume the rod ends (at $x = 0$ and $x = \ell$) to be simply supported and consider two variants of time conditions. In the first variant, the loading for $t = 0$ is applied to the rod at rest. Boundary and initial conditions have the form

$$w = \partial^2 w/\partial x^2 = 0 \,, \qquad (x = 0,\ x = \ell) \,, \tag{2.59}$$

$$w = \partial w/\partial t = 0 \,, \qquad (t = 0) \,. \tag{2.60}$$

In the second variant, steady vibrations take place under the conditions of stationary stochastic loading. The conditions

$$|w|, |\partial w/\partial t| < \infty , \qquad (t = \pm\infty) , \qquad (2.61)$$

are to be taken instead of the initial conditions (2.60). Assume that the external loading $q(x,t)$ is delta-correlated in time and in space, and satisfies the conditions

$$\langle q(x,t)\rangle = 0 ,$$

$$\langle q(x,t)q(x',t')\rangle = s\delta(x-x')\delta(t-t') , \qquad (2.62)$$

$$s = \text{const.}$$

Method of Differential Equations for Moment Functions

The method from Section 2.2 can be naturally extended to partial differential equations. The equations relating the moment functions of the fields $\underset{\sim}{u}(\underset{\sim}{x},t)$ and $\underset{\sim}{q}(\underset{\sim}{x},t)$ are obtained from (2.57) by multiplication and subsequent application of the operator of mathematical expectation. So, the equation for the mathematical expectation of the field $\underset{\sim}{u}(\underset{\sim}{x},t)$ has the form (2.6), and the equation for the moment functions of the second order has a form analogous to (2.7), i.e.,

$$\underset{\underset{\sim}{x},t}{L} \; \underset{\underset{\sim}{x}',t'}{L} \langle \underset{\sim}{u}(\underset{\sim}{x},t) \otimes \underset{\sim}{u}(\underset{\sim}{x}',t')\rangle = \langle \underset{\sim}{q}(\underset{\sim}{x},t) \otimes \underset{\sim}{q}(\underset{\sim}{x}',t')\rangle , \qquad (2.63)$$

etc. Here, as well as in (2.7), the subscript indices of the operator L indicate the set of arguments which are subjected to the action of the operator. If $\underset{\sim}{u}(\underset{\sim}{x},t)$ is an n-dimensional vector field, the product $\underset{\sim}{u}(\underset{\sim}{x},t) \otimes \underset{\sim}{u}(\underset{\sim}{x}',t')$ is to be treated as an ordered set of all admissible products of $u_j(\underset{\sim}{x},t)u_k(\underset{\sim}{x}',t')$, where $j,k = 1,2,\ldots,n$. This set characterizes a two-point tensor of the second order (see also Section 1.2). Some expressions given below relate to the case when $u(\underset{\sim}{x},t)$ and $q(\underset{\sim}{x},t)$ are scalar fields. The latter will be denoted by letters without tilde underlining to avoid any misunderstanding in the text.

The specific feature of operator relations of the type (2.63) is that they also include the averaged boundary conditions. We shall illustrate this by taking the equation (2.58) with additional conditions (2.59) to (2.61). Let us construct an equation relating correlation functions at the input and at the output, i.e.,

$$K_q(x,t;x',t') = \langle\tilde{q}(x,t)\tilde{q}(x',t')\rangle ,$$
$$K_w(x,t;x',t') = \langle\tilde{w}(x,t)\tilde{w}(x',t')\rangle . \tag{2.64}$$

Here, $\tilde{q}(x,t) = q(x,t) - \langle q(x,t)\rangle$ and $\tilde{w}(x,t) = w(x,t) - \langle w(x,t)\rangle$. The equation is obtained in the same way as the equation (2.63) and has the form

$$\left(EI\frac{\partial^4}{\partial x^4} + m\frac{\partial^2}{\partial t^2} + 2m\varepsilon\frac{\partial}{\partial t}\right)\left(EI\frac{\partial^4}{\partial x'^4} + m\frac{\partial^2}{\partial t'^2} + 2m\varepsilon\frac{\partial}{\partial t'}\right) \times$$
$$K_w(x,t;x',t') = K_q(x,t;x',t') . \tag{2.65}$$

To obtain the boundary conditions for the correlation function $K_w(x,t;x',t')$, corresponding to condition (2.59) for the displacement function $\tilde{w}(x,t)$, we multiply conditions (2.59) by $\tilde{w}(x',t')$ and average them over a set of realizations. Then, we perform the same operation for the conditions written for the primed variable x'. As a result, we obtain

$$K_w = \partial^2K_w/\partial x^2 = 0 , \qquad (x = 0,\ x = \ell) ,$$
$$K_w = \partial^2K_w/\partial x'^2 = 0 \qquad (x' = 0,\ x' = \ell) . \tag{2.66}$$

Instead of (2.66), we can use the conditions obtained from (2.66) by differentiating along the corresponding axis. The analogous averaging of the initial conditions (2.60) yields

$$K_w = \partial K_w/\partial t = 0 , \qquad (t = 0) ,$$
$$K_w = \partial K_w/\partial t' = 0 , \qquad (t' = 0) . \tag{2.67}$$

Now, let us consider the case of stationary stochastic vibrations. Since in this case $q(x,t)$ and $w(x,t)$ are processes stationary in time, the correlation functions (2.64) are dependent on $\tau = t'-t$ but independent of each time moment taken separately. Introducing the new variable τ, we obtain

$$\left(EI\frac{\partial^4}{\partial x^4} + m\frac{\partial^2}{\partial \tau^2} - 2m\varepsilon\frac{\partial}{\partial \tau}\right)\left(EI\frac{\partial^4}{\partial x'^4} + m\frac{\partial^2}{\partial \tau^2} + 2m\varepsilon\frac{\partial}{\partial \tau}\right) \times K_w(x,x',\tau) =$$
$$K_q(x,x',\tau) , \tag{2.68}$$

instead of (2.65). The boundary conditions (2.66) do not change. The initial conditions (2.67) must be replaced by

$$\partial K_w/\partial \tau = \partial^3 K_w/\partial \tau^3 = 0 , \quad (\tau = 0) , \qquad |K_w| < \infty , \quad (\tau \to \infty) . \tag{2.69}$$

It is assumed here that the derivatives written above exist for $\tau = 0$. Otherwise, the conditions (2.69) are substituted by the conditions of symmetry with respect to $\tau = 0$ and by the conditions of boundedness at $\tau \to \pm\infty$. The equations (2.65) and (2.68), with the boundary conditions (2.66), are solved by the method of separation of variables. We shall restrict ourselves to the solution for the process $w(x,t)$ which is stationary in time. This solution is

$$K_w(x,x',\tau) = \frac{s}{2m^2\varepsilon\ell} e^{-\varepsilon|\tau|} \sum_{k=1}^{\infty} \frac{1}{\omega_k^2} \left(\cos \omega_{k\varepsilon}\tau + \frac{\varepsilon}{\omega_{k\varepsilon}} \sin \omega_{k\varepsilon}|\tau|\right) \sin \frac{k\pi x}{\ell} \sin \frac{k\pi x'}{\ell} ,$$

$$\omega_k = \frac{k^2\pi^2}{\ell^2}\left(\frac{EI}{m}\right)^{1/2} , \qquad \omega_{k\varepsilon} = \sqrt{\omega_k^2 - \varepsilon^2} . \tag{2.70}$$

Method of Green's Functions for Continuous Systems

Let us introduce Green's function $G(\underset{\sim}{x},t;\underset{\sim}{\xi},\tau)$ for equation (2.57), where $u(\underset{\sim}{x},t)$ and $q(\underset{\sim}{x},t)$ are scalar fields. This function is determined as the solution of the equation:

$$L_{\underset{\sim}{x},t}\, G(\underset{\sim}{x},t;\underset{\sim}{\xi},\tau) = \delta(\underset{\sim}{x}-\underset{\sim}{\xi})\delta(t-\tau) , \tag{2.71}$$

satisfying the boundary conditions for $u(\underset{\sim}{x},t)$ and zero initial conditions at $t = \tau$. With respect to the variables $\underset{\sim}{x}$ and $\underset{\sim}{\xi}$ the function $G(\underset{\sim}{x},t;\underset{\sim}{\xi},\tau)$ is Green's function as commonly accepted, and with respect to the variables t and τ is an impulse response function (Section 2.2). The solution of the nonhomogeneous equation (2.57) is represented by means of Green's function as

$$u(\underset{\sim}{x},t) = \int_{-\infty}^{t} \int_V G(\underset{\sim}{x},t;\underset{\sim}{\xi},\tau) q(\underset{\sim}{\xi},\tau) d\underset{\sim}{\xi} d\tau . \tag{2.72}$$

The concept of Green's function can be extended to the case of vector and tensor fields. For example, let $\underset{\sim}{u}(\underset{\sim}{x},t)$ and $\underset{\sim}{q}(\underset{\sim}{x},t)$ be vector fields in the n-dimensional space. Equation (2.57) in its detailed form is

$$\sum_{k=1}^{n} L_{jk} u_k = q_j , \qquad (j = 1,2,\ldots,n) .$$

The Green's tensor $G_{jk}(\underset{\sim}{x},t;\underset{\sim}{\xi},\tau)$ is introduced as the solution of the system of equations

$$\sum_{k=1}^{n} \underset{\underset{\sim}{x},t}{L_{jk}} G_{k\ell}(\underset{\sim}{x},t;\underset{\sim}{\xi},\tau) = \delta_{j\ell}\delta(\underset{\sim}{x}-\underset{\sim}{\xi})\delta(t-\tau) , \quad (j,\ell = 1,2,\ldots,n) ,$$

satisfying the boundary conditions for $u(x,t)$ and zero initial conditions at $t = \tau$. The solution of the nonhomogeneous equation (2.57) now may have the form

$$u_j(\underset{\sim}{x},t) = \sum_{k=1}^{n} \int_{-\infty}^{t} \int_{V} G_{jk}(\underset{\sim}{x},t;\underset{\sim}{\xi},\tau) q_k(\underset{\sim}{\xi},\tau) d\underset{\sim}{\xi} d\tau . \qquad (2.73)$$

Formulae (2.72) and (2.73) are written symbolically as

$$\underset{\sim}{u}(\underset{\sim}{x},t) = \tilde{H}_{\underset{\sim}{\xi},\tau}^{\underset{\sim}{x},t} \; \underset{\sim}{q}(\underset{\sim}{\xi},\tau) , \qquad (2.74)$$

where the first pair of arguments of Green's function (tensor) is denoted by upper indices, the second by lower indices. Formula (2.74) is analogous to relation (2.1). Averaging this formula relates (2.20) to the mathematical expectations of the input and the output. Multiplying relation (2.74) for two different pairs of arguments $\underset{\sim}{x},t$ and $\underset{\sim}{x}',t'$ and averaging over a set of realizations, we obtain a relation of the type (2.21), namely,

$$\langle \underset{\sim}{u}(\underset{\sim}{x},t) \otimes \underset{\sim}{u}(\underset{\sim}{x}',t') \rangle = \tilde{H}_{\underset{\sim}{\xi},\tau}^{\underset{\sim}{x},t} \; \tilde{H}_{\underset{\sim}{\xi}',\tau'}^{\underset{\sim}{x}',t'} \; \langle \underset{\sim}{q}(\underset{\sim}{\xi},\tau) \otimes \underset{\sim}{q}(\underset{\sim}{\xi}',\tau') \rangle , \qquad (2.75)$$

etc.

As an illustrative example for the application of the method of Green's function, we shall take a system described by equation (2.58) and the boundary conditions (2.59). Green's function for equation (2.58) satisfies the equation

$$EI \frac{\partial^4 G}{\partial x^4} + m \frac{\partial^2 G}{\partial t^2} + 2m\varepsilon \frac{\partial G}{\partial t} = \delta(x-\xi)\delta(t-\tau)$$

with the additional conditions

$$G = \partial^2 G/\partial x^2 = 0 , \quad (x=0,\ x=\ell) ; \qquad G = \partial G/\partial t = 0 , \quad (t=\tau) .$$

The method of separation of variables gives the solution in the form of an infinite series

$$G(x,t;\xi,\tau) = \begin{cases} 0 & \text{for } t \leq \tau \,, \\ \dfrac{2}{m\ell} \displaystyle\sum_{k=1}^{\infty} \dfrac{1}{\omega k_{\varepsilon}} e^{-\varepsilon(t-\tau)} \sin[\omega_{k\varepsilon}(t-\tau)] \sin \dfrac{k\pi x}{\ell} \sin \dfrac{k\pi\xi}{\ell} & \text{for } t > \tau \,. \end{cases}$$

Further calculations according to formulae of the type (2.75) lead to the correlation function (2.70).

Method of Spectral Representation for the Solution of Stochastic Boundary-Value Problems

For continuous systems, the spectral representations (2.28) and (2.29) have the form

$$\begin{aligned} \underset{\sim}{q}(\underset{\sim}{x},t) &= \sum_{k} Q_k \underset{\sim}{\phi}_k(\underset{\sim}{x},t) \,, \\ \underset{\sim}{u}(\underset{\sim}{x},t) &= \sum_{k} Q_k \underset{\sim}{\psi}_k(\underset{\sim}{x},t) \,, \end{aligned} \tag{2.76}$$

where Q_k are stochastic values, $\underset{\sim}{\phi}_k(\underset{\sim}{x},t)$ and $\underset{\sim}{\psi}_k(\underset{\sim}{x},t)$ are deterministic vector functions of coordinates and time. As before, the relation between these functions is given by equation (2.30). Analogously to (2.32) and (2.34), the continuous spectral representations for the fields $\underset{\sim}{q}(\underset{\sim}{x},t)$ and $\underset{\sim}{u}(\underset{\sim}{x},t)$ are:

$$\begin{aligned} \underset{\sim}{q}(\underset{\sim}{x},t) &= \langle \underset{\sim}{q}(\underset{\sim}{x},t)\rangle + \int_{\underline{R}^{n+1}} Q(\underset{\sim}{k},\omega)\underset{\sim}{\phi}(\underset{\sim}{x},t;\underset{\sim}{k},\omega)\,d\underset{\sim}{k}\,d\omega \,, \\ \underset{\sim}{u}(\underset{\sim}{x},t) &= \langle \underset{\sim}{u}(\underset{\sim}{x},t)\rangle + \int_{\underline{R}^{n+1}} Q(\underset{\sim}{k},\omega)\underset{\sim}{\psi}(\underset{\sim}{x},t;\underset{\sim}{k},\omega)\,d\underset{\sim}{k}\,d\omega \,, \end{aligned} \tag{2.77}$$

where $Q(\underset{\sim}{k},\omega)$ are stochastic functions of the vector $\underset{\sim}{k}$ and of the real parameter ω, $\underset{\sim}{\phi}(\underset{\sim}{x},t;\underset{\sim}{k},\omega)$ and $\underset{\sim}{\psi}(\underset{\sim}{x},t;\underset{\sim}{k},\omega)$ are deterministic vector functions of spatial coordinates and time with the components of the vector $\underset{\sim}{k}$ and ω as parameters. The dimension of the vector $\underset{\sim}{k}$ usually coincides with the dimension of the physical space. The basis functions in the integral representations (2.77) are interrelated by the equation

$$L_{\underset{\sim}{x},t}\ \psi(\underset{\sim}{x},t;\underset{\sim}{k},\omega) = \phi(\underset{\sim}{x},t;\underset{\sim}{k},\omega) \ . \tag{2.78}$$

Let the loading q(x,t) in equation (2.58) be a stochastic function stationary in time and homogeneous in the coordinate x, with the mathematical expectation equal to zero. Then, the loading can be represented as

$$q(x,t) = \int_{-\infty}^{\infty}\int_{-\infty}^{\infty} Q(k,\omega)e^{i(kx+\omega t)}dkd\omega \ ,$$

where k is a wave number, ω is the frequency. Thus,

$$\phi(x,t;k,\omega) = e^{i(kx+\omega t)} \ .$$

The spectrum $Q(k,\omega)$ satisfies the relation

$$\langle Q^*(k,\omega)Q(k',\omega')\rangle = S_q(k,\omega)\delta(k-k')\delta(\omega-\omega') \ ,$$

where $S_q(k,\omega)$ is a space-time spectral density. Following the method of spectral representation, we seek the solution of equation (2.58) in the form

$$w(x,t) = \int_{-\infty}^{\infty}\int_{-\infty}^{\infty} Q(k,\omega)\psi(x,t;k,\omega)dkd\omega \ .$$

Function $\psi(x,t;k,\omega)$ should satisfy the equation

$$EI\frac{\partial^4\psi}{\partial x^4} + m\frac{\partial^2\psi}{\partial t^2} + 2m\varepsilon\frac{\partial\psi}{\partial t} = e^{i(kx+\omega t)}$$

and additional conditions. In the case of (2.59) and (2.60), these conditions will be

$$\psi = \frac{\partial^2 \psi}{\partial x^2} = 0 , \quad (x=0,\ x=\ell) ; \quad \psi = \frac{\partial \psi}{\partial t} = 0 , \quad (t=0) .$$

In the case of (2.59) and (2.61), the conditions will have the form

$$\psi = \frac{\partial^2 \psi}{\partial x^2} = 0 , \quad (x=0,\ x=\ell) ; \quad |\psi|, \left|\frac{\partial \psi}{\partial t}\right| < \infty , \quad (t\to\pm\infty) .$$

In the latter case, for example

$$\psi(x,t;k,\omega) = \psi_0(x;k,\omega)e^{i\omega t} ,$$

where the following notations are used

$$\psi_0 = \frac{e^{ik\ell}\sinh\beta x+\sinh\beta(\ell-x)}{2EI\beta^2(\beta^2+k^2)\sinh\beta\ell} + \frac{e^{ik\ell}\sin\beta x+\sin\beta(\ell-x)}{2EI\beta^2(\beta^2-k^2)\sin\beta\ell} - \frac{e^{ikx}}{EI(\beta^4-k^4)} ,$$

$$\beta^4 = (m\omega^2-2m\varepsilon i\omega)/(EI) .$$

Then, let us calculate the correlation function of the displacement $w(x,t)$:

$$K_w(x,t;x',t') = \int_{-\infty}^{\infty}\int_{-\infty}^{\infty} S_q(k,\omega)\psi^*(x,t;k,\omega)\psi(x',t';k,\omega)dkd\omega$$

or, after separating the time factor,

$$K_w(x,x';\tau) = \int_{-\infty}^{\infty}\int_{-\infty}^{\infty} S_q(k,\omega)\psi_0^*(x;k,\omega)\psi_0(x';k,\omega)e^{i\omega\tau}dkd\omega .$$

Unlike formula (2.70), which provides the solution in the form of an infinite series, here the solution is given in terms of double integrals.

On One Type of Spectral Representation for Stationary Problems

Let $q(\underset{\sim}{x},t)$ and $u(\underset{\sim}{x},t)$ in equation (2.57) be stochastic functions stationary in time and arbitrary in the spatial coordinates. Henceforth, for simplicity, we assume these functions to be scalar. In a correlation approximation they are prescribed by mathematical expectations and correlation functions (2.64). Instead of the latter, we shall use their Fourier transformations in time,

$$\begin{aligned} S_q(\underset{\sim}{x},\underset{\sim}{x}';\omega) &= \frac{1}{2\pi}\int_{-\infty}^{\infty} K_q(\underset{\sim}{x},\underset{\sim}{x}';\tau)e^{-i\omega\tau}d\tau \ , \\ S_u(\underset{\sim}{x},\underset{\sim}{x}';\omega) &= \frac{1}{2\pi}\int_{-\infty}^{\infty} K_u(\underset{\sim}{x},\underset{\sim}{x}';\tau)e^{-i\omega\tau}d\tau \ . \end{aligned} \tag{2.79}$$

Functions (2.79) have the characteristics of correlation functions with respect to the spatial coordinates, and the characteristics of spectral densities with respect to the frequency. From now on, since it will cause no misunderstanding, we shall refer to them as time spectral densities.

To solve equation (2.57), we take

$$\begin{aligned} q(\underset{\sim}{x},t) &= \langle q(\underset{\sim}{x},t)\rangle + \int_{-\infty}^{\infty} Q(\underset{\sim}{x},\omega)e^{i\omega t}d\omega \ , \\ u(\underset{\sim}{x},t) &= \langle u(\underset{\sim}{x},t)\rangle + \int_{-\infty}^{\infty} U(\underset{\sim}{x},\omega)e^{i\omega t}d\omega . \end{aligned} \tag{2.80}$$

Here $Q(\underset{\sim}{x},\omega)$ and $U(\underset{\sim}{x},\omega)$ are arbitrary stochastic functions in $\underset{\sim}{x}$ and delta-correlated stochastic functions in ω, and

$$\begin{aligned} \langle Q^*(\underset{\sim}{x},\omega)Q(\underset{\sim}{x}',\omega')\rangle &= S_q(\underset{\sim}{x},\underset{\sim}{x}';\omega)\delta(\omega-\omega') \ , \\ \langle U^*(\underset{\sim}{x},\omega)U(\underset{\sim}{x}',\omega')\rangle &= S_u(\underset{\sim}{x},\underset{\sim}{x}';\omega)\delta(\omega-\omega') \ . \end{aligned} \tag{2.81}$$

Substituting expressions (2.80) into (2.57) yields

$$L_{\underset{\sim}{x}}(i\omega)U(\underset{\sim}{x},\omega) = Q(\underset{\sim}{x},\omega) \ ,$$

where $L_{\underset{\sim}{x}}(i\omega)$ is derived from the operator L by substituting $i\omega$ for the operator $\partial/\partial t$. Thus, $L_{\underset{\sim}{x}}(i\omega)$ is the image of the operator L in the Fourier space. Hence, averaging by a set of realizations and taking into account the relations (2.81) we obtain the relation between the time spectral densities [15]

$$L_{\underset{\sim}{x}}(-i\omega)L_{\underset{\sim}{x}'}(i\omega)S_u(\underset{\sim}{x},\underset{\sim}{x}';\omega) = S_q(\underset{\sim}{x},\underset{\sim}{x}';\omega) \ . \tag{2.82}$$

The relation (2.82) is the analog of formula (2.41). Unlike this formula, relation (2.82) is a partial differential equation with respect to the spatial coordinates. Frequency ω plays the part of a parameter. This equation can also be treated as equation (2.63) for moment functions of the second order mapped into the Fourier space.

In our example, i.e., for equation (2.58), equation (2.82) takes the form

$$\left(EI\frac{\partial^4}{\partial x^4} - m\omega^2 - 2m\varepsilon i\omega\right)\left(EI\frac{\partial^4}{\partial x'^4} - m\omega^2 + 2m\varepsilon i\omega\right)S_w(x,x';\omega) = S_q(x,x';\omega) \ ,$$

where with consideration for the formulae (2.62) and (2.79), we must take

$$S_q(x,x';\omega) = \frac{s}{2\pi}\,\delta(x-x') \ . \tag{2.83}$$

The correlation function for the flexure of a rod is obtained from the corresponding time spectral density by the Fourier transformation with respect to time, i.e.,

$$K_w(x,x';\tau) = \int_{-\infty}^{\infty} S_w(x,x';\omega)e^{i\omega\tau}d\omega \ . \tag{2.84}$$

Method of Generalized Coordinates in Problems of Stochastic Vibrations

As in deterministic problems, the method of expansion in terms of coordinate functions (method of generalized coordinates) opens the way for the effective analytical and numerical solution of various problems of stochastic vibrations of elastic systems. Here, this method can be treated as one of the realizations of the method of spectral representation. Actually, the method is based on specifying external actions and the response of the system to them in the form of series

$$
\begin{aligned}
\underset{\sim}{q}(\underset{\sim}{x},t) &= \sum_k Q_k(t)\underset{\sim}{\phi}_k(\underset{\sim}{x}) \;, \\
\underset{\sim}{u}(\underset{\sim}{x},t) &= \sum_k u_k(t)\underset{\sim}{\phi}_k(\underset{\sim}{x}) \;,
\end{aligned}
\tag{2.85}
$$

where $\underset{\sim}{\phi}_k(\underset{\sim}{x})$ are deterministic vector functions, $Q_k(t)$ are the generalized forces and $u_k(t)$ the generalized coordinates (all of them stochastic functions of time). Thus, the generalized forces and the generalized coordinates are of stochastic character here, and it is over these functions that the basic theoretical-probabilistic calculations are performed.

A more detailed consideration of the method of generalized coordinates and its modification, the method of integral estimates, will be given in subsequent chapters.

2.4 METHODS IN THE THEORY OF MARKOV PROCESSES

Concept of Markov Processes

A stochastic process is called a Markov process if the distribution of its values at time t_k can be expressed in terms of the corresponding distribution at the proceeding moment of time $t_{k-1} < t_k$, independently of the process history. In other words, the Markov process is a process without memory. Markov processes are idealizations of real processes. Yet, many real processes may approximately be treated as Markov processes, or, at least, may be considered as components of some multidimensional Markov processes.

For the theory of stochastic vibrations, Markov processes continuous in time and continuous with respect to a set of states are of greatest interest. Referring the reader for details to [80,86,96], we shall give here some information on this class of Markov processes.

Consider the stochastic vector-function x(t) taking the possible values $\underset{\sim}{x}_1, \underset{\sim}{x}_2, \cdots, x_k$ at successive instants of time $t_0 < t_1 < \cdots < t_k$. Introduce the conditional probability density $p(\underset{\sim}{x}_k, t_k | \underset{\sim}{x}_{k-1}, t_{k-1}; \cdots; \underset{\sim}{x}_0, t_0)$, characterizing the distribution of process values at time t_k, provided that the process values have been specified at the preceding moments of time $t_0, \cdots, t_{k-1}$. The function $p(\underset{\sim}{x}_k, t_k | \underset{\sim}{x}_{k-1}, t_{k-1})$ which is equal to the conditional probability density of transition from the state $\{\underset{\sim}{x}_{k-1}, t_{k-1}\}$ to the state $\{\underset{\sim}{x}_k, t_k\}$ is called the transition probability density. The transition probability density has the ordinary characteristics of probability density and, specifically, satisfies the condition of normalization

$$\int_{\underline{R}^n} p(\underset{\sim}{x}_k, t_k | \underset{\sim}{x}_{k-1}, t_{k-1}) d\underset{\sim}{x}_k = 1 \, .$$

The transition probability density $p(\underset{\sim}{x}_k, t_k | \underset{\sim}{x}_{k-1}, t_{k-1})$ together with the initial distribution density $p(\underset{\sim}{x}_0, t_0)$ fully characterizes the properties of the Markov process. The integral equation which the probability transition density must satisfy is called the Smolouhovsky-Kolmogorov equation. To obtain this equation, it is necessary to consider three successive moments of time $t_0 < t_1 < t_2$. Relating the states to one another at moments t_0 and t_2 by means of the density $p(\underset{\sim}{x}_k, t_k | \underset{\sim}{x}_{k-1}, t_{k-1})$, we arrive at the integral equation

$$p(\underset{\sim}{x}_2, t_2 | \underset{\sim}{x}_0, t_0) = \int_{\underline{R}^n} p(\underset{\sim}{x}_2, t_2 | \underset{\sim}{x}_1, t_1) p(\underset{\sim}{x}_1, t_1 | \underset{\sim}{x}_0, t_0) d\underset{\sim}{x}_1 \, . \qquad (2.86)$$

Diffusion Markov Processes. Kolmogorov Equations

Below, we shall consider such processes in which the stochastic function $\underset{\sim}{x}(t)$ assumes sufficiently small increments $\Delta\underset{\sim}{x}$ within small time intervals Δt. To be more exact, there should be finite limits

$$\chi_\alpha(\underset{\sim}{x},t) = \lim_{\Delta t\to o} \frac{\langle\Delta x_\alpha\rangle}{\Delta t} , \qquad \chi_{\alpha\beta}(\underset{\sim}{x},t) = \lim_{\Delta t\to o} \frac{\langle\Delta x_\alpha \Delta x_\beta\rangle}{\Delta t} , \tag{2.87}$$

and all other limits read

$$\lim_{\Delta t\to o} \frac{\langle\Delta x_\alpha \Delta x_\beta \Delta x_\gamma \cdots\rangle}{\Delta t} = 0 . \tag{2.88}$$

The Markov process satisfying these conditions is called a diffusion process. Here, the angular brackets denote conditional mathematical expectations, calculated for the specified value of the vector $\underset{\sim}{x}$ at the instant of time t. For example

$$\langle\Delta x_\alpha\rangle = \int_{\underline{R}^n} \Delta x_\alpha p(\underset{\sim}{x}',t'|\underset{\sim}{x},t)d\underset{\sim}{x}' , \tag{2.89}$$

where $\underset{\sim}{x}' = \underset{\sim}{x}+\Delta\underset{\sim}{x}$, $t' = t+\Delta t$. Functions $\chi_\alpha(\underset{\sim}{x},t)$ and $\chi_{\alpha\beta}(\underset{\sim}{x},t)$ are called intensities of a diffusion process. Functions $\chi_\alpha(\underset{\sim}{x},t)$, called coefficients of translation, characterize the average evolution of the process $\underset{\sim}{x}(t)$. Functions $\chi_{\alpha\beta}(\underset{\sim}{x},t)$, forming a matrix of dimension n × n, characterize the "dispersion" of the process in time and are called diffusion coefficients.

Integral equation (2.86) for the diffusion Markov process is equivalent to a partial differential equation for the transition probability density $p(\underset{\sim}{x},t|\underset{\sim}{x}_0,t_0)$

$$\frac{\partial p}{\partial t} = - \sum_{\alpha=1}^{n} \frac{\partial}{\partial x_\alpha} [\chi_\alpha(\underset{\sim}{x},t)p] + \frac{1}{2} \sum_{\alpha=1}^{n} \sum_{\beta=1}^{n} \frac{\partial^2}{\partial x_\alpha \partial x_\beta} [\chi_{\alpha\beta}(\underset{\sim}{x},t)p] , \tag{2.90}$$

with the initial condition $p = \delta(\underset{\sim}{x}-\underset{\sim}{x}_0)$ for $t = t_0$. This equation of parabolic type is called the Kolmogorov forward equation, and its special case with $n = 1$ is called the Fokker-Planck-Kolmogorov equation. The probability density $p(\underset{\sim}{x},t)$ with the arbitrary initial distribution $p_0(\underset{\sim}{x}_0,t_0)$ is also the solution of the equation (2.90), the initial condition having the form $p = p_0(\underset{\sim}{x}_0,t_0)$ for $t = t_0$. Apart from the initial conditions, the solution of the equation (2.90) should satisfy requirements of positivity and the normalization condition. If the probability of reaching some boundaries in the phase space is to be determined (such problems arise, for instance, in the theory of reliability) the corresponding conditions are given on these boundaries.

Considering the transition probability density $p(\underset{\sim}{x},t|\underset{\sim}{x}_0,t_0)$ as a function of the initial state $\{\underset{\sim}{x}_0,t_0\}$, we arrive at the differential equation

$$\frac{\partial p}{\partial t_0} = -\sum_{\alpha=1}^{n} \chi_\alpha(\underset{\sim}{x}_0,t_0)\frac{\partial p}{\partial x_\alpha^0} - \frac{1}{2}\sum_{\alpha=1}^{n}\sum_{\beta=1}^{n} \chi_{\alpha\beta}(\underset{\sim}{x}_0,t_0)\frac{\partial^2 p}{\partial x_\alpha^0 \partial x_\beta^0}, \tag{2.91}$$

which is conjugate with respect to equation (2.90). As distinguished from the latter, equation (2.91) is called the Kolmogorov backward equation. The transition probability density $p(\underset{\sim}{x},t|\underset{\sim}{x}_0,t_0)$ should satisfy the initial condition $p = \delta(\underset{\sim}{x}-\underset{\sim}{x}_0)$ for $t_0 = t$. With coefficients χ_α and $\chi_{\alpha\beta}$ not explicitly dependent of time, it is possible to substitute $\partial p/\partial t_0 = -\partial p/\partial t$ in equation (2.91). As a result, we obtain the equation

$$\frac{\partial p}{\partial t} = \sum_{\alpha=1}^{n} \chi_\alpha(\underset{\sim}{x}_0)\frac{\partial p}{\partial x_\alpha^0} + \frac{1}{2}\sum_{\alpha=1}^{n}\sum_{\beta=1}^{n} \chi_{\alpha\beta}(\underset{\sim}{x}_0)\frac{\partial^2 p}{\partial x_\alpha^0 \partial x_\beta^0}. \tag{2.92}$$

Introducing for the differential operator the designation

$$\Lambda p = -\sum_{\alpha=1}^{n} \frac{\partial}{\partial x_\alpha}\left(\frac{\partial}{\partial x_\alpha}(\chi_\alpha p)\right) + \frac{1}{2}\sum_{\alpha=1}^{n}\sum_{\beta=1}^{n} \frac{\partial^2}{\partial x_\alpha \partial x_\beta}(\chi_{\alpha\beta} p), \tag{2.93}$$

we shall write the Kolmogorov forward equation (2.90) in the form

$$\frac{\partial p}{\partial t} = \Lambda p \ . \tag{2.94}$$

Denote the conjugate operator (with respect to the operator (2.93)) by Λ^*. Then, the Kolmogorov backward equation (2.91) will have the form

$$\frac{\partial p}{\partial t_0} + \Lambda_0^* p = 0 \ , \tag{2.95}$$

where the index in Λ_0^* implies that the operator acts on functions of $\underset{\sim}{x}_0$, the coefficients χ_α and $\chi_{\alpha\beta}$ being also considered as functions of $\underset{\sim}{x}_0$ and t_0.

Stochastic Differential Equations

The Markov diffusion processes are closely related to processes in dynamic systems excited by normal white noise. The equation of such a system with respect to the n-dimensional phase vector $\underset{\sim}{x}(t)$ reads

$$\frac{d\underset{\sim}{x}}{dt} = \underset{\sim}{f}(\underset{\sim}{x},t) + G(\underset{\sim}{x},t)\underset{\sim}{\xi}(t) \ . \tag{2.96}$$

Here, $\underset{\sim}{f}(\underset{\sim}{x},t)$ is an n-dimensional vector, $G(\underset{\sim}{x},t)$ a matrix of dimension $n \times m$ (analytical functions in $\underset{\sim}{x}$ and continuous functions in t), $\underset{\sim}{\xi}(t)$ an m-dimensional vector, the components of which are independent stationary normal white noises of unit intensity. Since the left and the right sides of equation (2.96) are generalized stochastic functions, writing out (2.96) has a formal meaning only. This is an essential fact, since calculating with functions in this equation in analogy to classical analysis may lead to wrong results. In particular, this occurs because for an excitation of the white noise type, the modulus of the increment $||\Delta\underset{\sim}{x}||$ in equation (2.96) is of order $\sqrt{\Delta t}$, and not of order Δt as in the classical analysis.

In the theory of stochastic processes, instead of equation (2.96), relations which are the result of limit processes applied to certain stochastic difference schemes with independent normal increments are considered. Generally, Ito stochastic differential equations are used such as

$$d\underset{\sim}{x}(t) = \underset{\sim}{f}(\underset{\sim}{x},t)dt + G(\underset{\sim}{x},t)d\underset{\sim}{w} , \tag{2.97}$$

where $\underset{\sim}{w}(t)$ is the m-dimensional Wiener process (a normal process with independent increments and mathematical expectation equal to zero). The equivalent integral equation has the form

$$\underset{\sim}{x}(t) = \int_{t_0}^{t} \underset{\sim}{f}(\underset{\sim}{x},t)dt + \int_{t_0}^{t} G(\underset{\sim}{x},t)d\underset{\sim}{w}(t) + \underset{\sim}{x}(t_0) , \tag{2.98}$$

where the second integral on the right side is treated as the limit in the root-mean-square sense, i.e.,

$$\int_{t_0}^{t} G(\underset{\sim}{x},t)d\underset{\sim}{w}(t) = \lim_{N\to\infty} \sum_{k=o}^{N-1} G[\underset{\sim}{x}(t_k),t_k][\underset{\sim}{w}(t_{k+1}-\underset{\sim}{w}(t_k)]. \tag{2.99}$$

Formula (2.99) defines the Ito stochastic integral. There may be other representations on the right side, for example, that of Stratonovich [94] who suggested the symmetrized expression

$$\int_{t_0}^{t} G(\underset{\sim}{x},t)d\underset{\sim}{w}(t) =$$

$$\lim_{N\to\infty} \sum_{k=o}^{N-1} G\left[\frac{\underset{\sim}{x}(t_{k+1})+\underset{\sim}{x}(t_k)}{2} , \frac{t_{k+1}+t_k}{2}\right] [\underset{\sim}{w}(t_{k+1})-\underset{\sim}{w}(t_k)] . \tag{2.100}$$

In operations on smooth functions both methods of transition to the limit are equally valid. Generally, the method of limit transition as applied to generalized stochastic functions influences the final result.

Equation (2.98) corresponds to equation (2.96) if we formally assume that $\xi = d\underset{\sim}{w}/dt$, i.e., if we assume that white noise is a derivative of the Wiener process, the correlation matrix of the Wiener process corresponding to the independent white noise of unit intensity

$$\langle \underset{\sim}{w}(t) \otimes \underset{\sim}{w}(t') \rangle = E \min\{t,t'\} , \tag{2.101}$$

where E is the m-dimensional unit matrix. In applications, it is more convenient to speak of the interpretation of white noise than of the method of specifying the stochastic integral. If the stochastic integral is taken according to Ito, the equation is said to contain Ito's white noise (2.96). If the representation of a stochastic integral is used in the form (2.100), one speaks of Stratonovich's white noise. The Stratonovich white noise can be interpreted as a result of some limit operation on smooth stochastic processes. This noise is a more suitable model for the description of a process in real systems, when white noises are used for the approximation of processes with a finite dispersion and with a spectral density varying but slightly in the considered frequency range.

Calculation of Intensities of Markov Process

Using the formulae (2.87), (2.98), (2.99) and (2.100), we find the relation between the intensities of the Markov process $\chi_\alpha(\underset{\sim}{x},t)$ and $\chi_{\alpha\beta}(\underset{\sim}{x},t)$ and the coefficients of the stochastic differential equation (2.97). In the case of Ito's white noise, the calculations yield

$$\chi_\alpha = f_\alpha , \quad \chi_{\alpha\beta} = \sum_{\gamma=1}^{m} g_{\alpha\gamma} g_{\alpha\beta} , \quad (\alpha,\beta = 1,2,\ldots,n) , \tag{2.102}$$

where f_α are components of the vector $\underset{\sim}{f}(\underset{\sim}{x},t)$, and the $g_{\alpha\beta}$ are the elements of the matrix $G(\underset{\sim}{x},t)$. Analogously, for Stratonovich's white noise, we obtain:

$$\chi_\alpha = f_\alpha + \frac{1}{2}\sum_{\beta=1}^{m}\sum_{\gamma=1}^{m}\frac{\partial g_{\alpha\gamma}}{\partial x_\beta}g_{\beta\gamma}\,, \quad \chi_{\alpha\beta} = \sum_{\gamma=1}^{m} g_{\alpha\gamma}g_{\beta\gamma}\,. \tag{2.103}$$

So, the difference shows only in the translation coefficients. As an example, we shall calculate the intensity coefficients for the system

$$\frac{d^2u}{dt^2} + [2\varepsilon + \mu\xi(t)]\frac{du}{dt} + \omega_0^2 u = 0\,, \tag{2.104}$$

where $\varepsilon > 0$, $\mu \geq 0$, $\omega_0^2 > 0$, and $\xi(t)$ is a normal stationary white noise of intensity s. Equation (2.104) describes vibrations of a system with a fluctuating damping coefficient. Ito's equivalent system of equations has the form

$$dx_1 = x_2 dt\,, \qquad dx_2 = -[2\varepsilon x_2 + \omega_0^2 x_1]dt - \mu\sqrt{s}x_2 dw\,, \tag{2.105}$$

where $x_1 = u$, $x_2 = \dot{u}$. In the designations of (2.97), $f_1 = x_2$, $f_2 = -(2\varepsilon x_2 + \omega_0^2 x_1)$, and the matrix G of dimension 2×1 has one non-zero element $g_{21} = -\mu\sqrt{s}x_2$. From the formulae (2.102) we find that

$$\chi_1 = x_2\,, \quad \chi_2 = -(2\varepsilon x_2 + \omega_0^2 x_1)\,, \quad \chi_{22} = \mu^2 s x_2^2\,, \tag{2.106}$$

(other diffusion coefficients $\chi_{\alpha\beta}$ are equal to zero). Correspondingly, the formulae (2.103) yields

$$\chi_1 = x_2\,, \quad \chi_2 = -(2\varepsilon x_2 + \omega_0^2 x_1) + \frac{1}{2}\mu^2 s x_2\,, \quad \chi_{22} = \mu^2 s x_2^2\,. \tag{2.107}$$

An additional term in the formula for χ_2 can essentially affect the results, say, in stability problems (see Section 5.3).

However, in many problems, the method of the representation of white noise does not affect the values of intensity coefficients and, consequently, the form of the Kolmogorov equation. Thus, if white noises enter only in the free terms of the differential equation, the matrix G does not depend on $\tilde{x}$. Therefore, the last terms in the formulae (2.103) for χ_α are identically equal to zero. Even in systems with coefficients which are stochastic functions of time, additional terms appear only under certain conditions. Consider, for instance, the system which is described by the n-th order equation:

$$\frac{d^n u}{dt^n} + [a_1+b_1\xi_1(t)]\,\frac{d^{n-1}u}{dt^{n-1}} + \cdots + [a_{n-1}+b_{n-1}\xi_{n-1}(t)]\,\frac{du}{dt} +$$

$$[a_n+b_n\xi_n(t)]u = 0 \ ,$$

where a_k and b_k are constant coefficients, and the $\xi_k(t)$ white noises. Turning to the case of the n-dimensional phase space, it is not difficult to show that there would be a difference between the results using either the formulae (2.102) or the formulae (2.103) only if $b_1 \neq 0$.

Some Generalizations

The sphere of application of the theory of Markov processes also includes actions which can be treated as a result of the passage of normal white noises through certain linear filters, such as systems with a finite number of degrees of freedom. Expanding the phase space by adding coordinates which describe the processes in the filter, we shall again have a Markov system. If the input process is stationary with a rational spectral density, and if the degree of the polynomial in the denominator is equal to 2ν, the dimension of the phase space will increase by ν.

Let us consider this problem in more detail. Let the external action q(t) be a stationary normal process with zero mathematical expectation, and with spectral density

$$S_q(\omega) = \frac{s}{2\pi}\left|\frac{P(\omega)}{Q(\omega)}\right|^2 \ . \qquad (2.108)$$

Here, s is a positive constant, and $P(\omega)$ and $Q(\omega)$ are polynomials with real coefficients:

$$P(\omega) = \sum_{k=1}^{\mu} a_k \omega^k , \qquad Q(\omega) = \sum_{k=1}^{\nu} b_k \omega^k , \tag{2.109}$$

and $0 \leq \mu < \nu$, $a_\mu = b_\nu = 1$. All the zeros of polynomials $P(\omega)$ and $Q(\omega)$ are in the upper half-plane. The process $q(t)$ is related to the normal noise $\xi(t)$ of intensity s by the formal relation

$$Q(d/dt)q(t) = P(d/dt)\xi(t) , \tag{2.110}$$

which can be considered as the differential equation of a certain linear filter.

If $\mu = 0$, then, assuming $q = y_1$, we obtain the filter equations in the form

$$\frac{dy_1}{dt} = y_2, \ \cdots, \ \frac{dy_{\nu-1}}{dt} = y_\nu , \quad \frac{dy_\nu}{dt} = - \sum_{k=o}^{\nu-1} b_k y_{k+1} + \xi(t) . \tag{2.111}$$

Adding the phase variables of the filter $y_1, y_2, \ldots, y_n$ to the phase variables of the system $x_1, x_2, \ldots, x_n$, we obtain an expanded system in the phase space of dimensionality $n+\nu$. The evolution of the system will be a Markov process.

For $\mu \geq 1$, the filter equation (2.110) will contain the derivatives of white noise. To give the equation a more exact meaning, we shall introduce an auxiliary differentiable process $y(t)$ such that

$$q(t) = P(d/dt)y(t) . \tag{2.112}$$

Equations (2.110) and (2.112) will be compatible provided

$$Q(d/dt)y(t) = \xi(t). \tag{2.113}$$

Assuming $y = y_1$, note that equation (2.113) is equivalent to the system of ν differential equations of the first order (2.111). Let us express the external action $q(t)$ in terms of phase variables $y_k(t)$. Taking into account equations (2.111) and (2.112), we find that

$$q(t) = \sum_{k=o}^{\mu} a_k y_{k+1} \; . \tag{2.114}$$

Thus, the expanded system $n+\nu$ of phase variables again turns out to be a Markov system.

Let us illustrate this by taking as an example the system (2.104) at whose input, instead of white noise, the exponentially correlated process $q(t)$ is specified. The spectral density of this process has the form (1.17), i.e.,

$$S_q(\omega) = \frac{\sigma_0^2}{\pi} \frac{\alpha}{\alpha^2+\omega^2} \; .$$

In formula (2.108) we should assume $P(\omega) = 1$, $Q(\omega) = \alpha+i\omega$, and $s = 2\sigma_0^2\alpha$. The filter equation has the form:

$$\frac{dq}{dt} + \alpha q = \xi(t) \; . \tag{2.115}$$

The Ito stochastic equations for the expanded phase space, including the additional phase variable $x_3 \equiv q(t)$, are

$$\begin{aligned} dx_1 &= x_2 dt \; , \\ dx_2 &= -(2\varepsilon x_2+\omega_0^2 x_1+\mu x_2 x_3)dt \; , \\ dx_3 &= -\alpha x_3 dt+\sigma_0\sqrt{2\alpha}dw \; . \end{aligned} \tag{2.116}$$

Any analytical function describing the spectral density can be approximated by means of rational functions of the class (2.108). This permits the extension of the described method to a wider class of stationary normal processes. The larger the dimension of an auxiliary phase space, the closer the approximation.

Methods of Solution of Kolmogorov Equations

If problems in the theory of random vibrations are reduced to the analysis of the Kolmogorov equations (2.90) - (2.92), it becomes possible to apply the effective methods of mathematical physics to these problems. To find the solution, both classical methods (of Fourier, of integral transformations, etc.), and approximate methods (difference methods, variational methods, etc.) can be used. Many methods and techniques as applied to the Kolmogorov equations were considered in [96].

In contrast to the methods based on spectral or correlation description of stochastic processes, the methods of the theory of Markov processes make it possible to formulate and solve the problems of finding distribution laws as well as probabilities for reaching boundaries. The methods are universal in the sense that they are applicable both to stationary and nonstationary processes, and to linear, parametric and nonlinear systems. Some problems can be solved in exact formulation and closed form which permits the use of the obtained solutions in evaluating other methods. However, it is necessary to realize the difficulties involved in applications of the theory of Markov processes, which increase when the coefficients are nonanalytical, the boundaries complicated, and when the number of dimensions of the phase space increases.

The Kolmogorov equations are also convenient for obtaining equations that should be satisfied by other characteristics of a stochastic process: characteristic functions, moments, cumulative distributions, etc. Consider, for instance, the moments of phase variables related to the probability density $p(\underset{\sim}{x},t)$ by the expression

$$m_{jk\ell} \cdots = \int_{\underline{R}^n} x_j x_k x_\ell \cdots p(\underset{\sim}{x},t)d\underset{\sim}{x} \; .$$

The density $p(\underset{\sim}{x},t)$ satisfies the Kolmogorov equation (2.90). Let the coefficients of this equation, i.e., the intensities of the Markov process, be polynomials of phase variables. To obtain the equations for the moment $m_{jk\ell} \cdots$, we multiply each term of the equation (2.90) by $x_j x_k x_\ell \cdots$ and integrate over the entire phase space. The derivative of the moment $m_{jk\ell}$ with respect to time will be on the left side. Transforming the integrals on the right side according to the Gauss-Ostrogradski formula and taking into account the behaviour of the function $p(\underset{\sim}{x},t)$ at infinity, we shall express these integrals as functions of the moments with some other moments besides the moment $m_{jk\ell}$ on the right side. If we can obtain the closed system of equations for a certain set of moments by multiple repetition of the procedure, further analysis reduces to the integration of this system at certain initial conditions.

2.5 METHODS OF STATISTICAL SIMULATION

General Characteristics of These Methods

Statistical simulation is the modelling of stochastic phenomena for a numerical solution of various problems. Among them, there may be both stochastic and deterministic problems (for instance, approximate calculations of definite integrals). Methods based on this idea are called Monte-Carlo methods. The extensive use of these methods was brought about by the development of computers which made possible the multiple reproduction of stochastic values and stochastic processes in systems of different levels of complexity, and effective statistical processing of the obtained results.

Methods of statistical simulation as applied to problems in the theory of stochastic vibrations are a universal device for analyzing both linear and nonlinear systems, systems with a finite number of degrees of freedom, and continuous systems.

The solution of problems by the method of statistical simulation consists of three stages: simulating sufficiently representative samples of stochastic values and stochastic processes, describing properties of a system and inputs; multiple analysis of a corresponding deterministic model aimed at obtaining samples of output parameters and processes; statistical processing of the results including the estimation of probabilistic characteristics, verification of statistical hypotheses, etc. Since most of the methods require the use of computers, their use also implies selection of a discretization scheme to approximate the given model. In particular, discretization with respect to time is performed, and stochastic functions of continuous time are substituted by corresponding stochastic sequences. Below, referring to special books [22,42] for detail, we shall briefly dwell on two basic questions: on methods of simulating stochastic values and stochastic sequences with prescribed properties, and on statistical processing of the results.

Simulation of Stochastic Variables

A great number of algorithms has been suggested for generating by computer a sequence of sample values of a stochastic variable with a prescribed distribution. In each algorithm, a certain scheme is used in which this variable is naturally formed. Most algorithms are based on simpler (standard) algorithms; for example, a generator of independent random numbers uniformly distributed within the interval [0,1], or a generator of independent normal numbers with a standard distribution. Most computers have these generators in their software.

Having, for instance, the sequence $\{\xi_k\}$ of values in [0,1] of the uniformly distributed variable ξ, it is possible to construct the sequence $\{x_k\}$ of the values of the continuously distributed variable x with the prescribed distribution function F(x). Here, the following easily proved fact is used: the function $x = F^{-1}(\xi)$ of the uniformly distributed variable ξ within the interval [0,1] has the distribution F(x). Hence, to construct the function $x = f(\xi)$, which maps the sequence $\{\xi_k\}$ into the sequence $\{x_k\}$, it is sufficient to find the inverse function $F^{-1}(\xi)$. For instance, for the variable distributed according to Weibull's law we have

$$F(x) = 1 - \exp\left[-\left(\frac{x-x_0}{x_c}\right)^{\alpha}\right], \qquad (x \geq x_0), \tag{2.117}$$

where x_0 and x_c are positive constants, $1 \leq \alpha < \infty$. Solving the equation $F(x) = \xi$ for x, we obtain the algorithm of transformation

$$x_k = x_0 + x_c\left(\ln\frac{1}{1-\xi_k}\right)^{1/\alpha}. \tag{2.118}$$

This method can be extended to multidimensional stochastic variables and, consequently, to stochastic sequences, the conditional distributions $p(x_k|x_{k-1}, \ldots, x_1)$ being used first. By means of the above described algorithm, the realizations of x_1 with the distribution $p(x_1)$ are calculated. Then, with the values of the variable x_1 found, the realizations of the variable x_2 with the distribution $p(x_2|x_1)$, etc. are determined. However, this method requires a lot of calculations and extensive initial information about the modelled variable. Therefore, in statistical simulation, algorithms are used in which the special properties of the simulated variables and functions are taken into account.

Simulation of Stochastic Functions

A stochastic function whose values are normally distributed can be treated as the result of passing normal white noise through a certain linear filter. In the algorithms based on this idea, generators of independent normal variables and computer blocks modelling the corresponding filters are used. This method is good for both stationary processes and nonstationary processes and it is applicable to the simulation of both multidimensional processes and stochastic fields.

Let us illustrate this method by an example of a normal process with rational density. If the spectral density has the form (2.108), the filter equation is (2.110). In computer calculations, white noise $\xi(t)$ of the intensity s is substituted by the sequence

$$\xi_k = \zeta_k \sqrt{s}/\sqrt{\Delta t} \ ,$$

where ζ_k is the sequence of values of an independent normal variable with unit variance and mathematical expectation equal to zero, and Δt is the step of discretization. Equation (2.110) or the equivalent equation (2.111) are substituted by their discrete analogs, computational technique and programming methods being extensively used, which permits saving computer time, utilizing computer memory economically and minimizing discretization errors. For example, when simulating an exponentially-correlated process, the filter equation has the form (2.115). The algorithm

$$x_k = x_{k-1} e^{-\alpha\Delta t} + \sigma_0 \sqrt{1-e^{-2\alpha\Delta t}}\, \zeta_k \ ,$$

based on the exact integration of equation (2.115) on each interval Δt, allows minimization of the discretization error.

Another class of algorithms for simulating normal processes is based on spectral representation. The stationary normal process $x(t)$ with the spectral density $S(\omega)$ can take the form

$$x(t) = \langle x \rangle + \int_{-\infty}^{\infty} X(\omega) e^{i\omega t} d\omega ,$$

where $X(\omega)$ is the delta-correlated complex-valued stochastic function with normal distribution. Using the representation in terms of real spectral functions, we introduce the following approximation for the process $x(t)$:

$$x(t) \approx \langle x \rangle + \sum_{k=1}^{N} (U_k \cos \omega_k t + V_k \sin \omega_k t) .$$

Here, U_k and V_k are independent normal variables with mathematical expectations equal to zero and variances

$$\underset{\sim}{D}(U_k) = \underset{\sim}{D}(V_k) = 2S(\omega_k)\Delta\omega ,$$

where $\omega_k = k\Delta\omega$, $\Delta\omega = \omega_c/N$ and ω_c is the cut-off frequency.

A wide class of stochastic functions which are not normal can be obtained from normal functions by nonlinear transformations. The simplest example is the Rayleigh process, whose values are expressed in terms of the sum of squares of two normal processes. The method of spectral representation (specifically of canonic expansion) of stochastic functions can also be used for simulating processes of sufficiently general type.

Simulation of Stochastic Fields

The representation of a normal process as the result of linear filtering of normal white noise can be extended to stochastic fields. Consider, for example, the scalar centred field $u(\underset{\sim}{x})$ in $\underline{V} \subseteq \underline{R}^n$. It can be represented as

$$u(\underset{\sim}{x}) = \int_V G(\underset{\sim}{x},\underset{\sim}{x}')\xi(\underset{\sim}{x}')d\underset{\sim}{x}' , \tag{2.119}$$

where $\xi(\underset{\sim}{x})$ is a delta-correlated normal field in V, and Green's function $G(\underset{\sim}{x},\underset{\sim}{x}')$ is related to the correlation function $K(\underset{\sim}{x},\underset{\sim}{x}')$ of the field $u(\underset{\sim}{x})$ by the expression

$$\int_V G(\underset{\sim}{x},\underset{\sim}{x}'')G(\underset{\sim}{x}'',\underset{\sim}{x}')d\underset{\sim}{x}'' = K(\underset{\sim}{x},\underset{\sim}{x}') . \tag{2.120}$$

The problem is reduced to the analytical or numerical solution of the integral equation (2.120) for $G(\underset{\sim}{x},\underset{\sim}{x}')$ and to the numerical realization of the equation of the filter (2.119). Another method is based on the representation of the field $u(\underset{\sim}{x})$ in the form of a series in terms of the nonstochastic functions $\phi_k(\underset{\sim}{x})$

$$u(\underset{\sim}{x}) = \sum_k U_k\phi_k(\underset{\sim}{x})$$

with stochastic coefficients U_k.

Statistical Processing of Results

The realizations of the output parameters obtained by statistical simulation form samples which, in principle, do not differ from the set of experimental data obtained under similar conditions from a multiply repeated experiment; the results of the mathematical experiment being "purer" in the sense that conditions of similarity are almost ideal. The last stage of statistical simulation is the estimation of probability characteristics and the verification of the hypothesis regarding possible types of behaviour of a system. The estimates obtained by this method, for instance, statistical averages

$$\bar{x} = \frac{1}{n}\sum_{k=1}^{N} x_k , \tag{2.121}$$

will be the realizations of stochastic variables varying in each mathematical experiment. In the Monte-Carlo method, these realizations are treated as approximate values of the estimated probabilistic characteristics, and the boundary values of confidence intervals are taken as indications of the possible error, and confidence probabilities as a measure of the reliability of an approximate solution. Thus, powerful and well-developed techniques of mathematical statistics, including the methods of obtaining point and interval estimates, methods of verification of hypotheses and methods of statistical solutions are fully applicable for processing the results of statistical simulation.

Evidently, the error of the method consists of the error of discretization of the original stochastic model, of the error in calculations, and of the error due to the limited number of samples. The first two types of error can be reduced to a minimum by the choice of good difference schemes and algorithms. Therefore, the general error is principally determined by the number of samples, so in order to diminish it, it is necessary to substantially increase computer time. Thus, if we apply the central limit theorem to the estimate (2.121), we obtain the fact that for the given amount of samples N, the error will be of the order $\sigma/\sqrt{N}$, where σ^2 is the estimate for the variance of the value x. The more variable the value x, the more calculations it requires to ensure the prescribed accuracy. The requirements for the number of samples are especially strict in those problems where the values of distribution functions for sufficiently rare events are to be estimated. Therefore, methods of statistical modelling are hardly applicable to the estimate of small probabilities (for instance, to the estimate of reliability of highly-reliable systems). This drawback is partly compensated by the fact that simulation provides one with a set of realizations of a stochastic function and allows one to make qualitative conclusions about possible variants in the behaviour of a system.

Chapter 3
Random Vibrations of Linear Continuous Systems

3.1 GENERAL RELATIONS FOR LINEAR SYSTEMS

Application of the Method of Generalized Coordinates

In this and the following chapters, we shall consider forced vibrations of linear systems subjected to random loadings. These problems have been thoroughly investigated for systems with a finite number of degrees of freedom; all the necessary information and illustrative examples were given in Sections 2.1 and 2.2. Therefore, we shall limit ourselves here to systems with an infinite number of degrees of freedom (i.e., continuous systems).

Let us consider the behaviour of a linear continuous system subjected to the action of external forces which stochastically depend on spatial coordinates and time. Let the equation of motion for the system have the form

$$A \frac{\partial^2 \underset{\sim}{u}}{dt^2} + B \frac{\partial \underset{\sim}{u}}{\partial t} + C\underset{\sim}{u} = \underset{\sim}{q} , \tag{3.1}$$

where $\underset{\sim}{u}(\underset{\sim}{x},t)$ and $\underset{\sim}{q}(\underset{\sim}{x},t)$ are functions of the coordinates $\underset{\sim}{x}$ and of time t with the range of definition $\underset{\sim}{x} \in G \subset \underline{R}^n$, $t \in \underline{R}$. For fixed t, the functions $\underset{\sim}{u}(\underset{\sim}{x},t)$ and $\underset{\sim}{q}(\underset{\sim}{x},t)$ are considered as elements of Hilbert spaces $\underline{H}_1$ and $\underline{H}_2$, respectively; and A, B and C are deterministic linear operators mapping $\underline{H}_1$ into $\underline{H}_2$. We shall call the operator A inertial, the operator B dissipative, and the operator C quasielastic. Generally, all these operators are symmetrical and positive definite, their domains of definition being given by $\underline{D}(A) \supseteq \underline{D}(C)$, $\underline{D}(B) \supseteq \underline{D}(C)$. Henceforth, it will be assumed that all these restrictions on operators are satisfied.

Let us apply the method of generalized coordinates, briefly described in Section 2.3, to the equation (3.1). To construct a suitable system of coordinate functions, we shall consider a related homogeneous problem

$$C\underset{\sim}{\phi} - \omega^2 A\underset{\sim}{\phi} = 0 \ . \tag{3.2}$$

Its eigenfunctions $\underset{\sim}{\phi}_1$, $\underset{\sim}{\phi}_2$, ..., have the meaning of natural modes of vibrations of the corresponding elastic system. The eigenvalues ω_1^2, ω_2^2, ..., are equal to the square of the natural frequencies of this system, which are related to the natural modes by Rayleigh's equations:

$$\omega_k^2 = \frac{(C\underset{\sim}{\phi}_k, \underset{\sim}{\phi}_k)}{(A\underset{\sim}{\phi}_k, \underset{\sim}{\phi}_k)} \ . \tag{3.3}$$

The modes of vibrations are mutually orthogonal with the weight of the operator A and of the operator C, respectively, i.e.,

$$(A\underset{\sim}{\phi}_j, \underset{\sim}{\phi}_k) = (C\underset{\sim}{\phi}_j, \underset{\sim}{\phi}_k) = 0 \ , \qquad (j \neq k) \ . \tag{3.4}$$

Any element from the domain of definition of the quasielastic operator C can be expanded into a series of eigenfunctions $\underset{\sim}{\phi}_1$, $\underset{\sim}{\phi}_2$, ..., the series converging, at least in the norm of operator C. In particular, if $\underset{\sim}{u}(\underset{\sim}{x},t)$ is an element of the corresponding phase space, i.e., $\underset{\sim}{u} \in \underline{D}(C)$, with the parametric dependence on time t, then the expansion reads

$$\underset{\sim}{u}(\underset{\sim}{x},t) = \sum_k u_k(t)\underset{\sim}{\phi}_k(\underset{\sim}{x}) \ . \tag{3.5}$$

Here, $u_k(t)$ are time functions (generalized coordinates). The mentioned properties of the expansion are based on the assumption that the spectrum of the problem (3.2) is discrete. This assumption is satisfied if C^{-1} is a fully continuous operator. For elastic systems (rods, plates, shells, etc.) of finite dimensions, C^{-1} is a fully continuous operator.

We seek the solution of equation (3.1) in the form of the series (3.5). For $B \equiv 0$, the substitution of the series (3.5) into equation (3.1) results in a system of separable equations with respect to the generalized coordinates $u_k(t)$. If $B \not\equiv 0$, this separation takes place only for dissipative operators of a special kind; for instance, for $B = 2\varepsilon A$ (the so-called external friction), $B = \eta C$ (Voigt's friction) etc. Here, ε and η are some constants.

If the conditions of separability of the equations with respect to the generalized coordinates are satisfied, then, substituting the series (3.5) into the equation (3.1) and taking into account the relations (3.3) and (3.4), we obtain

$$\frac{d^2u_k}{dt^2} + 2\varepsilon_k \frac{du_k}{dt} + \omega_k^2 u_k = Q_k(t) \ , \qquad (k = 1,2,\ldots), \tag{3.6}$$

where ε_k are the eigenvalues of operator B with respect to operator A, and are determined by the formula of type (3.3)

$$\varepsilon_k = \frac{1}{2}\frac{(B\underset{\sim}{\phi}_k,\underset{\sim}{\phi}_k)}{(A\underset{\sim}{\phi}_k,\underset{\sim}{\phi}_k)} \ . \tag{3.7}$$

The generalized forces $Q_k(t)$ are given by the relation

$$Q_k = \frac{(\underset{\sim}{q},\underset{\sim}{\phi}_k)}{(A\underset{\sim}{\phi}_k,\underset{\sim}{\phi}_k)} \ . \tag{3.8}$$

The advantage of the method of generalized coordinates is that, instead of seeking the solutions of the operator equation (3.1) which may be of a complicated analytical nature (say, a system of partial differential equations) we use the solution of ordinary differential equations of a standard type. In order to calculate the characteristics of the desired functions $\underset{\sim}{u}(\underset{\sim}{x},t)$, we have only to perform operations over the series (3.5). If $\underset{\sim}{u}(\underset{\sim}{x},t)$ is a stochastic function of coordinates $\underset{\sim}{x}$ and time t, its probabilistic characteristics are determined by averaging the series (3.5), as well as the series which are obtained by multiplying these series taken at different values of $\underset{\sim}{x}$ and t. Thus, for mathematical expectations and moment functions of the second order, we have the formulae

$$
\begin{aligned}
\langle \underset{\sim}{u}(\underset{\sim}{x},t)\rangle &= \sum_k \langle u_k(t)\rangle \underset{\sim}{\phi}_k(\underset{\sim}{x}) \,, \\
\langle \underset{\sim}{u}(\underset{\sim}{x},t) \otimes \underset{\sim}{u}(\underset{\sim}{x}',t')\rangle &= \sum_j \sum_k \langle u_j(t) u_k(t')\rangle \underset{\sim}{\phi}_j(\underset{\sim}{x}) \otimes \underset{\sim}{\phi}_k(\underset{\sim}{x}') \,,
\end{aligned}
\tag{3.9}
$$

i.e., these characteristics are represented in terms of corresponding moment functions of generalized coordinates. Analogously, for the correlation vector-function at the output of the system (3.1), we have the formula

$$
K_u(\underset{\sim}{x},t;\underset{\sim}{x}',t') = \sum_j \sum_k K_{u_j u_k}(t,t') \underset{\sim}{\phi}_j(\underset{\sim}{x}) \otimes \underset{\sim}{\phi}_k(\underset{\sim}{x}') \,, \tag{3.10}
$$

where $K_{u_j u_k}(t,t')$ are the joint correlation functions of generalized coordinates. The probabilistic characteristics of generalized coordinates are in their turn expressed in terms of the corresponding characteristics of generalized forces. For example, for the system given by equations (3.6), relations (3.9) take the form

$$\langle \underset{\sim}{u}(\underset{\sim}{x},t)\rangle = \sum_k \underset{\sim}{\phi}_k(\underset{\sim}{x}) \int_{-\infty}^{t} h_k(t-\tau)\langle Q_k(\tau)\rangle d\tau \ ,$$

$$\langle \underset{\sim}{u}(\underset{\sim}{x},t) \otimes \underset{\sim}{u}(\underset{\sim}{x}',t')\rangle = \tag{3.11}$$

$$\sum_j \sum_k \underset{\sim}{\phi}_j(\underset{\sim}{x}) \otimes \underset{\sim}{\phi}_k(\underset{\sim}{x}') \int_{-\infty}^{t} \int_{-\infty}^{t'} h_j(t-\tau)h_k(t'-\tau')\langle Q_j(\tau)Q_k(\tau')\rangle d\tau d\tau' \ .$$

Here, analogously to (2.19), we denote

$$h_k(t-\tau) = \begin{cases} 0 & (t \leq \tau) \\ \dfrac{1}{\omega_{ke}} e^{-\varepsilon_k(t-\tau)} \sin \omega_{k\varepsilon}(t-\tau) & (t > \tau) \ , \end{cases} \tag{3.12}$$

$\omega_{k\varepsilon}$ being the natural frequencies of the system allowing for dissipation, i.e.,

$$\omega_{k\varepsilon} = (\omega_k^2 - \varepsilon_k^2)^{1/2} \ , \qquad (\varepsilon_k < \omega_k) \ . \tag{3.13}$$

Stationary Response to Stationary Action

Let $\underset{\sim}{q}(\underset{\sim}{x},t)$ and $\underset{\sim}{u}(\underset{\sim}{x},t)$ be stationary stochastic time functions. Applying the Fourier transformation to the relation (3.10) with respect to the variable $\tau = t'-t$, we obtain the formula

$$S_u(\underset{\sim}{x},\underset{\sim}{x}';\omega) = \sum_j \sum_k S_{u_j u_k}(\omega) \underset{\sim}{\phi}_j(\underset{\sim}{x}) \otimes \underset{\sim}{\phi}_k(\underset{\sim}{x}') \ . \tag{3.14}$$

This formula relates the time spectral output density $S_u(\underset{\sim}{x},\underset{\sim}{x}';\omega)$ to the matrix elements of the mutual spectral densities of generalized coordinates. If the latter are separable, the joint spectral densities $S_{u_j u_k}(\omega)$ are found from the formula

$$S_{u_j u_k}(\omega) = \frac{S_{Q_j Q_k}(\omega)}{(\omega_j^2 - 2i\varepsilon_j\omega - \omega^2)(\omega_k^2 + 2i\varepsilon_k\omega - \omega^2)} \ , \tag{3.15}$$

$S_{Q_jQ_k}(\omega)$ being the matrix elements of joint spectral densities of generalized forces, i.e.,

$$S_{Q_jQ_k}(\omega) = \frac{1}{2\pi} \int_{-\infty}^{\infty} \langle \tilde{Q}_j^*(t) \tilde{Q}_k(t+\tau) \rangle e^{-i\omega\tau} d\tau \ . \tag{3.16}$$

In the general case, when instead of the equations (3.6) we obtain an infinite system of non-separable equations, there arises a problem of reducing this system to a finite one. After the reduction, the joint spectral densities of generalized co-ordinates are determined by formulae of the type (2.54).

If $q(\underset{\sim}{x},t)$ and $u(\underset{\sim}{x},t)$ are scalar fields, it is easy to find the analytical relation between the spectral densities $S_{Q_jQ_k}(\omega)$ and the time spectral density of the field $q(\underset{\sim}{x},t)$. In this case,

$$(\underset{\sim}{q}, \underset{\sim}{\phi}_k) \equiv \int_G q(\underset{\sim}{x},t) \phi_k(\underset{\sim}{x}) d\underset{\sim}{x} \ .$$

Substituting this expression into (3.8), and the latter into (3.16), we find

$$S_{Q_jQ_k}(\omega) = \frac{1}{\nu_j^2 \nu_k^2} \int_G \int_G S_q(\underset{\sim}{x}, \underset{\sim}{x}'; \omega) \phi_j(\underset{\sim}{x}) \phi_k(\underset{\sim}{x}') d\underset{\sim}{x} d\underset{\sim}{x}' \ . \tag{3.17}$$

Here, the notation

$$\nu_k^2 = (A\underset{\sim}{\phi}_k, \underset{\sim}{\phi}_k) \tag{3.18}$$

is used. Let $v(\underset{\sim}{x},t)$ be some scalar field which can be represented by the series

$$v(\underset{\sim}{x},t) = \sum_k r_k(\underset{\sim}{x}) u_k(t) \ . \tag{3.19}$$

Here, the $u_k(t)$ are the generalized coordinates in expansion (3.5), and the $r_k(\underset{\sim}{x})$ are, generally, some complex-valued deterministic functions. Calculate the variance of the field $v(\underset{\sim}{x},t)$:

$$D[\underset{\sim}{v}(\underset{\sim}{x},t)] = \sum_j \sum_k r_j^*(\underset{\sim}{x}) r_k(\underset{\sim}{x}) \langle \tilde{u}_j(t) \tilde{u}_k(t) \rangle \ . \tag{3.20}$$

On the right side, we have elements of the correlation matrix of generalized coordinates at coinciding moments of time. Further below, for brevity, we shall omit the argument t. Using the relation between the elements of the correlation matrix and the joint spectral densities:

$$\langle \tilde{u}_j \tilde{u}_k \rangle = \int_{-\infty}^{\infty} S_{u_j u_k}(\omega) d\omega \ ,$$

and taking into account the relation (3.15), we obtain

$$\langle \tilde{u}_j \tilde{u}_k \rangle = \int_{-\infty}^{\infty} \frac{S_{Q_j Q_k}(\omega) d\omega}{(\omega_j^2 - 2i\varepsilon_j \omega - \omega^2)(\omega_k^2 + 2i\varepsilon_k \omega - \omega^2)} \ . \tag{3.21}$$

Thus, calculation of the variance (3.20) is reduced to calculation of the integrals (3.21) and to the summation of the double series.

Approximate Calculation of Integrals

If the spectral densities $S_{Q_j Q_k}(\omega)$ are functions varying sufficiently slowly as compared to the denominator of the integrand, the following method can be used for an approximate calculation of integrals.

Generally, the spectral densities $S_{Q_j Q_k}(\omega)$ are complex-valued functions of ω. Separating the real and imaginary parts we write them in the form

$$S_{Q_j Q_k}(\omega) = S^{(R)}_{Q_j Q_k}(\omega) + i S^{(I)}_{Q_j Q_k}(\omega) \ . \tag{3.22}$$

Since the functions $Q_k(t)$, $(k = 1,2,\ldots)$ are real, we have (see Section 1.1)

$$S^{(R)}_{Q_jQ_k}(-\omega) = S^{(R)}_{Q_jQ_k}(\omega) , \qquad S^{(I)}_{Q_jQ_k}(-\omega) = -S^{(I)}_{Q_jQ_k}(\omega) . \tag{3.23}$$

With the relations (3.22) and (3.23) taken into account, the formula (3.21) has the form

$$\langle \tilde{u}_j \tilde{u}_k \rangle = \int_{-\infty}^{\infty} \frac{[(\omega_j^2-\omega^2)(\omega_k^2-\omega^2)+4\varepsilon_j\varepsilon_k\omega^2]S^{(R)}_{Q_jQ_k}(\omega)d\omega}{[(\omega_j^2-\omega^2)^2+4\varepsilon_j^2\omega^2][(\omega_k^2-\omega^2)^2+4\varepsilon_k^2\omega^2]} + \int_{-\infty}^{\infty} \frac{2[(\omega_j^2-\omega^2)\varepsilon_k-(\omega_k^2-\omega^2)\varepsilon_j]\omega S^{(I)}_{Q_jQ_k}(\omega)d\omega}{[(\omega_j^2-\omega^2)^2+4\varepsilon_j^2\omega^2][(\omega_k^2-\omega^2)^2+4\varepsilon_k^2\omega^2]} . \tag{3.24}$$

The right side, as could be expected, is real and symmetric with respect to permutation of the indices.

Calculate the integrals in (3.24) on the assumption that damping is sufficiently small, i.e.,

$$\varepsilon_{jk} = \max\{\varepsilon_j,\varepsilon_k\} \ll \omega_{jk} , \tag{3.25}$$

and the frequencies ω_j and ω_k are sufficiently close:

$$\Delta\omega_{jk} = |\omega_j - \omega_k| \ll \omega_{jk} . \tag{3.26}$$

Here, ω_{jk} is a certain frequency from the interval $[\omega_j,\omega_k]$, for example,

$$\omega_{jk} = \frac{1}{2}(\omega_j+\omega_k) . \tag{3.27}$$

The spectral densities $S^{(R)}_{Q_jQ_k}(\omega)$ and $S^{(I)}_{Q_jQ_k}(\omega)$ are required to be slowly varying functions of ω so that their increments within the intervals of the order ε_{jk} and $\Delta\omega_{jk}$ can be neglected. This is shown in Figure 8, where the graphs of the spectral densities of the generalized forces $S^{(R)}_{Q_jQ_k}(\omega)$ and of the generalized displacements $S^{(I)}_{u_ju_k}(\omega)$ are given, and $j \neq k$.

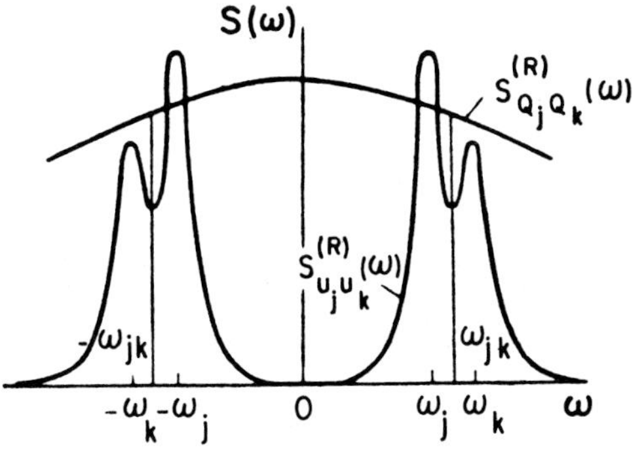

Figure 8

For an approximate calculation of the integrals in (3.24) we use the mean-value theorem, substituting the slowly varying expressions in the numerator by their values in the neighbourhood of sharp extremes of the transfer function and placing these values outside the integration sign. After the transformations, instead of (3.24), we obtain

$$\langle\tilde{u}_j\tilde{u}_k\rangle \approx S^{(R)}_{Q_jQ_k}(\omega_{jk})J^{(R)}_{jk} + \omega_{jk}S^{(I)}_{Q_jQ_k}(\omega_{jk})J^{(I)}_{jk} , \tag{3.28}$$

where the notations

$$J^{(R)}_{jk} = \int_{-\infty}^{\infty} \frac{F^{(R)}_{jk}(\omega)d\omega}{G_{jk}(\omega)} , \qquad J^{(I)}_{jk} = \int_{-\infty}^{\infty} \frac{F^{(I)}_{jk}(\omega)d\omega}{G_{jk}(\omega)} , \tag{3.29}$$

are being used. Under the integral sign are rational functions, and

$$F^{(R)}_{jk}(\omega) = (\omega_j^2-\omega^2)(\omega_k^2-\omega^2) + 4\varepsilon_j\varepsilon_k\omega^2 ,$$

$$F^{(I)}_{jk}(\omega) = 2\varepsilon_k(\omega_j^2-\omega^2) - 2\varepsilon_j(\omega_k^2-\omega^2) ,$$

$$G_{jk}(\omega) = [(\omega_j^2-\omega^2)^2+4\varepsilon_j^2\omega^2][(\omega_k^2-\omega^2)^2+4\varepsilon_k^2\omega^2] .$$

Using the residue theorem, we find that

$$J_{jk}^{(R)} = 2\pi i \sum_{\alpha=1}^{4} \frac{F_{jk}^{(R)}(\omega_\alpha)}{G'_{jk}(\omega_\alpha)} , \qquad J_{jk}^{(I)} = 2\pi i \sum_{\alpha=1}^{4} \frac{F_{jk}^{(I)}(\omega_\alpha)}{G'_{jk}(\omega_\alpha)} . \tag{3.30}$$

Here, ω_α are the poles of the integrand, located in the upper halfplane. These poles are $\pm\omega_{j\varepsilon}+i\varepsilon_j$, $\pm\omega_{k\varepsilon}+i\varepsilon_k$. The calculations according to the formulae (3.30) yield

$$\begin{aligned} J_{jk}^{(R)} &= \frac{4\pi(\varepsilon_j+\varepsilon_k)}{(\omega_j^2-\omega_k^2)^2+4(\varepsilon_j^2+\varepsilon_k^2)(\varepsilon_k\omega_j^2+\varepsilon_j\omega_k^2)} , \\ \omega_{jk}J_{jk}^{(I)} &= \frac{\pi\omega_{jk}}{\omega_j^2\omega_k^2} \frac{\omega^4{}_j-\omega_k^4+4(\varepsilon_j+\varepsilon_k)(\varepsilon_k\omega_j^2-\varepsilon_j\omega_k^2)}{(\omega_j^2-\omega_k^2)^2+4(\varepsilon_j+\varepsilon_k)(\varepsilon_k\omega_j^2+\varepsilon_j\omega_k^2)} . \end{aligned} \tag{3.31}$$

Formulae (3.28) and (3.31) are valid in the case when $j = k$. Here

$$\underset{\sim}{D}[u_k] \approx S_{Q_kQ_k}(\omega)J_{kk}^{(R)} . \tag{3.32}$$

The result coincides with (2.49) up to the notations used.

On Mutual Correlations of Generalized Coordinates

The application of the method of generalized coordinates is substantially simplified if the joint correlation of generalized coordinates can be neglected. To find the conditions under which such simplification is possible, we shall estimate the orders of the integrals $J_{jk}^{(R)}$ and $J_{jk}^{(I)}$ in the formula (3.28). Let

$$\begin{aligned} &\varepsilon_j \sim \varepsilon_k \sim \varepsilon_{jk} , \\ &S_{Q_jQ_j}(\omega_j) \sim S_{Q_kQ_k}(\omega_k) \sim S_{Q_jQ_k}^{(R)}(\omega_{jk}) \sim S_{Q_jQ_k}^{(I)}(\omega_{jk}) . \end{aligned} \tag{3.33}$$

The order of the variances of the generalized coordinates is determined from formula (3.32) by the value of the first integral (3.31) for $j = k$. The following estimate results:

$$J_{kk}^{(R)}(\omega_k) \sim \frac{1}{\varepsilon_k \omega_k^2} . \tag{3.34}$$

For $j \neq k$ there are three cases to be considered. If $|\omega_j - \omega_k|^2 \ll \varepsilon_{jk}^2$, i.e., if the distance between natural frequencies is substantially smaller than the bandwidth, then

$$J_{jk}^{(R)} \sim \frac{1}{\varepsilon_{jk}\omega_{jk}^2} , \qquad \omega_{jk} J_{jk}^{(I)} \sim \frac{\Delta\omega_{jk}}{\varepsilon_{jk}^2 \omega_{jk}^2} . \tag{3.35}$$

As the formulae (3.28), (3.34) and (3.35) show, the off-diagonal elements of the correlation matrix of the generalized coordinates under the above given conditions are of the same order as the diagonal elements. So, it is necessary here to take into account the joint correlation of the generalized coordinates. Here, the contribution of the imaginary part of the spectral density of the generalized forces is small as compared to the contribution of the real part (to be more exact it will give a correction of the order of $\Delta\omega_{jk}/\varepsilon_{jk}$).

If $|\omega_j - \omega_k|^2 \sim \varepsilon_{jk}^2$, the estimates following from the formulae (3.31) are

$$J_{jk}^{(R)} \sim \omega_{jk} J_{jk}^{(I)} \sim \frac{1}{\varepsilon_{jk}\omega_{jk}^2} . \tag{3.36}$$

For $|\omega_j - \omega_k|^2 \sim \varepsilon_{jk}^2$, the joint correlation of the generalized coordinates is to be taken into account. The contribution of the real and imaginary parts of the spectral densities of the generalized forces to the joint correlations of the generalized coordinates will be of the same order.

On satisfying the condition

$$|\omega_j-\omega_k|^2 \gg \varepsilon_{jk}^2 \tag{3.37}$$

for the integrals in the formula (3.28), we obtain the estimates

$$J_{jk}^{(R)} \sim \frac{\varepsilon_{jk}}{\Delta\omega_{jk}^2\omega_{jk}^2}, \qquad \omega_{jk}J_{jk}^{(I)} \sim \frac{1}{\Delta\omega_{jk}\omega_{jk}^2} . \tag{3.38}$$

The joint correlation moments of the generalized coordinates are of the order $\varepsilon_{jk}/\Delta\omega_{jk}$ or smaller as compared to the variances. The relative contribution of imaginary parts of the spectral densities of the generalized forces will be of the order $\Delta\omega_{jk}/\varepsilon_{jk}$ as compared to the contribution of real parts. Thus, as natural frequencies move farther away from one another, the role of imaginary parts of spectral densities increases (though the role of joint correlation of generalized coordinates decreases on the whole).

The above conclusions are valid in the case when the conditions (3.33) are satisfied. It is not difficult to modify these conclusions as applied to the case of essentially different coefficients of damping and to the case when principal and joint spectral densities of the generalized forces differ by one or more orders. The condition (3.37), and some other conclusions taking account of the joint correlation of generalized coordinates, were given in [13]. Some numerical examples will be considered in one of the following sections.

3.2 RANDOM VIBRATIONS IN LINEAR VISCOELASTIC SYSTEMS

Linear Viscoelastic Operators

The operation equation (3.1) is of a most general character. Yet, its application to real structures is not always possible, even if a structure can be treated as a linear system. An excessive idealization of dissipative forces is responsible for this. Taking into account dissipation by introducing terms proportional to the velocity into equation (3.1) results in such dependence of the relative dissipation on the frequency that this result does not agree with experimental data.

To get a more flexible method for describing energy dissipation in a system under deformation, the models of the linear theory of viscoelasticity should be used. Let us consider a material whose deformation follows the relations of the linear theory of viscoelasticity. Both elastic and dissipative properties of the system are described by means of some linear operator $\underline{C}$ analogous to the quasielastic operator C for a corresponding elastic system. Unlike the latter operator acting on the functions with respect to the spatial coordinates (time t is considered as a parameter), the viscoelastic operator acts on the functions with respect to spatial coordinates and time. It may be a differential operator with respect to t or an integral operator of the hereditary type.

In the problems of steady forced vibrations and of stationary random vibrations, it is sufficient to define the operator $\underline{C}$ as acting on the set of functions of the type $\underset{\sim}{U}(\underset{\sim}{x},\omega)e^{i\omega t}$. Here, ω is a real parameter (frequency), and $\underset{\sim}{U}(\underset{\sim}{x},\omega)$ is a deterministic or stochastic function of spatial coordinates and frequency. The operator $\underline{C}$ is introduced by the relation

$$\underline{C}[\underset{\sim}{U}(\underset{\sim}{x},\omega)e^{i\omega t}] = [C(\omega)\underset{\sim}{U}(\underset{\sim}{x},\omega)]e^{i\omega t} , \tag{3.39}$$

where $C(\omega)$ is a linear operator depending on the frequency ω as a parameter and acting on the functions with respect to the spatial coordinates. The domain of definition of operator $C(\omega)$, as a rule, coincides with the domain of definition of the quasielastic operator C for a corresponding elastic problem. This fact serves as a basis for one of the variants of the quasi-elastic analogy.

Separating the real and the imaginary parts in (3.39), we shall represent the Fourier transform of the viscoelastic operator as

$$C(\omega) = C_r(\omega) + iC_i(\omega) \ . \tag{3.40}$$

The real part $C_r(\omega)$ characterizes the elastic forces in the system, the imaginary part $C_i(\omega)$ represents the viscous forces.

Basic Relations

Let the equation of motion of the system have the form

$$A \frac{\partial^2 \underset{\sim}{u}}{\partial t^2} + \underset{\sim}{Cu} = \underset{\sim}{q} \ , \tag{3.41}$$

where $\underset{\sim}{u}(\underset{\sim}{x},t)$ and $\underset{\sim}{q}(\underset{\sim}{x},t)$ are stochastic functions of the coordinates and stationary stochastic functions of time with mathematical expectations equal to zero. Let us represent these functions as spectral expansions of the type (2.80),

$$\underset{\sim}{u}(\underset{\sim}{x},t) = \int_{-\infty}^{\infty} \underset{\sim}{U}(\underset{\sim}{x},\omega)e^{i\omega t}d\omega \ , \qquad \underset{\sim}{q}(\underset{\sim}{x},t) = \int_{-\infty}^{\infty} \underset{\sim}{Q}(\underset{\sim}{x},\omega)e^{i\omega t}d\omega \ . \tag{3.42}$$

For the stochastic functions $\underset{\sim}{U}(\underset{\sim}{x},\omega)$, we have the equation

$$[C(\omega)-\omega^2 A]\underset{\sim}{U}(\underset{\sim}{x},\omega) = \underset{\sim}{Q}(\underset{\sim}{x},\omega) \ , \tag{3.43}$$

which is the Fourier transform of equation (3.41). Further calculations are reduced to expanding the functions $\underset{\sim}{U}(\underset{\sim}{x},\omega)$ and $\underset{\sim}{Q}(\underset{\sim}{x},\omega)$ into series of the eigenvalue elements of the equation

$$[C_r(\omega)-\omega^2 A]\underset{\sim}{\phi} = \underset{\sim}{0} \ . \tag{3.44}$$

Frequency ω can enter the operator $C_r(\omega)$ in a complex way. Calculations are simplified in the case when the operator C_r

does not depend on ω. This corresponds to the assumption that dynamic elastic moduli in the entire considered range of frequencies are constant. This assumption is admissible for many engineering materials. The eigenvalue elements $\underset{\sim}{\phi}_k(\underset{\sim}{x})$ of equation (3.44) coincide with the modes of natural vibrations of the elastic system whose moduli are equal to the corresponding dynamic moduli of the viscoelastic system.

Let us consider the final results as applied to the problems for which the real part of operator $C(\omega)$ does not depend on ω, and the imaginary part is introduced by the relation

$$C_i(\omega) = \eta(\omega)C_r \, . \tag{3.45}$$

Here, $\eta(\omega)$ is some odd function of the frequency. This function represents the loss tangent $\chi(\omega)$ extended over the entire real axis of frequencies.* Thus,

$$\eta(\omega) = \chi(|\omega|)\,\mathrm{sign}\,\omega. \tag{3.46}$$

Without going into detailed calculations, we shall now write the formula for the spectral densities of the generalized coordinates:

$$S_{u_j u_k}(\omega) = \frac{S_{Q_j Q_k}(\omega)}{[\omega_j^2(1-i\eta)-\omega^2]\,[\omega_k^2(1+i\eta)-\omega^2]} \, . \tag{3.47}$$

Here, ω_j are natural frequencies of the elastic system with the operator C_r. The formula (3.47) is analogous to formula (3.15). Formulae (3.16), (3.17) and (3.20) remain unchanged. Instead of formula (3.21), we have

---

*In the theory of viscoelasticity (Rabotnov, Y., *Elements of Hereditary Mechanics of Solids*, Moscow, Nauka, 1977) the loss tangent is introduced as a non-negative function of the frequency ω, taking on values in the semi-infinite interval $[0,\infty)$.

$$\langle \tilde{u}_j \tilde{u}_k \rangle = \int_{-\infty}^{\infty} \frac{S_{Q_j Q_k}(\omega)\, d\omega}{[\omega_j^2(1-i\eta)-\omega^2][\omega_k^2(1+i\eta)-\omega^2]} \; . \tag{3.48}$$

Approximate Formulae for Joint Moments

The estimates for the integral (3.48) are derived analogously to those for the integral (3.21) in Section 3.1. Let $S^{(R)}_{Q_j Q_k}(\omega)$, $S^{(I)}_{Q_j Q_k}(\omega)$ and $\chi(\omega)$ be functions varying slowly within the intervals of the order $\eta\omega_{jk}$ and $\Delta\omega_{jk}$. Taking into account (3.22), we obtain instead of (3.28)

$$\langle \tilde{u}_j \tilde{u}_k \rangle \approx S^{(R)}_{Q_j Q_k}(\omega_{jk}) J^{(R)}_{jk} + \chi_{jk} S^{(I)}_{Q_j Q_k}(\omega_{jk}) J^{(I)}_{jk} \; , \tag{3.49}$$

where $\chi_{jk} = \chi(\omega_{jk})$, and the notation (3.27) is used. The integrals on the right side of formula (3.49) are

$$J^{(R)}_{jk} = \int_{-\infty}^{\infty} \frac{[(\omega_j^2-\omega^2)(\omega_k^2-\omega^2)+\chi^2_{jk}\omega_j^2\omega_k^2]\, d\omega}{[(\omega_j-\omega^2)^2+\chi^2_{jk}\omega_j^4][(\omega_k^2-\omega^2)^2+\chi^2_{jk}\omega_k^4]} \; ,$$

$$J^{(I)}_{jk} = \int_{-\infty}^{\infty} \frac{\omega^2(\omega_j^2-\omega_k^2)\, d\omega}{[(\omega_j^2-\omega^2)^2+\chi^2_{jk}\omega_j^4][(\omega_k^2-\omega^2)^2+\chi^2_{jk}\omega_k^4]} \; .$$

Using the residue theorem, we obtain

$$J^{(R)}_{jk} = \frac{\pi}{\sqrt{1+\chi^2_{jk}}} \, \frac{\omega_j+\omega_k}{\omega_j\omega_k} \, \frac{\alpha_{jk}\chi_{jk}(\omega_j^2+\omega_k^2)-\beta_{jk}(\omega_j-\omega_k)^2}{(\omega_j^2-\omega_k^2)^2+\chi^2_{jk}(\omega_j^2+\omega_k^2)^2} \; ,$$

$$\chi_{jk} J^{(I)}_{jk} = \frac{\pi}{\sqrt{1+\chi^2_{jk}}} \, \frac{\omega_j-\omega_k}{\omega_j\omega_k} \, \frac{\alpha_{jk}(\omega_j+\omega_k)^2+\beta_{jk}\chi_{jk}(\omega_j^2+\omega_k^2)}{(\omega_j^2-\omega_k^2)^2+\chi^2_{jk}(\omega_j^2+\omega_k^2)^2} \; , \tag{3.50}$$

and the notations used are

$$\alpha_{jk} = |\mathrm{Re}\sqrt{1+i\chi_{jk}}| \; , \qquad \beta_{jk} = |\mathrm{Im}\sqrt{1+i\chi_{jk}}|$$

Let us dwell briefly on taking account of joint correlations in problems of random vibrations in viscoelastic systems. The analysis of formulae (3.49) and (3.50) leads to conclusions analogous to the ones described above. In the estimates (3.33) to (3.38) it is sufficient to replace ε_{jk} by $\chi_{jk}\omega_{jk}$. Specifically, the condition to be satisfied for which the joint correlation of generalized coordinates can be neglected takes the form

$$|\omega_j - \omega_k|^2 \gg \chi_{jk}^2 \omega_{jk}^2 \ . \tag{3.52}$$

3.3 VIBRATIONS OF A PLATE IN A FIELD OF RANDOM PRESSURES

Statement of the Problem

Consider the application of the method of generalized coordinates to random vibrations of a plate of linear viscoelastic material. The equation of vibrations (3.41) has the form

$$\underline{D}\Delta\Delta w + \rho h \frac{\partial^2 w}{\partial t^2} = q \ , \tag{3.53}$$

where $w(x,y,t)$ is the deflection, $q(x,y,t)$ is the transversal load on the plate, ρ is the density of the plate material, h is the plate's thickness, $\underline{D}$ is the linear viscoelastic operator. Let the operator be specified by a relation of the type (3.39) on the set of solutions with the time factor $e^{i\omega t}$

$$\underline{D}[w_0(x,y)e^{i\omega t}] = D_r[1+i\eta(\omega)]\omega_0(x,y)e^{i\omega t} \ . \tag{3.54}$$

Here, D_r is the cylindrical stiffness not dependent on frequency and $\eta(\omega)$ is the loss tangent, defined on the range of frequency values $(-\infty,+\infty)$. For a material with full dissipation, the sign of the loss tangent coincides with the sign of ω. The modulus of the function $\eta(\omega)$ is denoted by $\chi(\omega)$ as in formula (3.46).

For a numerical example, we assume that the loading $q(x,y,t)$ in equation (3.53) is a stationary time function and a homogeneous function of the coordinates with a mathematical expectation equal to zero. For the time spectral density of pressure we take the expression

$$S_q(\xi,\zeta,\omega) = \Psi(\omega)\exp(-\alpha|\xi|-\beta|\zeta|) \ . \tag{3.55}$$

Here, $\Psi(\omega)$ is a real, non-negative frequency function, α and β are real, non-negative numbers or frequency functions, $\xi = x'-x$, and $\zeta = y'-y$. We seek the solution of the problem as the series (3.5), i.e.,

$$w(x,y,t) = \sum_k w_k(t)\phi_k(x,y) \ , \tag{3.56}$$

where $\phi_k(x,y)$ are the modes of free vibrations of an elastic plate. Further calculations are based on the formulae (3.14), (3.17), (3.48) and (3.49), ω_k being natural frequencies of an elastic plate with cylindrical stiffness D_r.

Rectangular Plate Supported Along Its Perimeter

The joint spectral densities of the generalized forces, $S_{Q_jQ_k}(\omega)$, enter in the formulae (3.48) and (3.49). According to relation (3.17), these spectral densities are expressed in terms of the time spectral density of loading and natural modes. For a rectangular plate simply supported along its perimeter, the modes of vibrations are ordered by two integers m and n, equal to the number of half-waves along the two sides of the plate. Therefore, it is expedient to replace the indices $j \to \{m,n\}$, $k \to \{r,s\}$ in the formulae (3.14), (3.17), etc. For example, instead of (3.14) we obtain

$$S_w(x,y;x',y';\omega) =$$

$$\sum_{m=1}^{\infty} \sum_{n=1}^{\infty} \sum_{r=1}^{\infty} \sum_{s=1}^{\infty} S_{u_{mn}u_{rs}}(\omega) \sin\frac{m\pi x}{a} \sin\frac{n\pi y}{b} \sin\frac{r\pi x'}{a} \sin\frac{s\pi y'}{b} , \tag{3.57}$$

and instead of formula (3.47)

$$S_{u_{mn}u_{rs}}(\omega) = \frac{S_{Q_{mn}Q_{rs}}(\omega)}{[\omega_{mn}^2(1-i\eta)-\omega^2][\omega_{rs}^2(1+i\eta)-\omega^2]} ; \tag{3.58}$$

and here,

$$\omega_{mn} = \left(\frac{m^2\pi^2}{a^2} + \frac{n^2\pi^2}{b^2}\right)\left(\frac{D_r}{\rho h}\right)^{1/2} . \qquad 3.59)$$

Substituting the expression (3.55) into formula (3.17), we obtain

$$S_{Q_{mn}Q_{rs}}(\omega) = \frac{\Psi(\omega)}{\rho^2 h^2} I_{mr}(\omega) J_{ns}(\omega) , \tag{3.60}$$

where the notations

$$I_{mr}(\omega) = \frac{4}{a^2} \int_0^a \int_0^a \exp(-\alpha|x'-x|) \sin\frac{m\pi x}{a} \sin\frac{r\pi x'}{a} dx dx' ,$$

$$J_{ns}(\omega) = \frac{4}{b^2} \int_0^b \int_0^b \exp(-\beta|y'-y|) \sin\frac{n\pi y}{b} \sin\frac{s\pi y'}{b} dy dy' ,$$

are used.

Without going into detailed calculations we write the final result

$$I_{mr}(\omega) = \begin{cases} \dfrac{4}{m^2\pi^2}\left\{\dfrac{a\alpha}{1+\alpha_m^2} + \dfrac{2[1-(-1)^m e^{-\alpha a}]}{(1+\alpha_m^2)^2}\right\} & (m = r) , \\[2ex] \dfrac{8}{mr\pi^2} \dfrac{1-(-1)^m e^{-\alpha a}}{(1+\alpha_m^2)(1+\alpha_r^2)} & \begin{pmatrix} m \neq r \\ |m \pm r| \text{ - even} \end{pmatrix} , \\[2ex] 0 & \begin{pmatrix} m \neq r \\ |m \pm r| \text{ - odd} \end{pmatrix} , \end{cases}$$

$$\alpha_m = \alpha a/(m\pi) . \tag{3.61}$$

We obtain analogous formulae for the integrals $J_{rs}(\omega)$. The result can be obtained from the relations (3.61) if we replace: $m \to n$, $r \to s$, $a \to b$, $\alpha \to \beta$, $\alpha_m \to \beta_n = \beta b/(n\pi)$, and $\alpha_r \to \beta_s = \beta b/(s\pi)$.

Discussion of Numerical Results

The results of some calculations for $a = b$, $\alpha = \beta$, χ = const., and $\Psi(\omega) = \Psi_0$ = const. are given below. To express the results in a nondimensional form, the following typical parameter values are used: wave number k_0, frequency ω_0, deflection w_0 and stress σ_0. These parameters are defined by

$$k_0 = \frac{\pi}{a}, \qquad \omega_0 = \frac{\pi^2}{a^2}\left(\frac{D_r}{\rho h}\right)^{1/2},$$

$$\omega_0 = \frac{\Psi_0}{\rho^2 h^2 \omega_0^2 \chi} = \frac{\Psi_0 \omega_0}{D_r^2 k_0^6 \chi}\left(\frac{D_r}{\rho h}\right)^{1/2}, \tag{3.62}$$

$$\sigma_0^2 = \frac{36\Psi_0(1+\chi^2)\omega_0}{k_0^4 h^4 \chi} = \frac{36\Psi_0(1+\chi^2)}{k_0^2 h^4 \chi}\left(\frac{D_r}{\rho h}\right)^{1/2}.$$

In Figure 9, the graphs of the spectral density $S_w(\omega)$ for the deflection in the centre of the plate are given. The graphs are constructed according to formulae (3.57) to (3.61) for the case $\chi = 0.1$ for two values of the correlation scale: $\alpha a = \pi$ and $\alpha a = 10\pi$. The values of the spectral density are shown in logarithmic scale. Each maximum of the function $S_w(\omega)$ corresponds to one of the natural frequencies of the plate. The variation of the functions in the vicinity of these maxima, as can also be seen from formula (3.58), is of the order of magnitude χ^{-2}. The analogous graph for the spectral density $S_\sigma(\omega)$ of the normal stress

$$\sigma = \pm \frac{6D_r(1+i\eta)}{h^2}\left(\frac{\partial^2 w}{\partial x^2} + \nu \frac{\partial^2 w}{\partial y^2}\right) \tag{3.63}$$

in the centre of the plate (in points most distant from its middle plane), is given in Figure 10.

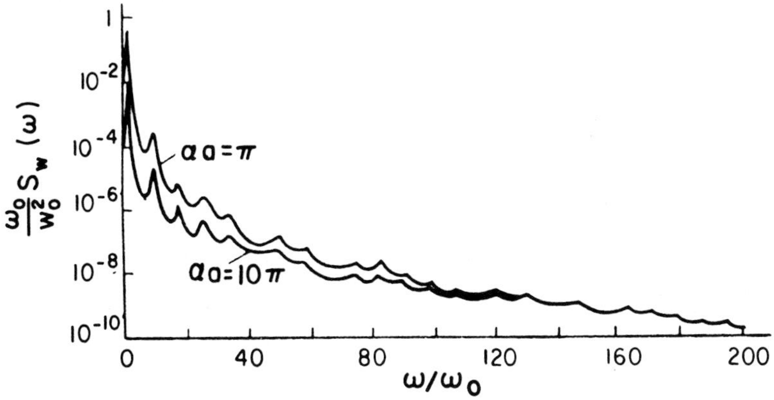

Figure 9

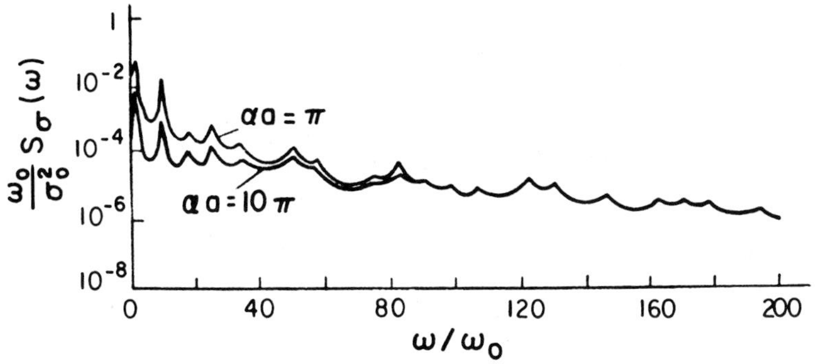

Figure 10

In Figure 11, the graph of the integral of the spectral density

$$\underset{\sim}{D}[w|\omega_*] = \int_{-\omega_*}^{\omega_*} S_W(\omega)d\omega \ , \qquad (3.64)$$

is given. This integral can be treated as the deflection variance for the loading of the type of truncated white noise with spectral density Ψ_0 and frequency bound ω_*. The graph is constructed for

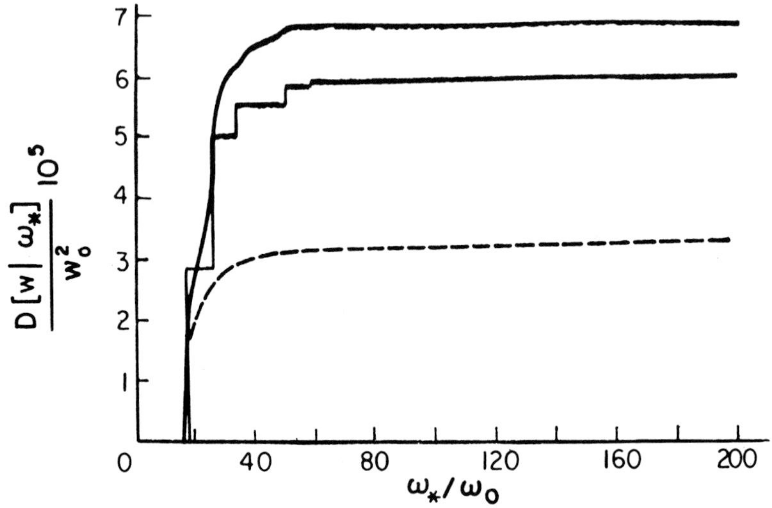

Figure 11

$\chi = 0.1$, $\alpha a = 1$. The continuous curve is obtained by a numerical integration of the spectral density $S_w(\omega)$. The stepped line represents the result of the summation of the series (3.20), with the moments of generalized coordinates substituted into it according to the approximate formulae (3.49) and (3.50). The dashed line is obtained by the same formulae, but without taking account of the joint correlations of generalized coordinates. In Figure 11, there is a good agreement between the result of the numerical integration of the spectral density and the result of the summation of the series, whose terms have been obtained by an approximate analytical integration. The analogous graph for the integral

$$\underset{\sim}{D}[\sigma|\omega_*] = \int_{-\omega_*}^{\omega_*} S_\sigma(\omega)\,d\omega \tag{3.65}$$

is given in Figure 12. In all the examples given above, the error caused by neglecting the mutual correlations of generalized coordinates is of an order agreeing with the estimates from Section 3.2. Various problems regarding stationary random vibrations of elastic systems are considered in [23,38,78,99,125,138,152,167] and in other references.

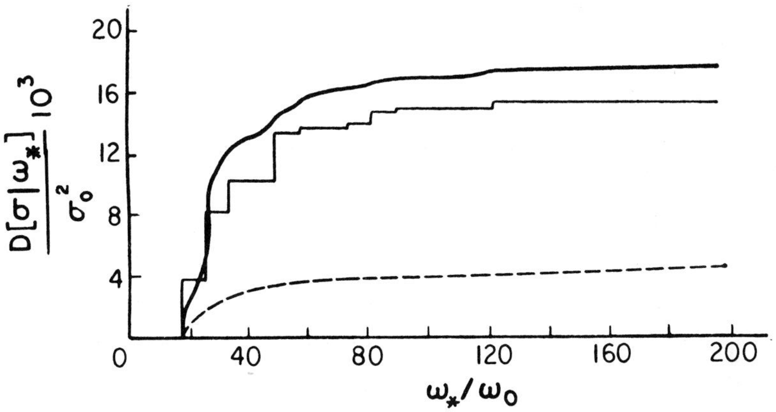

Figure 12

3.4 RANDOM VIBRATIONS OF SHELLS CONTAINING COMPRESSIBLE FLUID

Basic Assumptions

In this section, the method of generalized coordinates will be used for the analysis of random vibrations of thin elastic shells containing compressible fluid [16]. The task is to determine the probabilistic characteristics of the vibration field in a shell and in the fluid if probabilistic characteristics of the forces acting on the shell are known.

Consider a closed shell loaded by external forces normal to its surface. Let the intensity of these forces be a stationary random function of time, and, generally, an arbitrary random function of the coordinates of the median surface. The probabilistic characteristics of this function will be considered as given.

The shell consists of an external (carrying) layer and an internal layer. The objective of the latter is to damp the transmission of vibrations into the internal region. The internal shell may consist of several layers, is soft and therefore will be called insulation. The external shell is governed by the equations of the classical theory of thin shells, the only difference being that elastic constants are substituted by the corresponding linear viscoelastic operators. The insulating shell is made of some linear dissipative material. The normal displacements on the surfaces of the insulating shell are related to pressures on these surfaces by some operator equations. It is also assumed that perturbations in the insulating shell propagate in the normal direction, i.e., the relations between displacements and pressures are of a local character. The explicit operator expressions are not written down, since in the solution their Fourier transforms (vibration impedances) are used, which may be given in the form of numerical tables or graphs.

The interior of the shell is filled with a compressible fluid whose perturbations will be considered as small and having a potential. The losses in this medium and the radiation of the sound into the external medium can be neglected. Or, rather, we shall say that the losses due to the radiation are either insignificant (as, for instance, in the case when the external medium is of small density), or included in the general vibration losses of the carrying shell.

The above given scheme of assumptions corresponds to the design scheme for the fuselage structure of an aircraft whose surface is subjected to random pressure pulsations caused by a turbulent boundary layer. Although the frequencies of these pulsations are in the sonic range, they are actually of a pseudosonic character (see Section 1.3). We shall consider the probabilistic characteristics of pressure pulsations on the shell surface, which are determined by laboratory measurements or by measurements taken on the structure, as given. The losses due to radiation can be neglected as compared to the losses in the material of the carrying and insulating layers because of the great difference in the air densities inside and outside the cabin. A somewhat different approach to an analogous engineering problem can be found in [131].

Differential Equations and Boundary Conditions

First we shall formulate a corresponding deterministic problem. We introduce the curvilinear coordinates $\underset{\sim}{\xi} = \{\xi_1, \xi_2\}$ on the median surface of the carrying shell Ω. Inside the shell, we use the spatial curvilinear system $\underset{\sim}{x} = \{\xi_1, \xi_2, x_3\}$, adding the third coordinate x_3 evaluated along the external normal to Ω, to the coordinates $\{\xi_1, \xi_2\}$ (Figure 13).

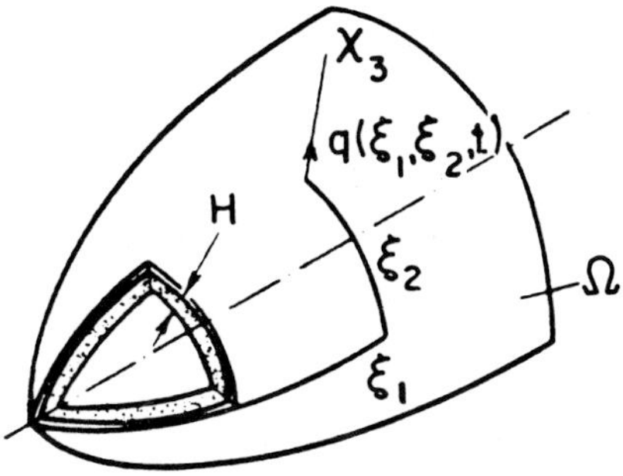

Figure 13

We now formulate the equations for the displacements $u_j(\underset{\sim}{\xi}, t)$ on the median surface of the carrying layer Ω and the velocity potential $\phi(\underset{\sim}{x}, t)$ in the volume V filled with compressible fluid. The shell equations take the form:

$$\sum_{k=1}^{3} \ell_{jk} u_k + \rho_0 h_0 \frac{\partial^2 u_j}{\partial t^2} = (q + p_0)\delta_{j3} , \quad (j = 1,2,3) . \quad (3.66)$$

Here, ℓ_{jk} are the shell theory operators extended to the viscoelastic case, ρ_0 is the material density of the carrying shell, h_0 is its thickness, q is the intensity of the external pressure on the carrying shell, and p_0 is the contact pressure on the boundary between the carrying and the insulating shell. Some of the boundary conditions of shell theory should be added to equation (3.66). The potential $\phi(\underset{\sim}{x},t)$ inside V satisfies the wave equation

$$\Delta\phi - \frac{1}{c^2}\frac{\partial^2\phi}{\partial t^2} = 0 , \tag{3.67}$$

where c is the acoustic velocity in the medium inside the shell.

We shall denote the normal displacements on the surfaces of the insulating shell, $x_3 = 0$ and $x_3 = -H$, by w and w_H. The kinematic conditions on these surfaces have the form

$$w = u_3 , \quad (x_3 = 0) , \quad \partial w_H/\partial t = \partial\phi/\partial x_3 , \quad (x_3 = -H) . \tag{3.68}$$

The dynamic condition for $x_3 = -H$ should be added to conditions (3.68); also, the expression for p_0 is included in equation (3.66). The relation between the pressures p_0 and p_H on the surfaces $x_3 = 0$ and $x_3 = -H$, and the normal displacements of the corresponding points, will be given as

$$p_0 = -(m_{11}w+m_{12}w_H) , \quad p_H = m_{21}w + m_{22}w_H . \tag{3.69}$$

Here, m_{jk} are some linear operators. Formulae (3.69) impose restrictions on the choice of a model for the insulating shell. In particular, models accounting for shear are to be excluded from consideration. However, most of the models of sound insulation theory can be described by (3.69). With consideration of (3.69), the additional boundary condition will be

$$-\rho\frac{\partial\phi}{\partial t} = m_{21}w + m_{22}w_H , \quad (x_3 = -H) , \tag{3.70}$$

(ρ is the fluid density).

It is often possible to exclude the tangential displacements u_1 and u_2 from equation (3.66). The approximation of the engineering theory of shells neglecting the tangential forces of inertia may serve as an example. Thus, we arrive at the following equation for the normal displacement

$$\ell w = q - (m_{11}w + m_{12}w_H) \ . \tag{3.71}$$

Here, ℓ is some linear operator. Henceforth, we make use of equation (3.71).

The boundary conditions for the carrying shell and, if necessary, the conditions for the potential $\phi(\underset{\sim}{x},t)$ are added to equations (3.67) and (3.71) and to the boundary conditions (3.68) and (3.70). For instance, if part Ω_0 of the surface bounding V is acoustically "stiff" or acoustically "soft", then one of the following two conditions, respectively, is prescribed on this part:

$$\left.\frac{\partial\phi}{\partial n}\right|_{\Omega_0} = 0 \ , \qquad \text{or} \qquad \left.\frac{\partial\phi}{\partial t}\right|_{\Omega_0} = 0 \ . \tag{3.72}$$

In a general case, conditions (3.72) are substituted by conditions of the impedance type.

Formulation of the Stochastic Boundary Value Problem

Let the loading $q(\underset{\sim}{\xi},t)$ form a random space-time field, stationary in time, and generally nonuniform on Ω. Since the problem is linear, and the mathematical expectations of the fields $q(\underset{\sim}{\xi},t)$, $w(\underset{\sim}{\xi},t)$, $w_H(\underset{\sim}{\xi},t)$ and $\phi(\underset{\sim}{x},t)$ are independent of time, we shall consider the case when

$$\langle q(\underset{\sim}{\xi},t)\rangle = \langle w(\underset{\sim}{\xi},t)\rangle = \langle w_H(\underset{\sim}{\xi},t)\rangle = \langle\phi(\underset{\sim}{x},t)\rangle = 0 \ .$$

We solve the stochastic boundary problem in correlational approximation, assuming the correlation function of the external pressure as given. We shall also need the Fourier transformation of this function with respect to time, which is

$$S_q(\underset{\sim}{\xi},\underset{\sim}{\xi}';\omega) = \frac{1}{2\pi}\int_{-\infty}^{\infty} \langle q^*(\underset{\sim}{\xi},t)q(\underset{\sim}{\xi}',t+\tau)\rangle e^{-i\omega\tau}d\tau \ . \qquad (3.73)$$

The task is to find an analogous function for the pulsation component of the pressure inside the shell

$$p = -\rho\ \partial\phi/\partial t \ . \qquad (3.74)$$

This function is introduced by the relation

$$S_p(\underset{\sim}{x},\underset{\sim}{x}';\omega) = \frac{1}{2\pi}\int_{-\infty}^{\infty} \langle p^*(\underset{\sim}{x},t)p(\underset{\sim}{x}',t+\tau)\rangle e^{-i\omega\tau}d\tau \ . \qquad (3.75)$$

Having this function, we can calculate the contribution of pulsations with frequencies from the interval $[\omega_*,\omega_{**}]$ to the value of the mean square of the pulsation pressure

$$\underset{\sim}{D}[p(\underset{\sim}{x};\omega_*,\omega_{**})] = 2\int_{\omega_*}^{\omega_{**}} S_p(\underset{\sim}{x},\underset{\sim}{x};\omega)d\omega \ . \qquad (3.76)$$

The final result of the calculation is the noise level in decibels

$$L(\underset{\sim}{x};\omega_*,\omega_{**}) = 10\log\{\underset{\sim}{D}[p(\underset{\sim}{x};\omega_*,\omega_{**})]/p_*^2\} \ , \qquad (3.77)$$

where p_* is the threshold pressure (Section 1.3).

Application of the Method of Spectral Expansions

Let $\psi_\alpha(\underset{\sim}{\xi})$ be the natural modes of vibrations of a corresponding elastic shell in a vacuum. We shall represent the displacement functions $w(\underset{\sim}{\xi},t)$ and $w_H(\underset{\sim}{\xi},t)$ as Fourier series in terms of the functions $\psi_\alpha(\underset{\sim}{\xi})$ and as stochastic Fourier integrals with respect to time,

$$w(\underset{\sim}{\xi},t) = \sum_{\alpha} \int_{-\infty}^{\infty} W_{\alpha}(\omega)\psi_{\alpha}(\underset{\sim}{\xi})e^{i\omega t}d\omega ,$$

$$w_H(\underset{\sim}{\xi},t) = \sum_{\alpha} \int_{-\infty}^{\infty} W_{H\alpha}(\omega)\psi_{\alpha}(\underset{\sim}{\xi})e^{i\omega t}d\omega . \qquad (3.78)$$

Here, $W_{\alpha}(\omega)$ and $W_{H\alpha}(\omega)$ are generalized stochastic functions (the spectra of the corresponding displacements). The system of functions $\psi_{\alpha}(\underset{\sim}{\xi})$ is complete and orthonormal on Ω and satisfies all the necessary boundary conditions. Let us represent the external loading in an analogous form:

$$q(\underset{\sim}{\xi},t) = \sum_{\alpha} \int_{-\infty}^{\infty} Q_{\alpha}(\omega)\psi_{\alpha}(\underset{\sim}{\xi})e^{i\omega t}d\omega . \qquad (3.79)$$

Let us expand the velocity potential $\phi(\underset{\sim}{x},t)$ into a Fourier series of the solutions $\chi_{\alpha}(\underset{\sim}{x};\omega)$ of the Helmholtz equation

$$\Delta\chi + \frac{\omega^2}{c^2}\chi = 0 , \qquad (3.80)$$

the solutions satisfying conditions of the type (3.72) and the conditions of boundedness inside V. Let the region V satisfy the condition that the normal coordinate x_3 can be separated in the solutions $\chi_{\alpha}(\underset{\sim}{x};\omega)$, i.e.,

$$\chi_{\alpha}(\underset{\sim}{x};\omega) = \phi_{\alpha}(\underset{\sim}{\xi})Z_{\alpha}(x_3,\omega)/Z_{\alpha}(-H,\omega) . \qquad (3.81)$$

Here, $Z_{\alpha}(x_3,\omega)$ is a certain function of the normal coordinate and the frequency, and the normalizing factor is introduced for the sake of a certain simplification in the subsequent representation of formulae. As before, we shall use the stochastic Fourier integral expansion with respect to time. Thus,

$$\phi(\underset{\sim}{x},t) = \sum_{\infty} \int_{-\infty}^{\infty} \Phi_{\alpha}(\omega)\chi_{\alpha}(\underset{\sim}{x};\omega)e^{i\omega t}d\omega . \qquad (3.82)$$

The substitution of expressions (3.78), (3.79) and (3.82) into the equations and boundary conditions leads to a certain system of linear algebraic equations for the spectra $W_\alpha(\omega)$, $W_{H\alpha}(\omega)$ and $\Phi_\alpha(\omega)$. Then, the probabilistic characteristics of the vibration fields in the system can be calculated by conventional methods.

Separation of Variables

Let us consider the case when a system of equations for the spectra decomposes into separate independent groups, containing spectra with identical index α. This is evidently possible if

$$\phi_\alpha(\underset{\sim}{\xi}) \equiv \psi_\alpha(\underset{\sim}{\xi}) \ , \tag{3.83}$$

i.e., if the surface modes in the solution of the Helmholtz equations (3.80) coincide with the modes of natural vibrations of the carrying shell in vacuum. Such separation is possible in a number of cases which are of interest. For instance, it is possible in the case of a closed spherical shell, or an infinitely long circular cylindrical shell, or an infinite plate bounding a layer of constant thickness, etc. In all these cases, the parameters are assumed to be constant on the entire median surface, and also certain restrictions on operators ℓ and m_{jk} are to be imposed.

Let condition (3.83) be satisfied. Performing the Fourier transformation with respect to time on the equations and the boundary conditions, we obtain the system of equations for the spectra

$$\begin{gathered}(L_\alpha+M_{11})W_\alpha + M_{12}W_{H\alpha} = Q_\alpha \ , \\ \Phi_\alpha \frac{\zeta'_\alpha}{\zeta_\alpha} = i\omega W_{H\alpha} \ , \qquad -i\rho\omega\Phi_\alpha = M_{12}W_\alpha + M_{22}W_{H\alpha} \ .\end{gathered} \tag{3.84}$$

Here, L_α and M_{jk} are the Fourier transforms of the corresponding operators, i.e., functions of frequency ω satisfying the relations

$$\ell(\psi_\alpha e^{i\omega t}) = L_\alpha(\omega)\psi_\alpha e^{i\omega t} ,$$
$$m_{jk}(\psi_\alpha e^{i\omega t}) = M_{jk}(\omega)\psi_\alpha e^{i\omega t} . \tag{3.85}$$

In equations (3.84), the notations

$$\zeta_\alpha(\omega) = Z_\alpha(-H,\omega) , \quad \zeta'_\alpha(\omega) = \left.\frac{\partial Z_\alpha(x_3,\omega)}{\partial x_3}\right|_{x_3=-H} \tag{3.86}$$

are introduced. Solving the equations, we obtain

$$W_\alpha(\omega) = F_\alpha(i\omega)Q_\alpha(\omega) , \qquad \Phi_\alpha(\omega) = G_\alpha(i\omega)Q_\alpha(\omega) , \tag{3.87}$$

where $F_\alpha(i\omega)$ and $G_\alpha(i\omega)$ are transfer functions:

$$F_\alpha(i\omega) = [L_\alpha+M_{11}-M_{12}M_{21}M_\alpha^{-1}\zeta'_\alpha(\omega)]^{-1} ,$$
$$G_\alpha(i\omega) = -i\omega\zeta_\alpha(\omega)M_{21}M_\alpha^{-1}F_\alpha , \tag{3.88}$$
$$M_\alpha = M_{22}\zeta'_\alpha(\omega) - \rho\omega^2\zeta_\alpha(\omega) .$$

The formula for the pressure in the fluid is given below. Taking account of relation (3.74), we find

$$p(\underset{\sim}{x},t) = \sum_\alpha \int_{-\infty}^{\infty} P_\alpha(\omega)\chi_\alpha(\underset{\sim}{x},\omega)e^{i\omega t}d\omega ,$$
$$P_\alpha(\omega) = H_\alpha(i\omega)Q_\alpha(\omega) , \qquad H_\alpha(i\omega) = -\rho\omega^2\zeta_\alpha(\omega)M_{21}M_\alpha^{-1}F_\alpha . \tag{3.89}$$

Let us introduce the following probabilistic characteristics of the loading $q(\underset{\sim}{\xi},t)$,

$$S_{Q_\alpha Q_\beta}(\omega) = \frac{1}{\nu_\alpha^2 \nu_\beta^2} \int_\Omega \int_\Omega S_q(\underset{\sim}{\xi}, \underset{\sim}{\xi}';\omega)\psi_\alpha^*(\underset{\sim}{\xi})\psi_\beta(\underset{\sim}{\xi}')d\underset{\sim}{\xi}d\underset{\sim}{\xi}' .$$

Here, $S_q(\underset{\sim}{\xi}, \underset{\sim}{\xi}';\omega)$ is the time spectral density of the loading (3.73), ν_α is the norm of the function $\psi_\alpha(\underset{\sim}{\xi})$, analogous to (3.18). The characteristics $S_{Q_\alpha Q_\beta}(\omega)$ are the joint spectral densities of generalized forces, their relations to the spectra of the loading $Q_\alpha(\omega)$ being given by the formula

$$\langle Q_\alpha^*(\omega)Q_\beta(\omega')\rangle = S_{Q_\alpha Q_\beta}(\omega)\delta(\omega-\omega') . \tag{3.90}$$

Let us calculate the correlation function of the displacement $w(\underset{\sim}{\xi},t)$ in the carrying shell. Using the formulae (3.78) and (3.87), and taking account of (3.90), we obtain

$$\langle w^*(\underset{\sim}{\xi},t)w(\underset{\sim}{\xi}',t+\tau)\rangle = \int_{-\infty}^{\infty} \sum_\alpha \sum_\beta F_\alpha^*(i\omega)F_\beta(i\omega)S_{Q_\alpha Q_\beta}(\omega)\psi_\alpha^*(\underset{\sim}{\xi})\psi_\beta(\underset{\sim}{\xi}')e^{i\omega\tau}d\omega . \tag{3.91}$$

Analogously, according to the formulae (3.89),

$$\langle p^*(\underset{\sim}{x},t)p(\underset{\sim}{x}',t+\tau)\rangle = \int_{-\infty}^{\infty} \sum_\alpha \sum_\beta H_\alpha^*(i\omega)H_\beta(i\omega)S_{Q_\alpha Q_\beta}(\omega)\chi_\alpha^*(\underset{\sim}{x};\omega)\chi_\beta(\underset{\sim}{x}';\omega)e^{i\omega\tau}d\omega \tag{3.92}$$

Using the formulae (3.91) and (3.92), it is not difficult to estimate other probabilistic characteristics. For instance, we shall obtain the formulae for variances from (3.91) and (3.92), assuming in them that $\underset{\sim}{\xi} = \underset{\sim}{\xi}'$, $\underset{\sim}{x} = \underset{\sim}{x}'$, $\tau = 0$. The time spectral densities for displacements and pressures are determined as

$$\begin{aligned} S_w(\underset{\sim}{\xi},\underset{\sim}{\xi}';\omega) &= \sum_\alpha \sum_\beta S_{W_\alpha W_\beta}(\omega)\psi_\alpha^*(\underset{\sim}{\xi})\psi_\beta(\underset{\sim}{\xi}') , \\ S_p(\underset{\sim}{x},\underset{\sim}{x}';\omega) &= \sum_\alpha \sum_\beta S_{P_\alpha P_\beta}(\omega)\chi_\alpha^*(\underset{\sim}{x};\omega)\chi_\beta(\underset{\sim}{x}';\omega) , \end{aligned} \tag{3.93}$$

where the following notations for the corresponding joint spectral densities were used:

$$S_{W_\alpha W_\beta}(\omega) = F^*_\alpha(i\omega)F_\beta(i\omega)S_{Q_\alpha Q_\beta}(\omega) ,$$
$$S_{P_\alpha P_\beta}(\omega) = H^*_\alpha(i\omega)H_\beta(i\omega)S_{Q_\alpha Q_\beta}(\omega) . \qquad (3.94)$$

Discussion of Qualitative Results

Joint spectral densities are functions of the parameter ω. Let us consider, for example, the spectral density $S_{W_\alpha W_\beta}(\omega)$. Its poles coincide with the poles of the transfer function $F_\alpha(i\omega)$. In the case of sufficiently small damping, these poles are close to the real axis. Each pole with $\omega > 0$ corresponds to one of the frequencies of free vibrations of the simultaneous system. Therefore, the equation Re $F_\alpha^{-1}(i\omega) = 0$ can be considered as the equation of the natural frequencies of this system. The degree of interdependence between elastic and acoustic oscillations is connected with the relations between the frequencies of the corresponding partial systems. The relation between the frequency ω_α of natural vibrations of the carrying shell, defined as a root of the equation Re $L_\alpha(\omega) = 0$, and the frequencies ω_{α_1}, ω_{α_2}, ..., of acoustic vibrations in the absolutely stiff carrying shell is most essential. These frequencies are determined from the equation Re $M_\alpha(\omega) = 0$, i.e.,

$$\mathrm{Re}[M_{22}\zeta'_\alpha(\omega) - \rho\omega^2\zeta_\alpha(\omega)] = 0 . \qquad (3.95)$$

If the insulation is sufficiently stiff ($w_H \approx 0$), the roots of the equation (3.95) are close to the roots of the equation $\zeta'_\alpha(\omega) = 0$. In the case of a soft insulation ($p_H \approx 0$) they are close to the roots of the equation $\zeta_\alpha(\omega) = 0$.

The function $S_{W_\alpha W_\alpha}(\omega)$ is shown in Figure 14 (curve 1). The influence of the acoustic medium is usually neglected when elastic vibrations of the shell are analyzed, assuming approximately that

$$F_\alpha \approx (L_\alpha + M_{11} - M_{12}M_{21}M_{22}^{-1})^{-1}$$

or even $F_\alpha \approx L_\alpha^{-1}$. The function corresponding to the latter approximation is shown by a dashed line in Figure 14.

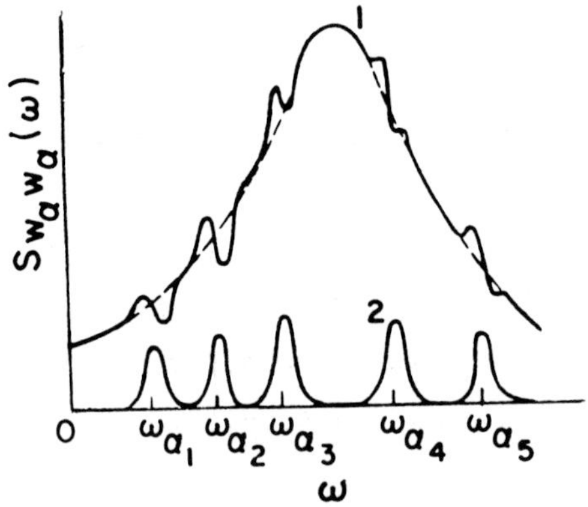

Figure 14

The poles of the spectral density $S_{p_\alpha p_\alpha}(\omega)$ coincide with the poles of the transfer function $H_\alpha(i\omega)$. Since the stiffness of the shell is substantially greater than the stiffness of the acoustic medium, these poles will be located close to the zeros of the function $M_\alpha(\omega)$. The condition of coincidence for the roots of the equations $L_\alpha(\omega) = 0$ and $M_\alpha(\omega) = 0$ is of the greatest interest. The function $S_{p_\alpha p_\alpha}(\omega)$ is also shown in Figure 14 (curve 2).

3.5 APPROXIMATE METHOD OF ANALYSIS

Basic Idea of the Method

The characteristics of the field inside the shell are determined by the formulae (3.76), (3.77), (3.93) and (3.94). The immediate application of these formulae presents difficulties, since the transfer functions $H_\alpha(i\omega)$ turn out to be complicated transcendental frequency functions. Besides, these functions vary very rapidly in the vicinity of the points on the real axis adjacent to the poles. When damping is small, and the density of natural frequencies high, the increment of the functions within a very small frequency interval can be of several orders of magnitude, thus making the numerical determination of the integrals in the formulae (3.76) practically impossible. To overcome these difficulties, it is possible to apply an approximation method similar to the saddle-point method. The essence of the method is that the rapidly varying part of the integrand in the neighbourhood of each pole is represented as a power series. After the application of the mean value theorem and the extension of integration limits to the entire real axis, the integrals are estimated analytically. Then, the contributions of each pole are summed up.

Derivation of the Basic Formulae

Consider the expression for the transfer function $H_\alpha(i\omega)$. Due to the high quality of the system, the rapidly varying functions in these expressions will be $\zeta'_\alpha(\omega)$ and $M_\alpha(\omega)$. Note that when the insulating layer is sufficiently stiff, the roots $\omega_{\alpha\gamma}$ of the equation $\mathrm{Re}\, M_\alpha(\omega) = 0$ are close to the corresponding roots $\Omega_{\alpha\gamma}$ of the equation $\zeta'_\alpha(\omega) = 0$. Therefore, let us expand the latter expression into a power series in the neighbourhood of the frequency $\Omega^2_{\alpha\gamma}$:

$$\zeta'_\alpha(\omega) = \left[\frac{\partial \zeta'_\alpha(\omega)}{\partial(\omega^2)}\right]_{\omega=\Omega_{\alpha\gamma}} (\omega^2 - \Omega^2_{\alpha\gamma}) + \cdots \; . \tag{3.96}$$

Taking only the first term of the series from equation (3.95), we find the following relation between the frequencies $\omega_{\alpha\gamma}$ and $\Omega_{\alpha\gamma}$:

$$\omega^2_{\alpha\gamma} \approx \Omega^2_{\alpha\gamma}(1 - b_{\alpha\gamma}) \; ,$$
$$b_{\alpha\gamma} = -\frac{\rho}{\mathrm{Re}\, M_{22}} \left[\frac{\zeta_\alpha(\omega)}{\partial \zeta_\alpha / \partial(\omega^2)}\right]_{\omega=\Omega_{\alpha\gamma}} \tag{3.97}$$

Then, taking account of the formulae (3.96) and (3.97), we write

$$\frac{\operatorname{Re} M_\alpha}{\zeta'_\alpha(\omega)} \approx \operatorname{Re} M_{22} \frac{\omega^2-\omega^2_{\alpha\gamma}}{\omega^2-\Omega^2_{\alpha\gamma}} ,$$

$$\frac{\rho\omega^2\zeta_\alpha(\omega)}{\zeta'_\alpha(\omega)} \approx -\operatorname{Re} M_{22} \frac{\Omega^2_{\alpha\gamma}-\omega^2_{\alpha\gamma}}{\omega^2-\Omega^2_{\alpha\gamma}} . \tag{3.98}$$

Substituting (3.98) into the formula (3.88) and (3.89), we obtain

$$F_\alpha(i\omega|\Omega_{\alpha\gamma}) \approx \left[L_\alpha+M_{11} - \frac{M_{12}M_{21}}{\operatorname{Re} M_{22}} \frac{\omega^2-\Omega^2_{\alpha\gamma}}{\omega^2-\omega^2_{\alpha\gamma}+ig_\alpha(\omega^2-\Omega^2_{\alpha\gamma})}\right]^{-1} ,$$

$$H_\alpha(i\omega|\Omega_{\alpha\gamma}) \approx \frac{M_{21}F_\alpha(i\omega|\Omega_{\alpha\gamma})(\Omega^2_{\alpha\gamma}-\omega^2_{\alpha\gamma})}{\omega^2-\omega^2_{\alpha\gamma}+ig_\alpha(\omega^2-\Omega^2_{\alpha\gamma})} , \tag{3.99}$$

$$g_\alpha = \operatorname{Im} M_{22}/\operatorname{Re} M_{22} .$$

Formulae (3.99) are applicable only in the neighbourhood of the frequency $\Omega_{\alpha\gamma} \approx \omega_{\alpha\gamma}$. The approximate expressions coincide with the exact ones for $\omega = \Omega_{\alpha\gamma}$ and for $\omega = \omega_{\alpha\gamma}$. Since the functions L_α and M_{jk} vary little when the frequency in the neighbourhood of $\Omega_{\alpha\gamma}$ varies little, they can be substituted by constants in the formulae (3.99). Then, on the right sides of the formulae there will be rational functions.

We now estimate contributions to the pressure variance of the frequencies in the neighbourhood of $\Omega_{\alpha\gamma}$. Let us calculate, for example, the integral

$$A_{\alpha\gamma} = \int_{-\omega_{\alpha\gamma}-\varepsilon}^{-\omega_{\alpha\gamma}+\varepsilon} |H_\alpha(i\omega|\Omega_{\alpha\gamma})|^2 d\omega + \int_{\omega_{\alpha\gamma}-\varepsilon}^{\omega_{\alpha\gamma}+\varepsilon} |H_\alpha(i\omega|\Omega_{\alpha\gamma})|^2 d\omega ,$$

where 2ε is a certain integral of frequencies outside of which the integrand is negligibly small. Approximately,

$$A_{\alpha\gamma} \approx \int_{-\infty}^{\infty} |H_\alpha(i\omega|\Omega_{\alpha\gamma})|^2 d\omega . \tag{3.100}$$

Substituting expressions (3.99) here, and introducing the notations

$$g_{\alpha\gamma} = \operatorname{Im}\left(M_{22} - \frac{M_{12}M_{21}}{L_\alpha + M_{11}} \right) \frac{1}{\operatorname{Re} M_{22}} ,$$
$$h_{\alpha\gamma} = -\operatorname{Re}\left(\frac{M_{12}M_{21}}{L_\alpha + M_{11}} \right) \frac{1}{\operatorname{Re} M_{22}} , \tag{3.101}$$

we shall represent formula (3.100) as

$$A_{\alpha\gamma} \approx \left| \frac{M_{21}(\Omega_{\alpha\gamma})}{L_\alpha(\Omega_{\alpha\gamma}) + M_{11}(\Omega_{\alpha\gamma})} \right|^2 \times$$

$$b^2_{\alpha\gamma}\Omega^4_{\alpha\gamma} \int_{-\infty}^{\infty} \frac{d\omega}{|\omega^2 - \omega^2_{\alpha\gamma} + (h_{\alpha\gamma} + ig_{\alpha\gamma})(\omega^2 - \Omega^2_{\alpha\gamma})|^2} . \tag{3.102}$$

The parameter $g_{\alpha\gamma}$ has the sense of the loss tangent at the frequency $\Omega_{\alpha\gamma}$. The parameter $h_{\alpha\gamma}$, as a non-dimensional correction of the frequency $\omega_{\alpha\gamma}$, is found without taking account of deformations of the carrying shell. The integral in the formula (3.102) is calculated using the residue theorem. Here, we shall only write the approximate result for $b_{\alpha\gamma} \ll 1$, $h_{\alpha\gamma} \ll 1$, $g_{\alpha\gamma} \ll 1$, i.e.,

$$A_{\alpha\gamma} \approx \left| \frac{M_{21}(\Omega_{\alpha\gamma})}{L_\alpha(\Omega_{\alpha\gamma}) + M_{11}(\Omega_{\alpha\gamma})} \right|^2 \frac{\pi b_{\alpha\gamma}\Omega_{\alpha\gamma}}{g_{\alpha\gamma}} . \tag{3.103}$$

Let the spectral densities of the external forces $S_{Q_\alpha Q_\beta}(\omega)$ be slowly varying functions of the frequency. Then, the approximate estimates for the spectral densities, variances and correlation functions of the pressure can be obtained from formulae (3.92), (3.93), (3.94) and (3.103). If, for example, the joint correlation of the modes of vibrations is neglected, the pressure variance defined as a function of the frequency interval ω_* and ω_{**} can be estimated as

$$D[p(\underset{\sim}{x};\omega_*,\omega_{**})] \approx \sum_\alpha \sum_\gamma \left| \frac{M_{21}(\Omega_{\alpha\gamma})\chi(\underset{\sim}{x},\Omega_{\alpha\gamma})}{L_\alpha(\Omega_{\alpha\gamma}) + M_{11}(\Omega_{\alpha\gamma})} \right|^2 \frac{\pi b_{\alpha\gamma}\Omega_{\alpha\gamma}}{g_{\alpha\gamma}} S_{Q_\alpha Q_\alpha}(\Omega_{\alpha\gamma}) . \tag{3.104}$$

Summation is performed over all the indices α, γ for which $\omega_* \le \Omega_{\alpha\gamma} \le \omega_{**}$. The formula which takes account of the joint correlations of the generalized coordinates is much more complicated, and we do not consider it here.

Let the density of the natural frequencies of the system be sufficiently high. Then, assuming that $\omega_* = 0$, smoothing the dependence of variance on the frequency interval ω_{**}, and differentiating the result with respect to this frequency, we shall obtain the approximate estimate for the smoothed spectral density of the pressure.

Examples

Consider a closed circular cylindrical shell. We take the equations of motion of the shell in the form

$$\begin{aligned} &\underline{D}\Delta_2\Delta_2 w - \frac{1}{R}\frac{\partial^2 f}{\partial x^2} + \rho_0 h \frac{\partial^2 w}{\partial t^2} = q - m_{11}w - m_{12}w_H ,\\ &\Delta_2\Delta_2 f + \frac{\underline{E}h}{R}\frac{\partial^2 w}{\partial x^2} = 0 . \end{aligned} \tag{3.105}$$

Here, f is a function of the loading, $\underline{D}$ and $\underline{E}$ are viscoelastic operators, R is the radius of the median surface, and Δ_2 is the Laplace operator on this surface. The interval volume will be related to the cylindrical system of coordinates (x,r,Θ). Therefore,

$$\Delta_2 = \frac{\partial^2}{\partial x^2} + \frac{1}{R^2}\frac{\partial^2}{\partial\Theta^2} , \qquad \Delta = \frac{\partial^2}{\partial r^2} + \frac{1}{r}\frac{\partial}{\partial r} + \frac{1}{r^2}\frac{\partial^2}{\partial\Theta^2} + \frac{\partial^2}{\partial x^2} .$$

Let Vlasov's boundary conditions be prescribed at the end of the shell, i.e.,

$$w = \frac{\partial^2 w}{\partial x^2} = \frac{\partial^2 f}{\partial x^2} = \frac{\partial^2 f}{\partial\Theta^2} = 0 , \qquad (x = 0,\ x = \ell) . \tag{3.106}$$

We shall consider the end walls to be acoustically soft, i.e., the pressure perturbations on them will be equal to zero. Then,

$$\partial\phi/\partial t = 0 , \qquad (x = 0, x = \ell) . \tag{3.107}$$

Equations (3.67) and (3.105) with conditions (3.106) and (3.107) will be satisfied if we take

$$\psi_{mn}(x,\Theta) = \sin\frac{m\pi x}{\ell}\, e^{in\Theta} ,$$

$$\chi_{mn}(x,r,\Theta;\omega) = \sin\frac{m\pi x}{\ell}\, e^{in\Theta} J_n(\kappa_m r) , \tag{3.108}$$

$$(m = 1,2,\ldots,; \ n = 0,\pm 1,\pm 2,\ldots) .$$

The pair of integers m and n are equivalent to the index α in formula (3.78) and correspond to the order of the modes of the shell vibration. Also, the following notation has been used in formulae (3.108):

$$\kappa_m^2 = \frac{\omega^2}{c^2} - \frac{m^2\pi^2}{\ell^2} . \tag{3.109}$$

Thus, the surface vibration modes of the fluid coincide with the modes of the shell's natural vibration modes, i.e., condition (3.83) is satisfied.

The closed spherical shell containing a compressible fluid is another example. In the approximation of shell theory, analogous to the approximations (3.105), the natural modes of the shell vibrations are surface spherical functions. The solutions of the corresponding Helmholtz equation (3.80) are spherical functions, i.e., condition (3.83) is also satisfied here. Another example is an infinitely long circular cylindrical shell. Here, instead of the conditions at the ends and end walls, the boundedness of the solutions at infinity is required. The solution of this problem is complicated by the fact that the spectrum of the elastic and acoustic subsystems is continuous. Therefore, in the expressions (3.78), the summation over the index α is substituted by integration with respect to a certain wave number.

The approximate relation between the natural frequencies of the simultaneous system $\omega_{\alpha\gamma}$ and the frequencies of the partial system $\Omega_{\alpha\gamma}$ is given by the formulae (3.97). Let us show how to calculate parameter $b_{\alpha\gamma}$, which occurs in these formulae, in the case of a closed cylindrical shell. It will be recalled that the index α is substituted by the pair of indices $\{m,n\}$; while the index γ relates, as before, to radial wave numbers. Consider the expression

$$\frac{\partial^2 J_n(z)}{\partial r \partial(\omega^2)} = \frac{1}{2\kappa c^2} \left[z J_n''(z) + J_n'(z) \right] .$$

Using the differential equation for Bessel functions, we find

$$\left[\frac{\zeta_\alpha(\omega)}{\partial \zeta_\alpha(\omega)/\partial \omega^2} \right]_{\omega=\Omega_{\alpha\gamma}} = - \frac{2c^2}{R_H} \frac{1}{1-n^2(\kappa_{mn\gamma} R_H)^{-2}} .$$

Here, $R_H = R\text{-}H$ and $\kappa_{mn\gamma}$ are the roots of the frequency equation $J_n'(\kappa_m R_H) = 0$ for the partial system ($\gamma = 1,2,\ldots$). Hence, using the second formula (3.97), we obtain

$$b_{mn\gamma} = \frac{2\kappa p_0}{R_H \operatorname{Re} M_{22}} \frac{1}{1-n^2(\kappa_{mn\gamma} R_H)^{-2}} , \tag{3.110}$$

where p_0 is the unperturbed pressure in the acoustic medium and κ is the polytropic index.

Parameter $b_{mn\gamma}$ characterizes the displacement of the natural frequencies of the simultaneous system with respect to the partial frequencies $\Omega_{mn\gamma}$. Note that under certain conditions this displacement is quite small. It follows from the formula for the natural frequencies of the fluid inside the cylinder that

$n^2 < \kappa^2_{mn\gamma} R^2_H$. The value $\tilde{E} = \text{Re}\, M_{22}/H$ can be treated as some typical stiffness (elasticity) modulus of the insulating layer. The smallness of frequency correction for $\Omega_{mn\gamma}$ follows from the conditions $p_0 \ll \tilde{E}$, $H \ll R_H$.

The Case of Non-Separable Variables

In the preceding sections, we considered the special case that the vibration modes of the carrying shell coincide with the surface vibration modes of the fluid. In the general case, the solution is sought in the form (3.78) and (3.82), where the functions $\phi_\alpha(\underset{\sim}{\xi})$ and $\psi_\alpha(\underset{\sim}{\xi})$ do not coincide identically. It is necessary to find the relation between the spectra of the pressure $P_\alpha(\omega)$ and the spectra of loading $Q_\alpha(\omega)$, i.e.,

$$P_\alpha(\omega) = \sum_\beta H_{\alpha\beta}(i\omega) Q_\beta(\omega) \ . \tag{3.111}$$

Here, $H_{\alpha\beta}(i\omega)$ are transfer functions. Substituting formulae (3.78), (3.79), and (3.82) into equation (3.71) as well as the conditions (3.68) and (3.70), we obtain the relations (3.111). Substitution into (3.71) yields

$$[L_\beta(\omega) + M_{11}(\omega)] W_\beta(\omega) + M_{12}(\omega) W_{H\beta}(\omega) = Q_\beta(\omega) \ , \tag{3.112}$$

where the notations of the preceding sections are used. We shall obtain two other groups of equations from (3.68) and (3.70) projecting the results of the substitution of formulae (3.78) and (3.82) onto the axes of the functional bases $\phi_\alpha(\underset{\sim}{\xi})$ or $\psi_\beta(\underset{\sim}{\xi})$. If the functions $\phi_\alpha(\underset{\sim}{\xi})$ are orthogonal on Ω, we obtain

$$\begin{aligned} \frac{\zeta'_\alpha(\omega)}{\zeta_\alpha(\omega)} \Phi_\alpha(\omega) &= i\omega \sum_\beta a_{\alpha\beta} W_{H\beta}(\omega) \ , \\ -i\omega\rho\Phi_\alpha(\omega) &= \sum_\beta a_{\alpha\beta} [M_{21}(\omega) W_\beta(\omega) + M_{22}(\omega) W_{H\beta}(\omega)] \ . \end{aligned} \tag{3.113}$$

Here, $a_{\alpha\beta}$ are the coefficients of re-expansion of the basis functions, i.e.,

$$a_{\alpha\beta} = \frac{(\psi_\beta, \phi_\alpha)}{(\phi_\alpha, \phi_\alpha)} . \tag{3.114}$$

If the functions $\phi_\alpha(\underset{\sim}{\xi})$ do not satisfy the orthogonality conditions on Ω, re-expansion is performed with respect to the basis $\psi_\beta(\underset{\sim}{\xi})$. Below, we shall consider the first case.

The infinite systems of equations (3.112) and (3.113) extend the equations (3.84) to the case when the condition (3.83) is not satisfied. It is not difficult to obtain a system of equations in which only the spectra $W_\beta(\omega)$ are involved:

$$\sum_\beta [L_\beta(\omega) + M_{11}(\omega) - M_{12}(\omega) M_{21}(\omega) M_\alpha^{-1}(\omega) \zeta'_\alpha(\omega)] a_{\alpha\beta} W_\beta(\omega) = \sum_\beta a_{\alpha\beta} Q_\beta(\omega) . \tag{3.115}$$

Approximate Calculation of Transfer Functions

The transfer functions in relations (3.111) are determined from the infinite systems of equations (3.112) and (3.113). Without considering the problem of convergence, we note that even the solution for truncated systems is very difficult because the frequency ω must be treated as a parameter taking the values from $(-\infty, \infty)$. However, if from the point of view of mechanics it is possible to break the feedback between the elastic system and the acoustic one, the transfer functions have explicit expressions. This can be done by means of several methods. If an insulating layer is such that its influence on the vibrations of a carrying layer can be neglected, we obtain, instead of the equation (3.71), an equation in which $m_{11} = m_{12} = 0$. If an insulating shell consists of a soft layer, bounded on the inside by a stiff wall, then, in considering the vibrations of a carrying shell, we can take $w_H = 0$. Then, we arrive at an equation (3.71) in which $m_{12} = 0$. And, finally, if the fluctuations of the pressure on the inside wall can be neglected, we shall find the relation between the displacements w and w_H by assuming in the second expression (3.69) that $p_H = 0$.

For the three cases listed above, we obtain, respectively, the following relations between the spectra:

$$W_\beta(\omega) = L_\beta^{-1}(\omega) Q_\beta(\omega) ,$$

$$W_\beta(\omega) = [L_\beta(\omega) + M_{11}(\omega)]^{-1} Q_\beta(\omega) , \qquad (3.116)$$

$$W_\beta(\omega) = [L_\beta(\omega) + M_{11}(\omega) - M_{12}(\omega) M_{21}(\omega) \; M_{22}^{-1}(\omega)]^{-1} Q_\beta(\omega) .$$

Hence, we shall find explicit expressions for the transfer functions in (3.111). These expressions have the form

$$H_{\alpha\beta}(i\omega) = -\rho\omega^2 \zeta_\alpha(\omega) M_{21}(\omega) M_\alpha^{-1}(\omega) F_\beta(i\omega) a_{\alpha\beta} , \qquad (3.117)$$

where $F_\beta(i\omega)$ is a transfer function corresponding to one of the relations (3.116).

Calculation of the Pressure's Spectral Density

The formula for the spectral density of the pressure is

$$S_p(\underset{\sim}{x}, \underset{\sim}{x}'; \omega) = \sum_\alpha \sum_\beta \sum_\gamma \sum_\delta \; H^*_{\alpha\beta}(i\omega) H_{\gamma\delta}(i\omega) S_{Q_\beta Q_\delta}(\omega) \chi^*_\alpha(\underset{\sim}{x}, \omega) \chi_\gamma(\underset{\sim}{x}', \omega) . \qquad (3.118)$$

This formula is analogous to the second formula (3.93). Here, unlike the latter, summation is performed over a greater number of indices: the indices α and γ characterize oscillation modes of the fluid, the indices β and δ are vibration modes of the shell.

As before, it is expedient to introduce the variance of the pressure in the frequency range $[\omega_*, \omega_{**}]$. The corresponding formula is obtained by integrating the right side of formula (3.118) with respect to the frequency ω for $\underset{\sim}{x} = \underset{\sim}{x}'$. An approximate estimate of integrals can be performed by the above described method. With joint correlations of generalized forces ignored, the final result for $w_H = 0$ takes the form

$$D[p(\underset{\sim}{x};\omega_*,\omega_{**})] \approx \sum_\alpha \sum_\beta \sum_\mu \left| \frac{M_{21}(\Omega_{\alpha\mu})\chi_\alpha(\underset{\sim}{x},\Omega_{\alpha\mu})}{L_\beta(\Omega_{\alpha\mu})+M_{11}(\Omega_{\alpha\mu})} \right|^2 \frac{\pi b_{\alpha\mu}\Omega_{\alpha\mu}a^2_{\alpha\beta}}{g_{\alpha\mu}} S_{Q_\beta Q_\beta}(\Omega\alpha\mu) . \tag{3.119}$$

Summation is performed over all α and μ for which $\omega_* \leq \Omega_{\alpha\mu} \leq \omega_{**}$. And here, the same notations are used as in the formula (3.104). However, in calculating the parameters $b_{\alpha\beta}$, $g_{\alpha\mu}$ and $\Omega_{\alpha\mu}$, the simplifications used when deriving approximate relations (3.116) should be taken into account.

Circular Cylindrical Shell with Acoustically Stiff End Walls

Let us again consider a circular cylindrical shell, the difference being that instead of the boundary conditions (3.107) on the end walls, here, these conditions are

$$\partial\phi/\partial x = 0 , \qquad (x = 0,\ x = \ell) . \tag{3.120}$$

Equations (3.105) and the conditions (3.106) will be taken in the same form. To satisfy the conditions (3.120), we assume that

$$\chi_{mn}(\underset{\sim}{x},\omega) = \cos\frac{m\pi x}{\ell}\, e^{in\theta} J_n(\kappa_m r) ,$$
$$(m = 1,2,\ldots;\ n = 0,\pm 1,\pm 2,\ldots) . \tag{3.121}$$

The analysis requires the use of the formula (3.119), the coefficients (3.114) of the re-expansion of the functions $\psi_\beta(\underset{\sim}{x})$ with respect to the basis $\phi_\alpha(\underset{\sim}{x})$ being determined for $m \neq \mu$ as

$$a_{(mn)(\mu\nu)} = \eta_\mu m[1-(-1)^{m+\mu}]\delta_{n\nu}/[m^2+\mu^2)\pi] . \tag{3.122}$$

The coefficient η_μ is equal to unity for $\mu = 0$ and is equal to two in other cases. If $m = \mu$, then $a_{(mn)(\mu\nu)} = 0$.

A shell reinforced by a system of stringers and frames (Figure 15) is of great practical interest. Let the reinforcement be sufficiently stiff so that the deformations of stringers and frames can be neglected when estimating the shell vibrations. The typical space scale of correlation of an external loading is assumed to be small as compared to the size of the panel. The shell will be treated as a set of independent panels attached to the frame.* We shall consider simply supported panels as an example. Let the frames form a regular rectangular grid on Ω, its sides being a_1 and a_2 long. The end walls $x = 0$ and $x = \ell$ will be considered as acoustically stiff. We shall denote the median surface of the j-th panel by Ω_j, and the local coordinates for the panel by $\xi_1^{(j)}$ and $\xi_2^{(j)}$. The index β in the formulae of the type (3.119) will be substituted by three indices m_1, m_2, and j, so that we have

$$\psi_{m_1 m_2 j}(\underset{\sim}{\xi}) = \begin{cases} \sin \dfrac{m_1 \pi \xi_1^{(j)}}{a_1} \sin \dfrac{m_2 \pi \xi_2^{(j)}}{a_2} & (\underset{\sim}{\xi} \in \Omega_j) , \\ 0 & (\underset{\sim}{\xi} \notin \Omega_j) . \end{cases} \tag{3.123}$$

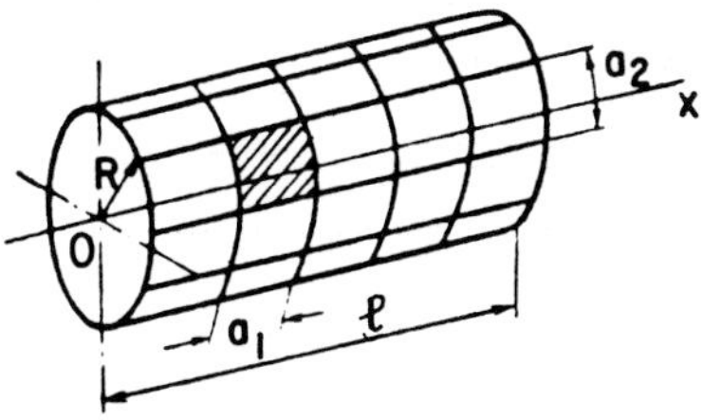

Figure 15

*Bolotin, V.V., Moskalenko, V.N., and others, "Methods of Calculation of Random Acoustic Fields Inside Reinforced Shells", *Dokladi Nauchnotekhnicheskoi Conference of Moscow Energy Institute, Dynamics and Strength of Machines*, Moscow, M.E.I. Publishing House, 1969.

Characterizing the panel's location on the shell by two numbers N_1 and N_2, we assume that $\xi_1^{(j)} = x_1 - N_1 a_1$, and $\xi_2^{(j)} = R\theta - N_2 a_2$ $(N_1, N_2 = 0,1,2,\ldots,)$. We calculate the coefficients (3.114) as

$$a_{(mn)(m_1 m_2 j)} = \frac{2}{\pi R \ell} \int_0^{a_1} \int_0^{a_2} \sin \frac{m_1 \pi \xi_1}{a_1} \sin \frac{m_2 \pi \xi_2}{a_2} \cos \frac{m\pi (N_1 a_1 + \xi_1)}{\ell} \times$$

$$\exp\left[\frac{jn}{R} (N_2 a_2 + \xi_2)\right] d\xi_1 d\xi_2 \ . \tag{3.124}$$

We do not write down the explicit expression for these coefficients. If the typical scale of correlation of the external loading is of the same order as the sizes of the assumed panel, then the joint correlation between the identical modes of vibrations of all the panels should be taken into account in the formulae of type (3.118). In the formula (3.119), summation should be performed over the indices $\alpha = (m,n)$, $\beta = (m_1, m_2)$, $j = (N_1, N_2)$, $j' = (N_1, N_2)$ and μ, i.e.,

$$D[p(\underset{\sim}{x}; \omega_*, \omega_{**})] \approx$$

$$\sum_\alpha \sum_\beta \sum_j \sum_{j'} \sum_\mu \frac{|M_{21}(\Omega_{\alpha\mu}) \chi_\alpha(\underset{\sim}{x}, \Omega_{\alpha\mu})|^2}{[L_{j\beta}(\Omega_{\alpha\mu}) + M_{11}(\Omega_{\alpha\mu})]^* [L_{j'\beta}(\Omega_{\alpha\mu}) + M_{11}(\Omega_{\alpha\mu})]} \times$$

$$\frac{\pi b_{\alpha\mu} \Omega_{\alpha\mu} a^*_{\alpha j \beta} a_{\alpha j' \beta}}{g_{\alpha\mu}} S_{Q_{j\beta} Q_{j'\beta}}(\Omega_{\alpha\mu}) \ . \tag{3.125}$$

If the field of external forces is uniform, the spectral densities of the generalized forces corresponding to the identical modes of vibrations of each panel are independent of the index j. The impedance $L_{j\beta}$ is independent of j since all the panels are assumed to be identical. Instead of formula (3.125), the following formula is used:

$$D[\underset{\sim}{p}(\underset{\sim}{x};\omega_*,\omega_{**})] \approx \sum_\alpha \sum_\beta \sum_\mu \left| \frac{M_{21}(\Omega_{\alpha\mu})\chi_\alpha(\underset{\sim}{x},\Omega_{\alpha\mu})}{L_\beta(\Omega_{\alpha\mu})+M_{11}(\Omega_{\alpha\mu})} \right|^2 \frac{\pi b_{\alpha\mu}\Omega_{\alpha\mu}}{g_{\alpha\mu}} \times$$

$$\left[\sum_j \left| a_{(mn)(m_1m_2j)} \right|^2 S_{Q_\beta Q_\beta}(\Omega_{\alpha\mu}) + \right.$$

$$\left. \sum_{j\neq j'} \sum a^*_{(mn)(m_1m_2j)} a_{(mn)(m_1m_2j')} S_{Q_{j\beta}Q_{j'\beta}}(\Omega_{\alpha\mu}) \right] . \qquad (3.126)$$

Note that the second term in square brackets takes account of the mutual correlations between identical modes of vibration of different panels. With formula (3.124) taken into consideration, it is possible to find the sum of the first of the series

$$\sum_j |a_{(mn)(m_1m_2j)}|^2 = \frac{2\eta_m^2\eta_n^2}{\pi R\ell a_1^3 a_2^3} \frac{1-(-1)^{m_1}\cos(m\pi a_1/\ell)}{(m_1/a_1)^2-(m/\ell)^2} \frac{1-(-1)^{m_2}\cos(na_2/R)}{(m_2\pi/a_2)^2-(n/R)^2} .$$

The cases when $m_1/a_1 = m/\ell$ or $m_2\pi/a_2 = n/R$ can be obtained by the limiting transition.

Some numerical results for the reinforced shell subjected to pulsations from the turbulent boundary layer are described in the paper quoted in the footnote above. A model corresponding to a correlation function of the type (1.45) is taken for pressure pulsations. Table 3.1 shows the results of calculations; it gives the number of natural frequencies within the standard octave ranges. The number of natural frequencies of one panel are given; the number of natural frequencies of the acoustic medium inside the shall are given in the last column. As can be seen in the table, more than $5\cdot10^3$ series terms had to be summed up to take into account the spectral components of the pressure up to 1500 hertz.

The pressure field inside the shell turns out to be non-uniform. Closer to the shell axis, the full noise level decreases. Along the shell axis, we find that close to the end walls (which were assumed to be acoustically stiff) the full noise level is higher than in the middle.

Table 3.1

Octave No.	Limit Frequencies		Number of Natural Frequencies	
	Hz	rad.sec^{-1}	Elastic Panel	Acoustic Panel
1	22.3 - 44.6	140 - 280	-	-
2	44.6 - 89.2	280 - 560	-	3
3	88.5 - 177.0	555 - 1111	1	12
4	177.0 - 354.0	1111 - 2222	2	77
5	354.0 - 707.5	2222 - 4445	5	564
6	707.5 - 1415.0	4445 - 8890	8	$4.5\cdot 10^3$

3.6 APPLICATION OF THE METHOD OF SPECTRAL REPRESENTATION

Vibrations of an Infinite Viscoelastic Plate in a Field of Stochastic Forces

The method of integral spectral representation is convenient for the analysis of random vibrations of infinite or semi-infinite elastic bodies, especially if the loading they are subjected to is a uniform one with respect to the spatial coordinates and stationary in time. As an example, let us consider random vibrations of an infinite plate of constant thickness made of linear viscoelastic material. The equation of motion of the plate will have the form (3.53). Let $q(\underset{\sim}{x},t)$ be a stationary random field. Then, it admits the representation

$$q(\underset{\sim}{x},t) = \int_{-\infty}^{\infty}\int_{\underline{R}^2} Q(\underset{\sim}{k},\omega)e^{i(\underset{\sim}{k}\underset{\sim}{x}+\omega t)}d\underset{\sim}{k}d\omega \ , \tag{3.127}$$

where $Q(\underset{\sim}{k},\omega)$ is a generalized random function of the wave vector k and of frequency ω (the spectrum of external loadings). For the deflection $w(\underset{\sim}{x},t)$, an analogous representation will be taken:

$$w(\underset{\sim}{x},t) = \int_{-\infty}^{\infty}\int_{\underline{R}^2} W(\underset{\sim}{k},\omega)e^{i(\underset{\sim}{k}\underset{\sim}{x}+\omega t)}dkd\omega \ . \tag{3.128}$$

The relations between the spectrum of the deflection $W(\underset{\sim}{k},\omega)$ and the spectrum of loading $Q(\underset{\sim}{k},\omega)$ is given by the formula

$$W(\underset{\sim}{k},\omega) = \frac{Q(\underset{\sim}{k},\omega)}{k^4(D_r+iD_i)-\rho h\omega^2}, \tag{3.129}$$

where $k = |\underset{\sim}{k}|$. This formula follows from equation (3.53), if the Fourier transformation with respect to coordinates and time is applied and the formulae (3.127) and (3.128) are used. Here, D_r+iD_i is a complex cylindrical stiffness (the Fourier transform of the viscoelastic operator $\underline{D}$).

To find the relation between the space-time spectral density of the deflection $S_w(\underset{\sim}{k},\omega)$ and the corresponding spectral density of the loading $S_q(\underset{\sim}{k},\omega)$, we apply the formulae:

$$\langle Q^*(\underset{\sim}{k},\omega)Q(\underset{\sim}{k}',\omega')\rangle = S_q(\underset{\sim}{k},\omega)\delta(\underset{\sim}{k}-\underset{\sim}{k}')\delta(\omega-\omega'),$$

$$\langle W^*(\underset{\sim}{k},\omega)W(\underset{\sim}{k}',\omega')\rangle = S_w(\underset{\sim}{k},\omega)\delta(\underset{\sim}{k}-\underset{\sim}{k}')\delta(\omega-\omega').$$

Substituting the expression (3.129) into the second formula and taking account of the first one, we obtain the desired relation:

$$S_w(\underset{\sim}{k},\omega) = \frac{S_q(\underset{\sim}{k},\omega)}{(k^4D_r-\rho h\omega^2)^2+(\chi k^4D_r)^2}. \tag{3.130}$$

Here, $\chi(\omega)$ is the loss tangent corresponding to the operator $\underline{D}$.

Having the relation (3.130), it is not difficult to derive the formulae for variations, correlation functions and spectral densities of displacements, velocities, accelerations, stresses, etc. in the plate. For instance, for the variation of the deflection, the formula is

$$\underset{\sim}{D}[w(\underset{\sim}{x},t)] = \int_{-\infty}^{\infty}\int_{\underline{R}^2} \frac{S_q(\underset{\sim}{k},\omega)\,d\underset{\sim}{k}\,d\omega}{(k^4D_r-\rho h\omega^2)^2+(\chi k^4D_r)^2}. \tag{3.131}$$

As another example, we shall take the formula for the time spectral density of the normal component of acceleration. It reads

$$S_{\ddot{w}}(\omega) = \omega^4 \int_{\underline{R}^2} \frac{S_q(\underset{\sim}{k},\omega)d\underset{\sim}{k}}{(k^4 D_r - \rho h\omega^2)^2 + (\chi k^4 D_r)^2} .$$

Some other examples of the applications of this method can be found in references [45,74,78] and others. The method of spectral representation as applied to infinite homogeneous plates is actually a modification of the method of generalized coordinates for systems with a continuous spectrum of natural frequencies. For example, representation of the deflection (3.128) as the Fourier integral over the wave vector $\underset{\sim}{k}$ is equivalent to the expansion of the deflection into the modes of natural vibrations of an infinite plate. However, in the examples below, the method of integral representations does not admit of such interpretation.

Vibrations of a Multispan Regular Plate in a Field of Stochastic Forces

This problem was investigated by V. Moskalenko [68]. Consider the vibrations of an infinite plate resting on two equidistant rigid supports and subjected to normal forces homogeneous in coordinates and stationary in time. The equation of motion has the form (3.53). Its right side has the form (3.127). Unlike the problem considered above, here, the solution is sought in the form of the canonical integral

$$w(\underset{\sim}{x},t) = \int_{-\infty}^{\infty} \int_{\underline{R}^2} Q(\underset{\sim}{k},\omega)\psi(\underset{\sim}{x}|\underset{\sim}{k},\omega)e^{i\omega t}d\underset{\sim}{k}d\omega . \tag{3.132}$$

Substituting the expression (3.132) into equation (3.53) and the appropriate boundary conditions, we arrive at the boundary value problem for the auxiliary function $\psi(\underset{\sim}{x}|\underset{\sim}{k},\omega)$. Having solved this problem, we can calculate the probabilistic characteristics of the plate's vibration field. For example, for the spectral density of the deflection $w(\underset{\sim}{x},t)$ at a fixed point of the plate, we have the formula

$$S_w(\underset{\sim}{x},\omega) = \int_{\underline{R}^2} S_q(\underset{\sim}{k},\omega)\,|\psi(\underset{\sim}{x}|\underset{\sim}{k},\omega)|^2 d\underset{\sim}{k} \; . \tag{3.133}$$

The calculations were made for the case of cylindrical flexure. The spectral density of the loading $S_q(\underset{\sim}{k},\omega)$ was of the type (1.42). The integral (3.133) was calculated by the method of contour integration using the assumption of the smallness of damping. The graph for the nondimensional spectral density (3.133) in the middle of the span of length ℓ is given in Figure 16. As is generally known, the spectrum of the natural frequencies of a regular plate or beam of infinite span is a band spectrum. The spectrum bands are shown in Figure 16 as solid lines.

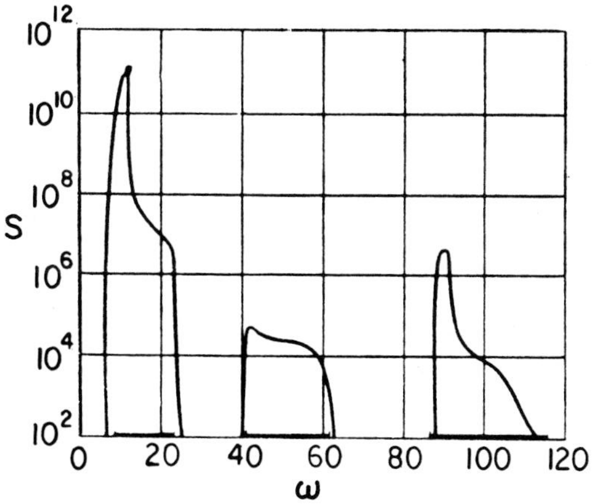

Figure 16

Propagation of Random Vibrations in a Viscoelastic Rod

The problem of the propagation of vibrations in elastic bodies leads naturally to the consideration of semi-infinite models. One of the problems of this type was considered by V. Palmov [76].

Let the axial force p(t), which is a stationary random function of time, be applied to the free end of a semi-infinite prismatic linear viscoelastic rod. The equation of the rod's vibration has the form

$$\underline{E}F \frac{\partial^2 u}{\partial x^2} - \rho F \frac{\partial^2 u}{\partial t^2} = 0 \ , \qquad (3.134)$$

where $\underline{E}$ is a viscoelastic operator corresponding to the elastic modulus. On the loaded end, the following condition should be satisfied

$$\underline{E}F \frac{\partial u}{\partial x} = p(t) \ , \qquad (x = 0) \ . \qquad (3.135)$$

The requirement of the boundedness of the solution at infinity must be added. Following the general method, we represent the force p(t) as a Fourier stochastic integral

$$p(t) = \int_{-\infty}^{\infty} P(\omega) e^{i\omega t} d\omega \ . \qquad (3.136)$$

Here, $P(\omega)$ is a generalized random function of the frequency ω related to the spectral density of the force p(t) by the expression

$$\langle P^*(\omega) P(\omega') \rangle = S_p(\omega) \delta(\omega - \omega') \ . \qquad (3.137)$$

We look for the solution of equation (3.134) in the form

$$u(x,t) = \int_{-\infty}^{\infty} U(x,\omega) e^{i\omega t} d\omega \ .$$

For the random function $U(x,\omega)$, we obtain the equation

$$\frac{d^2 U}{dx^2} + \lambda^2 U = 0 \ , \qquad (3.138)$$

where the notations

$$\lambda^2 = \omega^2/[c^2(1+i\eta)] \ , \qquad c^2 = E_r/\rho \ , \qquad \eta = E_i/E_r$$

are used. The boundary conditions (3.135) can be transformed into the following expression

$$\frac{dU}{dx} = \frac{P(\omega)}{E_r F(1+i\eta)} \ , \qquad (x = 0) \ . \tag{3.139}$$

Taking the solution of the equation (3.138), bounded at infinity, and imposing the condition (3.139), we find

$$U(x,\omega) = \frac{P(\omega)e^{i\lambda x}}{iE_r F(1+i\eta)\lambda} \ ,$$

where $i\lambda$ is a characteristic index having a negative real part. Hence, using the relation (3.137), we calculate spectral densities, correlation functions and variances of displacements, velocities, accelerations, stresses, etc. in the rod. For instance, for the correlation function of the acceleration, we have

$$K_{\ddot{u}}(x_1,x_2,\tau) = \frac{1}{(\rho F)^2} \int_{-\infty}^{\infty} \exp\left[i\omega\tau - \frac{i\omega\alpha}{c}(x_1-x_2) - \right.$$

$$\left. - \frac{\omega\beta}{c}(x_1+x_2)\right] |\lambda(\omega)|^2 S_p(\omega)d\omega \ ,$$

where

$$\alpha = \frac{1}{\sqrt{2}}\left[\frac{1}{1+\chi^2} + \frac{1}{(1+\chi^2)^{1/2}}\right]^{1/2} \ , \qquad \beta = \frac{\chi}{2\alpha(1+\chi^2)} \ .$$

In [77], the analogous problem of the propagation of longitudinal vibrations in a rod with damping of the dry friction type is considered. After statistical linearization has been applied, the problem is reduced to the one considered above. The results obtained permit one to estimate the length beyond which (due to the nonlinear type of friction) vibration does not propagate.

Propagation of Random Vibrations in Viscoelastic Shells

The problem of vibrations of semi-infinite cylindrical shells subjected to random kinematic or static loadings at the end of the shell was considered by V. Radin and V. Chirkov [83,84]. As an example, we may take the results of the calculations for the case of axisymmetric random vibrations. The graphs in Figures 17 and 18 show the non-dimensional spectral densities of a normal displacement for two cases: when the normal displacement $w_0(t)$ is specified at the end $x_1 = 0$, and when the angle of rotation of the normal $\phi_0(t)$ is specified at this end, $w_0(t)$ and $\phi_0(t)$ being stationary random time functions. The non-dimensional spectral densities are introduced as

$$s_1 = S_w(x_1,\omega)/S_{w_0}(\omega) , \qquad s_2 = S_w(x_1,\omega)/[\lambda_0^2 S_{\phi_0}(\omega)] ,$$

where λ_0 is the characteristic length of the zone of boundary effect. On the horizontal axis, the dimensionless frequency ω/ω_0 is plotted, where ω_0 represents the eigenfrequencies of the moment-free radial oscillations.

$$\lambda_0 = (Rh)^{1/2} , \qquad \omega_0 = \frac{1}{R}(E_r/\rho)^{1/2} . \qquad (3.140)$$

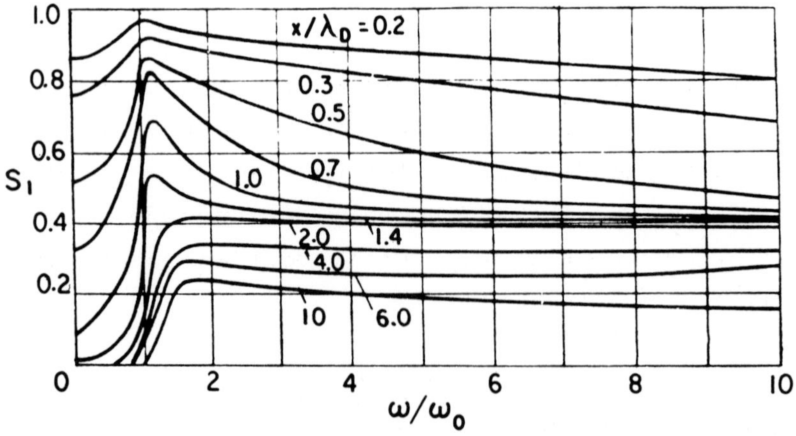

Figure 17

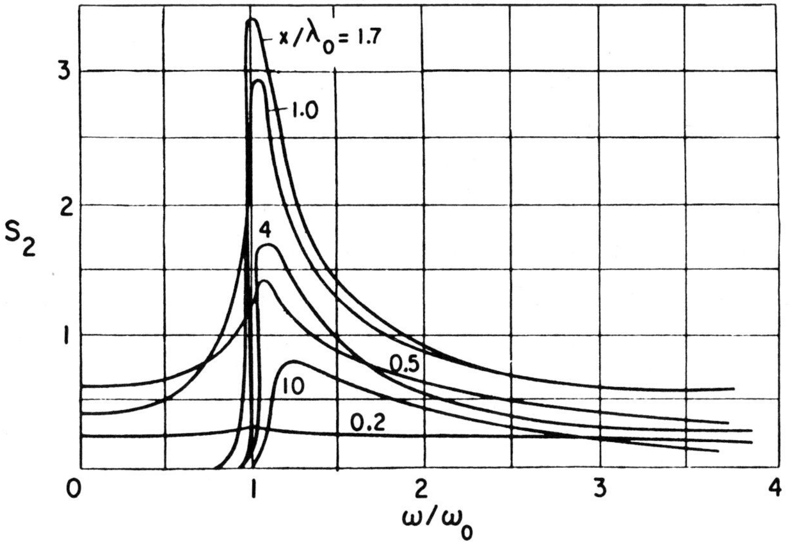

Figure 18

The spectral densities have a pronounced maximum close to ω_0. Variation of the functions $s_\alpha(x_1,\omega)$ is of a different character within the zone of boundary effects and outside this zone. For $x_1 < \lambda_0$, the high frequency vibration components damp faster, and for $x_1 > \lambda_0$ the low frequency ones damp faster.

Vibrations of the Elastic Half-Space Subjected to Moving Random Loading

This problem was considered in connection with the estimate of stresses originating in a wall due to the high-frequency cavitation of a rapidly moving fluid.* Small vibrations of the elastic half-space $x_3 > 0$ subjected to the normal pressure $q(x_1,x_2,t)$ are described by the Lamé equation for the displacement vector (notations are those universally accepted)

---

*Bolotin, V.V., Volokhovski, V.Yu., Chirkov, V.P., "Vibrations of the Elastic Half-Space under Dynamic Random Loads", *Izvestia Ac. of Sciences USSR*, MTT (Mekhanika Tverdovo Tela), 1975, No. 5.

$$(\lambda+\mu)\ \text{grad div}\ \underset{\sim}{u} + \mu\Delta\underset{\sim}{u} = \rho\frac{\partial^2\underset{\sim}{u}}{\partial t^2}$$

and the boundary conditions at $x_3 = 0$ for the components σ_{jk} of the stress tensor: $\sigma_{13} = \sigma_{23} = 0$ and $\sigma_{33} = -q$. In addition, the condition of boundedness of the solution for $x_3 \to +\infty$ and the radiation condition of wave propagation into the half-space should be satisfied.

Let the external loading q in the moving system of coordinates x_1-vt, x_2, x_3 be a homogeneous random function of the variables x_1-vt, x_2, and t, with the mathematical expectation equal to zero. Henceforth, we shall represent it as the Fourier stochastic integral

$$q(\underset{\sim}{x},t) = \int_{-\infty}^{\infty}\int_{\underline{R^2}} Q(\underset{\sim}{k},\omega)\exp\{i[\underset{\sim}{k}\underset{\sim}{x}+(\omega-k_1v)t]\}d\underset{\sim}{k}d\omega\ , \qquad (3.141)$$

where $\underset{\sim}{x} = \{x_1,x_2\}$ is a two-dimensional radius-vector, $\underset{\sim}{k} = \{k_1,k_2\}$ is the corresponding wave vector, ω is the frequency, and $Q(\underset{\sim}{k},\omega)$ is a generalized random function (loading spectrum). The task is to find the probabilistic characteristics of displacement fields and stresses for $x_3 > 0$. Introduce the vector-function $\underset{\sim}{w}(\underset{\sim}{x},t)$ whose elements include, besides the components of the displacement vector and the stress tensor, the first invariant of the stress tensor $\sigma = \sigma_{11}+\sigma_{22}+\sigma_{33}$. We shall seek the components $w_\alpha(\underset{\sim}{x},t)$ of the vector-function $\underset{\sim}{w}(\underset{\sim}{x},t)$ in a form analogous to (3.141), as

$$w_\alpha(\underset{\sim}{x},t) = \int_{-\infty}^{\infty}\int_{\underline{R^2}} W_\alpha(\underset{\sim}{k},\omega,x_3)\exp\{i[\underset{\sim}{k}\underset{\sim}{x}+(\omega-k_1v)t]\}d\underset{\sim}{k}d\omega\ . \qquad (3.142)$$

Here, $W_\alpha(\underset{\sim}{k},\omega,x_3)$ are generalized random functions of the variables k_1, k_2 and ω, dependent also on the coordinate x_3. Having found the relation between $W_\alpha(\underset{\sim}{k},\omega,x_3)$ and the spectral density of the loading $S_q(\underset{\sim}{k},\omega)$, we shall be able to find spectral densities, correlation functions and variances for the components of the vector-function $\underset{\sim}{w}(\underset{\sim}{x},t)$.

The character of the solution depends substantially on the relation between the frequency ω and the wave number k_1 on the one hand, and the velocity of motion of the loading v, on the other hand. If even one of the conditions

$$k_1 v = \omega , \qquad |k_1 v - \omega| = kc_R , \qquad (k = |\underset{\sim}{k}|) \tag{3.143}$$

(c_R is the rate of propagation of Raleigh waves) is satisfied, then nonintegrable singularities appear in the expressions for the probabilistic characteristics of displacements and stresses. From the viewpoint of physics, this implies the appearance of such resonances (in both frequencies and modes of excited vibrations) that radiation of energy into the half-space cannot provide stationary regimes. To obtain in this case physically sound results, energy dissipation in the material and, possibly, energy radiation and dissipation in the adjacent medium must be taken into consideration.

The straight lines specified by equations (3.143) divide the first quadrant of the parameter plane $\{v,\omega\}$ into six open regions. These regions are shown in Figure 19 and denoted by Roman numerals. Stationary solutions for the elastic half-space considered here exist only for those loadings whose spectral characteristics and velocity of motion are such that they belong to one of the six mentioned classes.

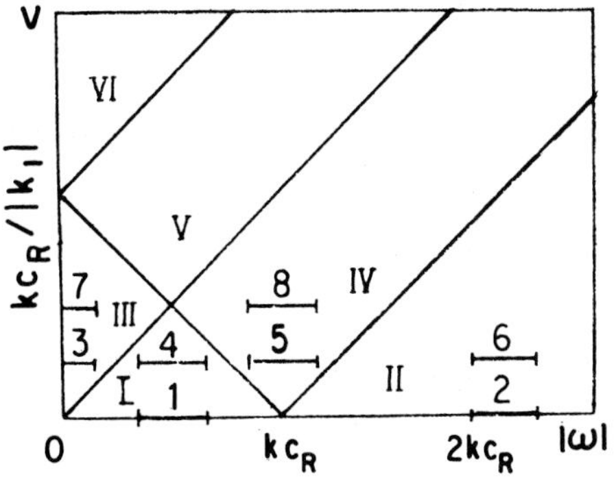

Figure 19

Some numerical results are shown in Figure 20. For calculations, the spectral density of loading is taken in the form

$$S_q(\underset{\sim}{k},\omega) = \begin{cases} \frac{s_0}{2\pi}\left[\delta(\underset{\sim}{k}-\underset{\sim}{\kappa})+\delta(\underset{\sim}{k}+\underset{\sim}{\kappa})\right] & (\omega \in [\omega_*,\omega_{**}]) , \\ 0 & (\omega \notin [\omega_*,\omega_{**}]) , \end{cases}$$

where s_0 is the intensity of the truncated white noise with the frequency bounds ω_* and ω_{**} and $\underset{\sim}{\kappa}$ is a given wave vector with the components κ_1 and κ_2. The graphs in Figure 20 are plotted for $\kappa_1 = \kappa_2 = \kappa$. The curves 1 correspond to the values $v = 0$, $\omega_* = \kappa(0.5c_R-0.2c_2)$, and $\omega_{**} = \kappa(0.5c_R+0.2c_2)$, the curve 2, to the values $v = 0$, $\omega_* = \kappa(2c_R-0.2c_2)$, and $\omega_{**} = \kappa(2c_R+0.2c_2)$, etc. Here, c_2 is the velocity of propagation of shear waves. The variances of the displacements u_3 and the variance of the first invariant of the stress tensor σ decrease monotonically when moving into the half-space. Generally, this does not take place for the other components of the vector-function $\underset{\sim}{w}(\underset{\sim}{x},t)$. Of particular interest is the shear stress σ_{13}, for which the maximum variance is reached at depths of the order to the typical length of a loading wave.

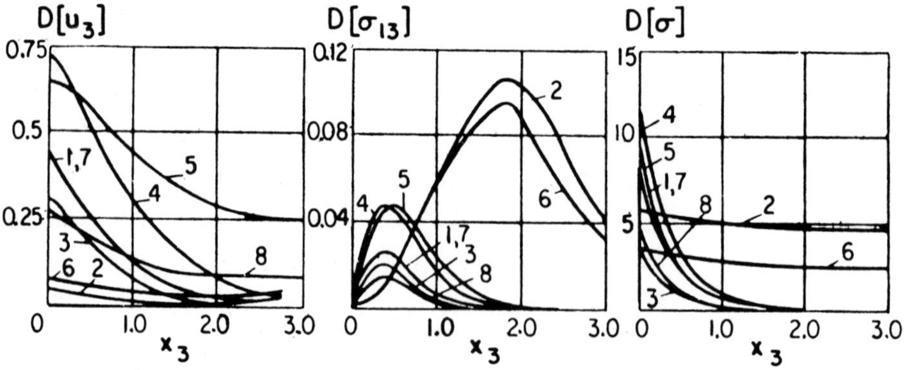

Figure 20

Chapter 4
The Asymptotic Method in the Theory of Random Vibrations of Continuous Systems

4.1 ASYMPTOTIC ESTIMATES FOR NATURAL FREQUENCIES AND NATURAL MODES

Preliminary Remarks

The method of generalized coordinates, which was considered in detail and illustrated by a number of examples in the preceding chapter, is effective for obtaining numerical results and qualitative conclusions. The application of this method requires summation of the contributions of all the modes of natural vibrations excited by a random loading. If the time spectrum of a loading is sufficiently wide, and the spectrum of the natural frequencies of the system is sufficiently dense, the number of these modes may be very large. As an example, consider the vibration of the covering panels of an aircraft subjected to the acoustic pressures of a jet stream. The spectrum of these pulsations actually occupies the entire acoustic range (from tens to 10,000 - 20,000 Hertz). On the other hand, the spectrum of frequencies of the natural vibrations of the covering panels is quite dense. The difference between adjacent natural frequencies may be 10 Hertz or less. For instance, in the numerical example of Section 3.5, there are hundreds of vibration modes of the surface (taking account of the large number of panels constituting the surface) and thousands of acoustic modes. Therefore, summation of a large number of terms is necessary when random vibrations of an aircraft surface are analyzed.

At the same time, the high density of the spectrum can be used for the construction of approximate asymptotic solutions. The method, in which the asymptotic behaviour of the high natural frequencies and corresponding modes and the high density of the spectrum are used, was suggested in [11]. The method is essentially based on substituting the sum of contributions of each mode by an integral over a certain region in the space of wave numbers. The wider the excitation spectrum and the higher the density of natural frequencies, the better is the approximation of the method using integral estimates. Unlike the direct summation over natural modes of vibrations, the integral method leads, in some cases, to results in closed form. Hence, there arises the possibility for systematic analysis of the influence of various factors on the random behaviour of elastic systems.

Referring the reader to the author's works (for instance, [9]) for details, we shall present here the essentials of the asymptotic method, which makes it possible to analyze natural modes and frequencies of vibrations of elastic systems at sufficiently high wave numbers.

Fundamentals of the Asymptotic Method

Let us consider some generalized rectangle in the m-dimensional space of the variables $x_1, x_2, \ldots, x_m$: $0 \leq x_\alpha \leq a_\alpha$ (Figure 21). Assume we are seeking functions $\phi_1, \phi_2, \ldots, \phi_n$ in this region which satisfy the system of differential equations

$$\sum_{\gamma=1}^{n} L_{\beta\gamma}(\phi_\gamma) - \lambda \sum_{\gamma=1}^{n} M_{\beta\gamma}(\phi_\gamma) = 0 \, , \qquad (\beta = 1,2,\ldots,n) \, , \tag{4.1}$$

and the conditions of each boundary of the region, say,

$$N_{\alpha\gamma}(\phi_1,\phi_2,\ldots,\phi_n|0) = 0 \, ,$$
$$N_{\alpha\gamma}(\phi_1,\phi_2,\ldots,\phi_n|a_\alpha) = 0 \, , \qquad (\alpha = 1,2,\ldots,n; \; \gamma = 1,2,\ldots,p). \tag{4.2}$$

Here, $L_{\beta\gamma}$, $M_{\beta\gamma}$, and $N_{\alpha\gamma}$ are linear differential operators, and 2p is the general order of the system (4.1). The boundary value problem is assumed to be self-adjoint; the problem is to find the eigenvalues λ and eigenfunctions ϕ_β of the given boundary value problem.

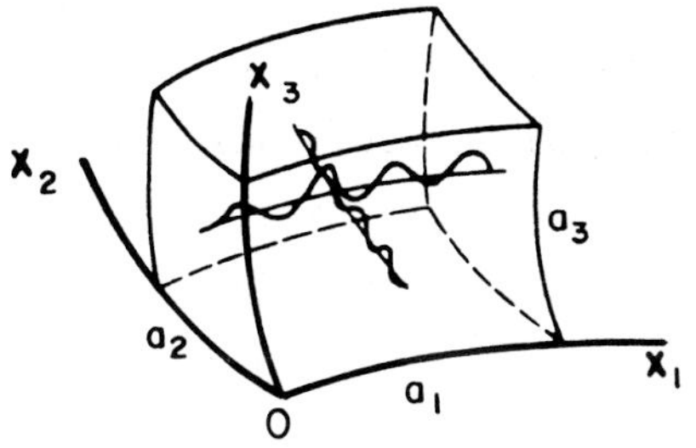

Figure 21

The following terminology will be used: The boundary value problem whose eigenfunctions can be represented as a product of functions of each separate argument:

$$\phi_\beta = \prod_{\alpha=1}^{m} \psi_{\beta\alpha}(x_\alpha) \, , \qquad (\beta = 1,2,\ldots,n) \, , \tag{4.3}$$

will be called a boundary value problem with separable variables. Vibrations of plates and shells with the Navier boundary conditions on the contour are examples.

The selfadjoint problem described by the system of equations (4.1) and boundary conditions (4.2) will be called a boundary value problem with quasi-separable variables if the following conditions are satisfied:

(a) The problem becomes a boundary value problem with separable variables if the boundary conditions (4.2) are appropriately chosen, i.e., if the system (4.1) allows a generating solution in the form (4.3).

(b) The system (4.1) allows the solution in the form

$$\phi_\beta = \Phi_{\beta\alpha_0}(x_{\alpha_0}) \prod_{\alpha \neq \alpha_0}^{m} \psi_{\beta\alpha}(x_\alpha) , \quad (\beta = 1,2,\ldots,n;\ \alpha_0 = 1,2,\ldots,m) , \tag{4.4}$$

where $\Phi_{\beta\alpha_0}$ are some functions of one variable x_{α_0}. In other words, the substitution of (4.4) converts (4.1) into a system of ordinary differential equations for the functions $\Phi_{\beta\alpha_0}$.

(c) The substitution of (4.4) into the boundary conditions (4.2), corresponding to $\alpha = \alpha_0$, converts them into conditions containing only $\Phi_{\beta\alpha_0}$.

A typical example of boundary problems with quasi-separable variables are problems described by systems of equations with constant coefficients and containing derivatives of even orders. The boundary conditions for each face of the region should contain an operation of differentiation of the same even order with respect to each of the coordinates tangential to the given face.

The asymptotic method of solution of boundary value problems in the theory of vibrations with quasi-separable variables is based on the following assumptions: The solution (4.3) has asymptotic properties of the eigenfunctions which are not affected by a change of boundary conditions. Therefore, it can be used as a generating solution also for those problems which do not admit an exact solution in the form (4.3). In this case, the solution (4.3) can be considered only as an approximate one for the inner region at a sufficient distance from the boundaries. Close to the boundaries, the exact solution will be different from the generating solution. If a solution can be constructed for each face that will satisfy its boundary conditions and will tend to the generating solution as the distance from the face increases, the asymptotic expressions for the eigenfunctions can be obtained by means of the operation of matching. It is easy to see that the asymptotic method is actually based on constructing approximate solutions for rectangular regions by means of exact solutions obtained for certain semi-infinite regions.

When constructing a solution, we should note that each function $\psi_{\beta\alpha}(x_\alpha)$ in the expression (4.3) will contain two constants if an appropriate choice of normalization conditions is made. We shall choose one of them (the wave number k_α) so that the value k_α^{-1} will characterize the scale of the variation of the function $\psi_{\beta\alpha}(x_\alpha)$. Then, the second constant will be a certain phase characteristic. These constants can be found only after matching the solutions. Substituting (4.3) into (4.4), we shall find the relation between the eigenvalue λ and the wave numbers $k_1, k_2, \ldots, k_m$ as:

$$\lambda = \lambda(k_1, k_2, \ldots, k_m) \ . \tag{4.5}$$

Consider, for example, the boundary $x_{\alpha_0} = 0$. Assuming that the parameter in the equations (4.1) is determined according to (4.5), we shall seek their solutions in the form (4.4). The obtained system of ordinary differential equations will be assumed to have a solution of the type $\psi_{\beta\alpha_0}(x_{\alpha_0})$ depending on two arbitrary constants. Assume that the system mentioned above admits p - 1 linearly independent solutions with boundary effect properties, i.e., solutions which damp down as they recede into the inner region. We shall call these solutions corrective. Their properties are analogous to the integrals of the boundary effect type in shell theory, or of the boundary layer type in hydromechanics of viscous fluids. Joining the generating solution and the corrective ones, we shall have p + 1 constants. With the help of these constants, we shall satisfy p boundary conditions and the normalization condition. The wave numbers k_1, $k_2, \ldots, k_m$ are assumed to be known at this stage. Analogously, we shall construct the solution for the opposite face $x_{\alpha_0} = a_{\alpha_0}$. Requiring from both equations that they should coincide in the inner region up to a small difference (caused by the existence of corrective solutions), we shall obtain the matching condition. This condition yields the equations for the wave numbers $k_1, k_2, \ldots, k_m$. The number of equations from which wave numbers can be obtained is m. Then, the eigenvalues can be determined from the formula (4.5).

The error of a matching operation is of the same order as the one that corrective solutions assume in the inner region. The faster the boundary effect damps down, the smaller the error is. Ordinarily, the error decreases rapidly as the wave numbers increase. However, in some cases, the solution of the boundary effect type cannot be constructed for a certain region of wave numbers. In these cases, we are faced with the degeneration of the dynamic boundary effect. Degeneration of the boundary effect indicate a strong influence of the boundary conditions on the behaviour of eigenfunctions in the inner region. The asymptotic method certainly can not be used if the boundary effect degenerates.

Solutions obtained by means of the asymptotic method can be considered as approximate expressions for eigenfunctions valid everywhere except in the vicinities of ribs and vertices of the region. Here, differential equations and boundary conditions are to be satisfied exactly. However, the uniqueness of the solution is achieved only by the operation of matching which is not rigorously justified and does not correspond to the spirit of classical analysis. From the practical viewpoint, it is better to consider the solutions for the inner region and the solutions for each boundary separately, as it is done in hydromechanics when viscous and non-viscous solutions are considered separately, or in the statics of shells when moment and non-moment solutions are considered.

Application of the Asymptotic Method to Differential Equations with Constant Coefficients

A very important field of application of the asymptotic method is that of partial differential equations with constant coefficients. Equations of this type generally describe vibrations of rectangular plates, panels with constant metrics of the median surface supported by a rectangular contour, and circular cylindrical shells. For simplicity, consider the case of a single equation of order 2n; most of the problems of the theory of elastic vibrations can be reduced to such a "resolving" equation.

Thus, we consider the homogeneous linear self-adjoint boundary value problem

$$L(\phi) - \lambda M(\phi) = 0 , \qquad (4.6)$$

for the rectangular region $0 \leq x_\alpha \leq a_\alpha$ $(\alpha = 1,2,\ldots,m)$ in the space of m variables $x_1,x_2,\ldots,x_m$. Here, L and M are linear differential operators with constant coefficients of the order 2n and 2(n-1), respectively, containing no derivatives of odd order:

$$L(\phi) = \sum_{|\beta|=0}^{n} a_\beta D^{2\beta}\phi , \quad M = \sum_{|\beta|=0}^{n-1} b_\beta D^{2\beta}\phi , \quad D^\beta = \frac{\partial^{|\beta|}}{\partial x_1^{\beta_1}\partial x_2^{\beta_2}\ldots\partial x_m^{\beta_m}} .$$

Here, $\beta = \{\beta_1,\beta_2,\ldots,\beta_m\}$ is a multi-index, $\beta_\alpha \geq 0$ are integers, a_β, b_β are constant real coefficients. Summation is performed over all the indices for which the value $|\beta| = \beta_1+\beta_2+\cdots+\beta_m$ varies within the mentioned intervals.

The functions ϕ should satisfy the boundary conditions, which, generally, are different for each face of the region. Assume that these conditions have the form $N_\alpha^{|\beta|}(\phi|0) = 0$,

$N_\alpha^{|\beta|}(\phi|a_\alpha) = 0$ ($\alpha = 1,2,\ldots,m$; $|\beta| = 0,1,\ldots,n$), where $N_\alpha^{|\beta|}$ are certain linear differential expressions with constant coefficients. We have to determine the eigenvalues λ and eigenfunctions ϕ of the considered boundary value problem.

Equation (4.6) will be satisfied if we assume

$$\phi = \sin k_1(x_1-\xi_1)\sin k_2(x_2-\xi_2) \cdots \sin k_m(x_m-\xi_m) \ , \tag{4.7}$$

where the $k_1,k_2,\ldots,k_m$, $\xi_1,\xi_2,\ldots,\xi_m$ are constants. The constants $k_1,k_2,\ldots,k_m$ are wave numbers. The corresponding value of the parameter λ is determined from a relation of the type (4.5)

$$\lambda = \frac{\sum\limits_{|\beta|=0}^{n} (-1)^{|\beta|} a_\beta k_1^{2\beta_1} k_2^{2\beta_2} \cdots k_m^{2\beta_m}}{\sum\limits_{|\beta|=0}^{n-1} (-1)^{|\beta|} b_\beta k_1^{2\beta_1} k_2^{2\beta_2} \cdots k_m^{2\beta_m}} \ . \tag{4.8}$$

The boundary conditions here will not generally be satisfied.

We shall find an approximate solution of the problem, considering the expression (4.7) as a generating solution, valid only in the inner region sufficiently distant from the boundaries. This will be possible if for the values of the parameter λ, determined from (4.8), we can construct n-1 independent solutions for each boundary which damp down as the distance from the boundary increases. Then, we shall have a sufficient number of arbitrary constants to satisfy all the boundary conditions. We shall construct 2m different solutions, satisfying both the differential equation (4.6) and the corresponding boundary conditions. All these solutions tend to the generating solution (4.7) as the distance from the boundary increases. Bringing the obtained solutions into agreement, we shall obtain m conditions for the determination of the wave numbers $k_1,k_2,\ldots,k_m$ of the generating solution. The error of matching can be estimated for each case. In usual circumstantes, the error decreases as the wave numbers

$k_1, k_2, \ldots, k_m$ increase. For example, consider the boundary $x_\alpha = 0$. We shall seek the solution for this boundary in the form

$$\phi_\alpha = \Phi_\alpha(x_\alpha) \prod_{\gamma \neq \alpha} \sin k_\gamma (x_\gamma - \xi_\gamma) \ . \tag{4.9}$$

The function Φ_α should satisfy the equation

$$L_\alpha(\Phi_\alpha) - \lambda M_\alpha(\Phi_\alpha) = 0 \ , \tag{4.10}$$

where

$$L_\alpha(\Phi_\alpha) = \sum_{|\beta|=0}^{n} a_\beta \frac{d^{2\beta_\alpha}\Phi_\alpha}{dx_\alpha^{2\beta_\alpha}} \prod_{\gamma \neq \alpha} (-1)^{\beta_\gamma} k_\gamma^{2\beta_\gamma} \ ,$$

$$M_\alpha(\Phi_\alpha) = \sum_{|\beta|=0}^{n-1} b_\beta \frac{d^{2\beta_\alpha}\Phi_\alpha}{dx_\alpha^{2\beta_\alpha}} \prod_{\gamma \neq \alpha} (-1)^{\beta_\gamma} k_\gamma^{2\beta_\gamma} \ .$$

The parameter λ is substituted into equation (4.10) from the formula (4.8). The corresponding characteristic equation obtained from (4.10) by substituting $\Phi_\alpha = Ce^{rx_\alpha}$ is of the order 2n; let us denote this equation by $\Delta_\alpha(r^2) = 0$.

It is obvious that equation $\Delta_\alpha(r^2) = 0$ has two imaginary roots $r_{1,2} = \pm ik_\alpha$. Separating these roots, which correspond to the generating solution (4.7), we obtain an equation of the order 2(n-1) which will be written, for brevity, as $\Delta_\alpha^*(r^2) = 0$. Then

$$\Delta_\alpha(r^2) = \Delta_\alpha^*(r^2)(r^2 + k_\alpha^2) \ . \tag{4.11}$$

The possibility of constructing an asymptotic solution depends on the properties of equation $\Delta_\alpha^*(r^2) = 0$ only. Let (n-1) roots of this equation have negative real parts. We shall denote them by $r_\gamma = -p_\gamma + iq_\gamma$, where $p_\gamma > 0$ ($\gamma = 1, 2, \ldots, n-1$). The expression

$$\Phi_\alpha(x_\alpha) = \sin k_\alpha(x_\alpha - \xi_\alpha) + \sum_{\gamma=1}^{n-1} C_{\alpha\gamma} \exp[(-p_\gamma + iq_\gamma)x_\alpha] \tag{4.12}$$

contains n constants, so that n conditions on the boundary $x_\alpha = 0$ can be satisfied. Then, for $x_\alpha = 0$, the solution (4.9) with the function Φ_α, determined from (4.12), will satisfy all the conditions $N_\alpha^{|\beta|}(0) = 0(|\beta| = 0, 1,\ldots,n)$, and for $x_\alpha \to \infty$ it tends to the generating solution (4.7).

The expression (4.12) is valid only in the case of simple characteristic roots. In the case of multiple roots, the expression will generally have terms of the type

$$x_\alpha^\ell \exp[-(p_\gamma+iq_\gamma)x_\alpha] \ ,$$

where the highest power ℓ equals the multiplicity of the root minus one; all the other properties of the solution (4.12) remain unchanged.

The corrective solution

$$\Psi_\alpha(x_\alpha|0) = \sum_{\gamma=1}^{n-1} C_{\alpha\gamma} \exp[(-p_\gamma+iq_\gamma)x_\alpha] \tag{4.13}$$

describes the deviation of the solution from the generating solution in the vicinity of the boundary $x_\alpha = 0$. If all $q_\gamma \equiv 0$, a nonoscillating boundary effect takes place; otherwise the effect is oscillating. If n-1 roots having negative real parts cannot be found among the roots r, then, degeneration of the dynamic boundary effect takes place. It is easy to find the sufficient conditions for nondegeneration of the boundary effect. Consider the equation

$$\Delta_\alpha^*(r^2) = g_0 r^{2(n-1)} + g_1 r^{2(n-2)} + \cdots + g_{n-2}r^2 + g_{n-1} = 0 \ . \tag{4.14}$$

For equation (4.14) to have n-1 roots with negative real parts, it is sufficient that the equation

$$\Delta_\alpha^*(\rho) = g_o\rho^{n-1}(-1)^{n-1} + g_1\rho^{n-2}(-1)^{n-2} + \cdots - g_{n-2}\rho + g_{n-1} = 0 \ , \tag{4.15}$$

where $\rho = -r^2$, has no positive or zero real roots. According to Descarte's theorem, equation (4.15) has no positive or zero roots if all its coefficients have the same sign. Thus, the boundary effect does not degenerate if $g_{n-1} > 0$ and the coefficients of equation (4.14) form an alternating series.

Let $h_0, h_1, \ldots, h_n$ be the coefficients of the equation

$$\Delta_\alpha(r^2) = h_0 r^{2n} + h_1 r^{2(n-1)} + \cdots + h_{n-1} r^2 + h_n = 0 .$$

Then, the coefficients of equation (4.15) are calculated by the following recurrence formulae:

$$g_0 = h_0 , \quad g_1 = h_1 - g_0 k_\alpha^2 , \quad g_2 = h_2 - g_1 k_\alpha^2, \ \ldots .$$

Matching of Solutions

Assume that solutions of the type (4.9) have been constructed for each boundary of the region $x_\alpha = 0$ and $x_\alpha = a_\alpha$ $(\alpha = 1,2,\ldots,m)$. Each solution satisfies exactly the differential equation (4.6) and the conditions on one of the faces; all of them tend to the generating solution (4.7) as the distance x_α from the boundary increases. We shall find the wave numbers $k_1, k_2, \ldots, k_m$, matching the solutions originating at the opposite boundaries of the region. The matching conditions can be obtained by various methods. The most convenient one is as follows.

Let us consider two solutions, say,

$$\begin{aligned} \Phi_\alpha(x_\alpha|0) &= \sin k_\alpha(x_\alpha - \xi_\alpha) + \Psi(x_\alpha|0) , \\ \Phi_\alpha(x_\alpha|a_\alpha) &= \pm[\sin k_\alpha(a_\alpha - x_\alpha - \eta_\alpha) + \Psi_\alpha(x_\alpha|a_\alpha)] . \end{aligned} \tag{4.16}$$

The second of them is constructed for the boundary $x_\alpha = a_\alpha$. Analogously to (4.13), we have

$$\Psi_\alpha(x_\alpha|a_\alpha) = \sum_{\gamma=1}^{n-1} D_{\alpha\gamma} \exp[(-p_\gamma+iq_\gamma)(a_\alpha-x_\alpha)] .$$

Eliminate the constants $C_{\alpha\gamma}$ ($\gamma = 1,2,\ldots,n-1$) using the boundary conditions for the solution $\Phi_\alpha(x_\alpha|0)$ and express the phase of the generating solution ξ_α in terms of wave numbers. It is easy to see that the formula for ξ_α has the form

$$\tan k_\alpha\xi_\alpha = F_\alpha(k_1,k_2,\ldots,k_m) , \tag{4.17}$$

where F_α is some rational function of wave numbers. Making the solution $\Phi_\alpha(x_\alpha|a_\alpha)$ satisfy the boundary conditions, we obtain analogously, the formula for the phase η_α:

$$\tan k_\alpha\eta_\alpha = G_\alpha(k_1,k_2,\ldots,k_m) . \tag{4.18}$$

Since in the inner region the functions $\Psi(x_\alpha|0)$ and $\Psi(x_\alpha|a_\alpha)$ are small as compared to the principal terms, which take into account the generating solution, the matching condition has the form

$$\sin\, k_\alpha(x_\alpha-\xi_\alpha) = \pm\sin\, k_\alpha(a_\alpha-x_\alpha-\eta_\alpha) . \tag{4.19}$$

And evidently, it is possible to have a discrepancy ε of the order

$$\varepsilon \sim \exp(-\frac{1}{2}\, a_\alpha \min\, p_\gamma) , \tag{4.20}$$

where $\min p_\gamma$ is the smallest value (in magnitude) among the negative real parts of the roots of the equation $\Delta^*_\alpha(r^2) = 0$. The estimate (4.20) corresponds to the assumption that matching is performed in the centre of the region (for $x_\alpha = \frac{1}{2} a_\alpha$). If the generating solution satisfies the boundary conditions on one of the boundaries exactly, then, instead of (4.20), we obtain the following estimate for the discrepancy:

$$\varepsilon \sim \exp(-a_\alpha \min\, p_\gamma) . \tag{4.21}$$

Let us consider the matching condition (4.19) in greater detail. It can have the form

$$k_\alpha a_\alpha = k_\alpha(\eta_\alpha + \xi_\alpha) + m_\alpha \pi \ , \tag{4.22}$$

where m_α may generally be an arbitrary integer or zero. Using (4.17) and (4.18), we obtain

$$k_\alpha a_\alpha = \tan^{-1} F_\alpha(k_1, k_2, \ldots, k_m) + \tan^{-1} G_\alpha(k_1, k_2, \ldots, k_m) + m_\alpha \pi \ . \tag{4.23}$$

The number of such conditions of matching solutions that can be constructed for all x_α ($\alpha = 1,2,\ldots,m$) is exactly m. Thus, we obtain a system of equations sufficient for finding the wave numbers.

The matching conditions in the form (4.23) have an advantage: the functions on their right sides are determined only by the conditions on the given boundary, and do not depend on which boundary we consider, $x_\alpha = 0$ or $x_\alpha = a$. Thus, the expression for most frequently encountered boundary conditions can be constructed a priori; it is then not difficult to obtain the matching conditions for various combinations of boundary conditions, by combining appropriate functions. In particular, if the boundary conditions on opposite faces are identical, the equation (4.23) takes the form

$$k_\alpha a_\alpha = 2 \tan^{-1} F_\alpha(k_1, k_2, \ldots, k_m) + m_\alpha \pi \ . \tag{4.24}$$

Equations (4.23) and (4.24) admit of a convenient geometric interpretation. Consider a two-dimensional case (Figure 22). Equations (4.23) describe two families of curves on the plane k_1, k_2; one of them depends on the parameter m_1, the other on the parameter m_2 (both parameters are integers). If the Navier conditions are satisfied on the contour, then $F_\alpha \equiv G_\alpha \equiv 0$, and the curves become an orthogonal mesh of straight lines spaced at π/a_1 and π/a_2. They are plotted as thin lines in Figure 22. For boundary conditions different from the Navier conditions these lines become curvilinear; points of intersection of the curves represent the wave numbers k_1, k_2. If the dimensionality is larger than two, the geometrical interpretation holds, though it is not so obvious as in the two-dimensional case.

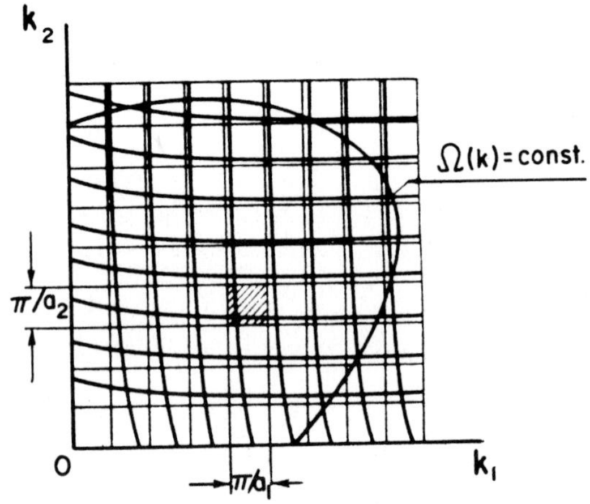

Figure 22

After the wave numbers have been found, the asymptotic expressions for the eigenvalues λ and for the eigenfunctions ϕ are determined from the formulae (4.7) and (4.8). The more exactly defined expressions for the eigenfunctions near the boundaries (except for the vicinities of vertices and edges) are determined from formulae of the type (4.9). It can be expected, and it is confirmed by calculations, that the error in determining eigenvalues is generally of the same order as the discrepancy ε. However, this assumption needs more consideration and proof.

Another problem connected with the matching operation which needs more consideration is ordering of the spectrum. It seems possible, at first sight, that if $\tan^{-1}$ is understood in the sense of its principal values, then the case when all $m_\alpha = 1$ will correspond to the principal frequency of natural vibrations. However, there are examples when this simple assumption is not valid: in these cases the frequencies for which all $m_\alpha = 1$ are either higher frequencies or are "extraneous", not corresponding to the given problem.

General considerations which should be kept in mind when ordering the spectrum are as follows. The asymptotic method has been developed to fit the highest frequencies of the spectrum; the lower the natural frequency is, the bigger, generally, is the error of the method. Although, as will be seen later, in a number of problems the asymptotic method gives very good results even for the principal frequency, determination of principal frequencies is not within the scope of application of this method. Thus, the first natural frequency cannot be identified by the asymptotic method. The first frequencies or modes of natural vibrations corresponding to them should be found by some other method. Comparing them to the estimates found by the asymptotic method, it is easy to determine the root of equations (4.23) and (4.24) corresponding to the principal frequency and to disregard extraneous roots. From the practical point of view, the first modes of natural vibrations can always be predicted by proceeding from assumptions of mechanics.

4.2 APPLICATION OF THE ASYMPTOTIC METHOD TO PLATES AND SHELLS

Application of the Asymptotic Method to Plates

As an example to illustrate this method, consider a rectangular elastic plate with the side lengths a_1 and a_2. We shall assume the cylindrical stiffness D and the surface density of the plate ρh to be constant. The equation for the modes $\phi(x_1,x_2)$ and the frequencies ω of the natural vibrations has the form

$$D\Delta\Delta\phi - \rho h\omega^2\phi = 0 . \qquad (4.25)$$

Following the asymptotic method, we seek a generating (inner) solution in the form (4.7), i.e.,

$$\phi(x_1,x_2) = \sin k_1(x_1-\xi_1)\sin k_2(x_2-\xi_2) \ , \tag{4.26}$$

where k_1 and k_2 are the unknown wave numbers, while ξ_1 and ξ_2 are the phases of the generating solution. This expression satisfies equation (4.25) and corresponds to the frequency:

$$\omega = (k_1^2+k_2^2)(D/\rho h)^{1/2} \ . \tag{4.27}$$

But expression (4.26) generally does not satisfy the boundary conditions. The Navier boundary conditions are an exception. For these conditions

$$k_1 = m_1\pi/a_1 \ , \qquad k_2 = m_2\pi/a_2 \ ,$$

and the formula (4.27) is exact.

In the case of nondegenerate boundary effects, the expression (4.26) can be considered as an asymptotic solution of the boundary value problem, valid in the region sufficiently distant from the plate contour. The solution close to the boundary $x_1 = 0$ will be sought in the form of an expression of the type (4.29):

$$\phi(x_1,x_2) = W(x_1)\sin k_2(x_2-\xi_2) \ . \tag{4.28}$$

By substituting expression (4.28) into equation (4.25), we obtain

$$W'''' - 2k_2^2W'' + (k_2^4-\rho h\omega^2/D)W = 0$$

or, after substituting here the expression (4.27) for the natural frequency,

$$W'''' - 2k_2^2 W'' - (2k_1^2 k_2^2 + k_1^4) W = 0 . \quad (4.29)$$

The characteristic equation corresponding to it,

$$r^4 - 2k_2^2 r^2 - (2k_1^2 k_2^2 + k_1^4) = 0 ,$$

has two imaginary roots and two real ones:

$$r_{1,2} = \pm i k_1 , \quad r_{3,4} = \pm (k_1^2 + 2k_2^2)^{1/2} .$$

The imaginary roots correspond to the generating solution (4.26), the real roots to the corrective solutions. Thus, a non-degenerated, nonoscillating dynamic boundary effect always occurs in plates.

The general integral of equation (4.29) has the form

$$W(x_1) = C_1 \sin k_1 x_1 + C_2 \cos k_1 x_1 + C_3 \exp[-x_1 (k_1^2 + 2k_2^2)^{1/2}] +$$

$$C_4 \exp[x_1 (k_1^2 + 2k_2^2)^{1/2}] .$$

If the boundary $x_1 = 0$ is considered, the last term should be omitted, since it increases infinitely with the increase of x_1. Out of the remaining terms, the first two fully coincide with the generating solution (4.26), and the first three terms taken together describe the dynamic boundary effect in the boundary zone:

$$W(x_1) = \sin k_1 (x_1 - \xi_1) + C \exp [-x_1 (k_1^2 + 2k_2^2)^{1/2}] . \quad (4.30)$$

Using the expression (4.30), it is not difficult to estimate the width of the region of the dynamic boundary effect. Since the constant C does not exceed unity in its order of magnitude, the influence of the last term in formula (4.30) can be estimated by the factor $\exp[-x_1(k_1^2+2k_2^2)^{1/2}]$. Let $x_1 = \lambda_1$, where $\lambda_1 = \pi/k_1$ is half of the wave length of the generating solution. Then, for $k_1 = k_2$, we have $e^{-\pi\sqrt{3}} = 0.0043$. Even in the most unfavourable case ($k_2 = 0$), we obtain $e^{-\pi} = 0.0432$. Thus, the width of the region of the dynamic boundary effect does not exceed half of the wavelength.

An equation of the type (4.30) can be constructed for each side of the rectangular plate. Satisfying the corresponding boundary conditions, we shall express the constant C and the phase ξ in terms of wave numbers k_1 and k_2. Then, we shall require that all the four solutions coincide to an accuracy of dynamic boundary effects. Hence, we arrive at the matching conditions of the type (4.23), i.e.,

$$\begin{aligned} k_1 a_1 &= \tan^{-1} F_1(k_1,k_2) + \tan^{-1} G_1(k_1,k_2) + m_1\pi , \\ k_2 a_2 &= \tan^{-1} F_2(k_1,k_2) + \tan^{-1} G_2(k_1,k_2) + m_2\pi , \end{aligned} \qquad (4.31)$$

where m_1, m_2 are integers or zero. The functions $F_\alpha(k_1,k_2)$ and $G_\alpha(k_1,k_2)$ are equal to the tangents of the phase constants ξ_α, η_α found from the boundary conditions for $x_\alpha = 0$ and $x_\alpha = a_\alpha$, respectively. Consequently, they depend on the boundary conditions only. If $w = 0$ on the plate contour, and if the functions $\tan^{-1} F_\alpha$ and $\tan^{-1} G_\alpha$ are understood in the sense of their principal values, the numbers m_1 and m_2 run through all the values of positive integer, thus ordering the spectrum of the natural frequencies. The values $m_1 = m_2 = 1$ correspond to the principal mode of vibrations. In the case when one or several plate sides are free or elastically supported, an additional investigation is necessary to determine the numbers m_1 and m_2 which correspond to the principal frequency. It should be noted that strictly speaking, the asymptotic method is applicable only to sufficiently high modes of vibration.

Rectangular Plate Clamped on its Contour

It is well known that there is no exact solution to this problem in a closed form. Let us apply the asymptotic method to it. Let the side $x_1 = 0$ be clamped. The conditions for the function $W(x_1)$ at $x_1 = 0$ have the form $W = W' = 0$. The substitution of the expression (4.30) leads here to the system of two equations

$$\sin k_1\xi_1 - C = 0 \ , \qquad k_1 \cos k_1\xi_1 - (k_1^2+2k_2^2)^{1/2}C = 0 \ ,$$

hence

$$\tan k_1\xi_1 = k_1/(k_1^2+2k_2^2)^{1/2} \ , \qquad C = k_1/[2(k_1^2+k_2^2)]^{1/2}$$

so

$$\begin{aligned} F_1(k_1,k_2) &= k_1/(k_1^2+2k_2^2)^{1/2} \ , \\ F_2(k_1,k_2) &= k_2/(k_2^2+2k_1^2)^{1/2} \ , \end{aligned} \tag{4.32}$$

(the second formula is obtained by the circular permutation of indices).

Expression (4.30) for the dynamic boundary effect takes the form

$$W(x_1) = \frac{k_1}{2^{1/2}(k_1^2+k_2^2)^{1/2}} \left\{ \frac{(k_1^2+2k_2^2)^{1/2}}{k_1} \sin k_1x_1 - \cos k_1x_1 + \exp[-x_1(k_1^2+2k_2^2)^{1/2}] \right\} . \tag{4.33}$$

Having the expression (4.33), we can easily find bending moments in the zone of the dynamic boundary effect. Let us determine the law which gives the change of moments

$$M_{11} = D\left(\frac{\partial^2 w}{\partial x_1^2} + \mu \frac{\partial^2 w}{\partial x_2^2}\right) \tag{4.34}$$

along the lines on which $\sin k_2(x_2-\xi_2) = 1$ (μ is the Poisson ratio). In other words, let us calculate the expression

$$M_{11}(x_1) = D(W''-\mu k_2^2 W) .$$

Substituting here the formula (4.33), we obtain

$$M_{11}(x_1) = \frac{k_1(k_1^2+\mu k_2^2)D}{2^{1/2}(k_1^2+k_2^2)^{1/2}} \left\{ -\frac{(k_1^2+2k_2^2)^{1/2}}{k_1} \sin k_1x_1 + \cos k_1x_1 + \right.$$

$$\left. \frac{k_1^2+k_2^2(2-\mu)}{k_1^2+\mu k_2^2} \exp[-x_1(k_1^2+2k_2^2)^{1/2}] \right\} . \qquad (4.35)$$

The matching conditions (4.31) with formula (4.32) taken into account have the form:

$$k_1a_1 = 2\tan^{-1}\frac{k_1}{(k_1^2+2k_2^2)^{1/2}} + m_1\pi ,$$
$$k_2a_2 = 2\tan^{-1}\frac{k_2}{(k_2^2+2k_1^2)^{1/2}} + m_2\pi . \qquad (4.36)$$

The $\tan^{-1}$ quantities in the equations (4.36) are to be understood in the sense of their principal values. The wave numbers k_1 and k_2 corresponding to the case $m_1 = m_2 = 1$ will be somewhat larger than π/a_1 and π/a_2. So, it is natural to expect the equations (4.36) to yield an approximate value of the principal frequency for $m_1 = m_2 = 1$. This is confirmed by the analysis of the modes of vibration. The roots of the equations (4.36) for $m_1 \leq 0$, $m_2 \leq 0$ are extraneous and should be dropped.

For some special cases, the system of equations (4.36) admits a solution in closed form. For example, for the vibrations of a rod of length a, or for the cylindrical deflection of a plate ($k_2 \to 0$), we easily obtain from the first equation

$$ka = (m+1/2)\pi , \qquad (m = 1,2,\ldots) .$$

Substituting (4.27) into the formula, we obtain

$$\omega = \frac{(m+1/2)^2\pi^2}{a^2}\left(\frac{D}{\rho h}\right)^{1/2} , \qquad (m = 1,2,\ldots) . \qquad (4.37)$$

This formula, giving an error of less than 1 percent even for the principal frequency, was obtained by Rayleigh from the analysis of the exact solution to the problem of vibrations of a clamped rod.

Consider a square plate ($a_1 = a_2 = a$). The exact solution of the equations (4.36) for those modes of vibrations to which $m_1 = m_2 = m$ correspond is

$$ka = (m+1/3)\pi \;, \qquad (m = 1,2,\ldots) \;.$$

Hence, using the formula (4.27), we obtain

$$\omega = \frac{2(m+1/3)^2\pi^2}{a^2}\left(\frac{D}{\rho h}\right)^{1/2} , \qquad (m = 1,2,\ldots) \;. \tag{4.38}$$

The accuracy of the asymptotic method has been investigated by many authors. The most recent data can be found in [166]. Some results are given in Table 4.1, where one can find the values of the frequency of a square plate

$$\bar{\omega}^2 = \frac{a^4}{\pi^4}\frac{\rho h}{D}\,\omega^2 \;, \tag{4.39}$$

for three types of natural modes: double-symmetric, symmetric-antisymmetric, and double-antisymmetric. The results obtained by the Ritz method, with a large number of the series terms retained (for instance, up to 36 terms for double-symmetric modes), were used for verification. These results yield slightly overestimated frequency values.

As can be seen from the table, the parameter $\bar{\omega}^2$ for the principal natural frequency ($m_1 = m_2 = 1$) is determined by means of the asymptotic method with an error not exceeding 5 percent. The resulting correction for the frequency is about 2.5 percent. This value corresponds to the discrepancy of matching which for the modes of vibrations at $m_1 = m_2 = m$ is of the order

$$\varepsilon = \exp[-\tfrac{1}{2}\pi\sqrt{3}(m+1/3)] \;. \tag{4.40}$$

Table 4.1 - Estimates for the Frequency Parameter $\bar{\omega}^2$

r	s	$m_1 = 2r-1, \quad m_2 = 2s-1$		$m_1 = 2r-1, \quad m_2 = 2s$		$m_1 = 2r, \quad m_2 = 2s$	
		Asymptotic	Control	Asymptotic	Control	Asymptotic	Control
1	1	12.6567	13.2939	54.5568	55.2999	118.568	120.225
2 1	1 2	177.870	178.587	277.418 454.291	279.500 454.987	602.196	604.463
2	2	493.827	497.043	897.631	901.544	1410.42	1415.71
3 1	1 3	979.756	980.724	1188.45 1874.24	1190.83 1874.89	2155.98	2158.37
3 2	2 3	1583.92	1588.33	2235.19 2672.16	2241.46 2676.90	3491.02	3498.09
3	3	3236.35	3243.98	4706.08	4715.25	6435.60	6446.77
4 1	1 4	3281.74	3282.53	3648.66 5370.25	3651.07 5371.05	5834.46	5836.83
4 2	2 4	4306.02	4310.93	5321.31 6653.30	5328.86 6658.33	7893.64	7901.71
3 4	4 3	6783.47	6797.80	9647.86 8822.31	9659.16 8834.50	12033.3	12047.3
5 1	1 5	8331.76	8332.40	8905.3 12382.3	8908.15 12382.9	13077.2	13079.7
5 2	2 5	9905.88	9910.97	11399.8 14279.7	11408.1 14284.8	16055.5	16064.0
4	4	11568.2	11583.2	15191.9	15208.6	19290.1	19309.3

At $m = 1$, formula (4.40) yields $\varepsilon = 2.66\%$. The error of the asymptotic method decreases very rapidly as the frequency increases. At least two or three correct significant digits can be obtained from the simple formula (4.38) for sufficiently high frequencies. It should be noted that at the end of the table, the accuracy of the Ritz results decreases somewhat and, due to this, the table does not exhibit an exponential convergence of the errors toward zero.

The asymptotic method has been applied to the analysis of rectangular plates [8,127,142], multispan plates [67,69], plate systems [126], multilayer plates and panels of medium thickness [66,114,149], etc. Recently, the following modification of the asymptotic method was developed [128,166]: the solution is sought in the form (4.9) with the functions $\phi(x)$ satisfying both the equation (4.10) and the conditions for $x = 0$ and $x = a$. Thus, the method can be extended to problems with a partial degeneration of the dynamic boundary effect. To find the natural frequencies and the asymptotic relations of the type (4.5), the Rayleigh-Ritz method is used, the asymptotic expressions for natural modes being used as basis functions. According to the comparative estimates [166], this modification permits a much more accurate determination of natural frequencies.

Application of the Asymptotic Method to Shells

Without dwelling on details (see [8]), we shall analyze some results concerning thin elastic shells. Consider a thin elastic shell undergoing free vibrations. Let the typical wave lengths be small as compared to the size of the shell. Then, for flexural modes of vibrations, the following equations can be taken

$$D\Delta\Delta w - \Delta_k\psi - \rho h\omega^2 w = 0 \,, \qquad \Delta\Delta\psi + Eh\Delta_k w = 0 \,, \tag{4.41}$$

where $w(x_1,x_2)$ denotes the normal displacement, $\psi(x_1,x_2)$ denotes the forces in the median surface, Δ is the Laplace operator on the median surface, and Δ_k is Vlasov's operator. If metric and curvature vary but slightly on the median surface, the coordinates x_1 and x_2 can be chosen in such a way that

$$\Delta = \frac{\partial^2}{\partial x_1^2} + \frac{\partial^2}{\partial x_2^2} \,, \qquad \Delta_k = \kappa_2 \frac{\partial^2}{\partial x_1^2} + \kappa_1 \frac{\partial^2}{\partial x_2^2} \,. \tag{4.42}$$

Here κ_1 = const., κ_2 = const. are the mean values of the principal curvatures of the median surface. The asymptotic formula of the type (4.27) for the frequencies of primarily flexural vibrations of the shell whose median surface is a rectangle with the sides a_1 and a_2 has the form

$$\omega^2 \sim \left[(k_1^2+k_2^2)^2 + \frac{k_0^4(k_1^2\gamma+k_2^2)^2}{(k_1^2+k_2^2)^2}\right]\frac{D}{\rho h} \,. \tag{4.43}$$

The notations used in formula (4.43) and further below are

$$k_0 = (Eh\kappa_1^2/D)^{1/4}, \qquad \gamma = \kappa_2/\kappa_1 \tag{4.44}$$

(it is assumed that $\kappa_1 \neq 0$). The wave numbers k_1 and k_2 are determined from the matching conditions (4.31), where the functions $F_\alpha(k_1,k_2)$ and $G_\alpha(k_1,k_2)$ are found from the boundary conditions on the sides $x_1 = 0$, $x_1 = a_1$, $x_2 = 0$, $x_2 = a_2$.

To make use of formula (4.43), it is necessary for the system (4.41) to have solutions of the type of the dynamic boundary effect, and a sufficient number of them to satisfy the boundary conditions on each side of the contour. Formula (4.43) is exact for a simply supported shell when $k_1a_1 = m_1\pi$, $k_2a_2 = m_2\pi$ $(m_1,m_2 = 1,2,\ldots)$. For a spherical shell, the dynamic boundary effect does not degenerate for any wave numbers. Formula (4.43) is valid as an asymptotic relation for spherical shells with an arbitrary contour, (Figure 23 (a) and (b)). Indeed, the median surface can be divided into small rectangular regions, and the solutions for the adjacent regions can be matched by the solution of the type of the dynamic boundary effect. For circular cylindrical shells, the dynamic boundary effect does not degenerate at the circular boundary for any wave numbers, but does at the rectilinear boundary, provided

$$k_1^2 + k_2^2 > k_0^2 . \tag{4.45}$$

In the case of a closed shell, no correction of solutions in the direction of the circular coordinate is necessary. So, the formula (4.43) is fully applicable to closed circular cylindrical shells with arbitrary boundary conditions being homogeneous along each circular boundary (Figure 23(c)). The formula is also applicable to circular cylindrical panels supported along rectilinear edges, since in this case, too, no correction in the direction of the circular coordinate is necessary (Figure 23(d)). Under analogous conditions, the formula extends to shells of positive and negative Gaussian curvature provided principal curvatures can be approximated by some mean values. This is the case, for example, with closed shells of revolution and corresponding panels, supported along generatrices, provided $R_1 \gg R_2$ (Figures 23(e), (f)). In other cases, it is necessary to verify the conditions of application of the asymptotic method. An exhaustive answer to this question is given in the diagram (Figure 24) taken from reference [8]. The region of degeneration of the dynamic boundary effect is shown by thick hatching. This diagram also shows that, if $R_1 \neq R_2$, the dynamic boundary effect degenerates at the boundary of smaller curvature for arbitrary boundary conditions and for sufficiently small wave numbers.

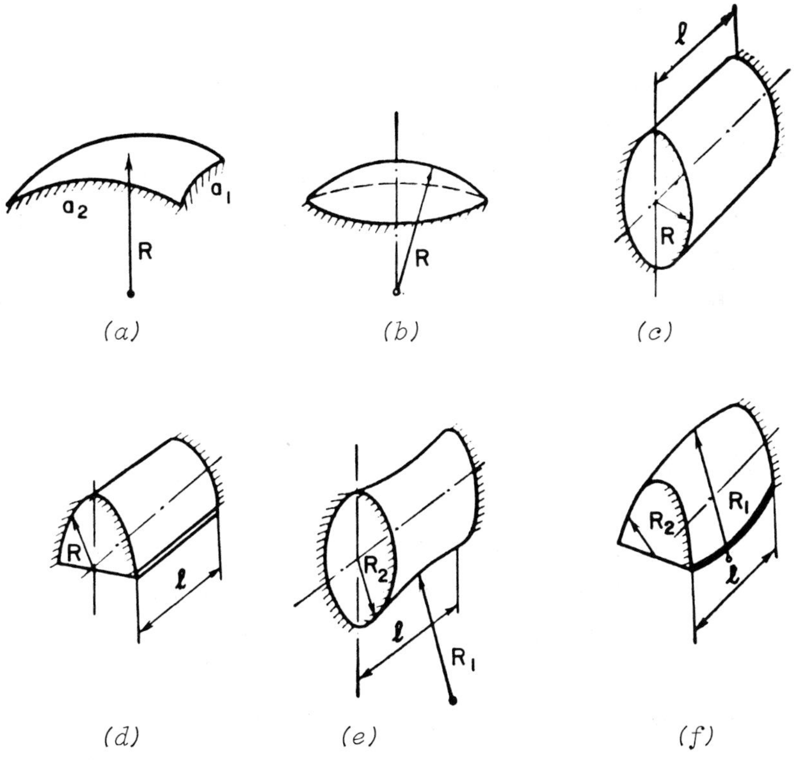

Figure 23

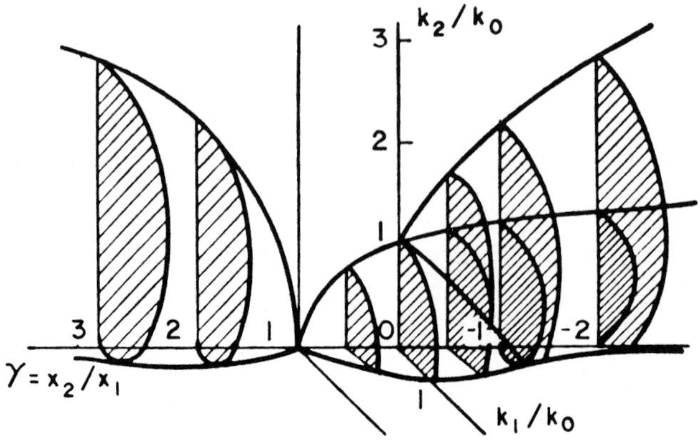

Figure 24

Other examples of applications of the asymptotic method to shells can be found in references [8,30,46,47,70]. A review of recent publications can be found in [166].

4.3 THEORY OF THE DISTRIBUTION OF NATURAL FREQUENCIES OF ELASTIC SYSTEMS

Basic Concepts

Problems concerning the distribution of natural frequencies of an elastic continuum were first formulated in connection with the statistical theory of the heat capacity of solids. Only in this decade have engineers become interested in these problems, mostly in connection with the design of structures subjected to broadband loadings and the interpretation of the corresponding experimental data. It was within this period that the main results concerning the distribution of natural frequencies in thin and thin-walled systems were obtained.

Consider a body whose behaviour is described within the scope of the classical theory of elasticity or its self-adjoint generalizations, or within the scope of one of the applied theories (for instance, the theory of elastic rods, plates, shells). Let the body undergo small free vibrations close to the state of stable equilibrium. These vibrations are described by linear equations with a self-adjoint operator. The body is assumed to be finite, and the Green operator corresponding to it completely continuous. With these limitations, the spectrum of natural frequencies will be discrete and the multiplicity of natural frequencies finite, there being no points of condensation in the spectrum of natural frequencies, except possibly at infinitely distant points. The ordered set of natural frequencies will be denoted by $\underset{\sim}{\Theta} = \{\omega_1, \omega_2, \ldots\}$, and the number of times each frequency is repeated corresponds to its multiplicity.

Let us introduce basic concepts in the theory of the distribution of natural frequencies. Consider the function $N(\omega)$ assuming integer values equal to the number of natural frequencies smaller than ω. We shall refer to this function, which is expressed analytically as

$$N(\omega) = \sum_{\omega_\alpha < \omega} \eta(\omega - \omega_\alpha) , \tag{4.46}$$

as the function of the distribution of natural frequencies. Here $\eta(\omega)$ is the Heaviside function. The derivative

$$\nu(\omega) = \frac{dN(\omega)}{d\omega} , \tag{4.47}$$

is the generalized function

$$\nu(\omega) = \sum_{\omega_\alpha < \omega} \delta(\omega - \omega_\alpha) , \tag{4.48}$$

where $\delta(\omega)$ is the Dirac function. The function $\nu(\omega)$ will be described as the density of natural frequencies. It is evident that both $N(\omega)$ and $\nu(\omega)$ prescribe uniquely the set $\underset{\sim}{\Theta}$. These functions are characteristics of the given elastic body. To distinguish them from the characteristics introduced further below they will be referred to as the exact function of the distribution of natural frequencies and the exact density of natural frequencies, respectively.

Assume now that the spectrum of natural frequencies is sufficiently dense in the sense that the range we are interested in (for instance, the resolving range of measuring equipment) contains a sufficiently large number of frequencies. Then, it is not necessary to know the exact distribution. It is expedient to use approximate distributions instead of the exact one, approximating the piecewise-constant function (4.46) by means of some piecewise-smooth functions. Further below, we shall dwell on asymptotic distributions which are introduced in the following way [12,20]. Let the parameters characterizing the exact distribution of the spectrum be non-dimensional parameters β_1, β_2,... that under certain conditions take on values which are small as compared to unity. For thin and thin-walled systems, relative thickness may for example be such a parameter. We shall refer to the function $\overline{N}(\omega)$ as the asymptotic function of the distribution of natural frequencies with respect to the parameters β_1, β_2,..., if it is related to the exact function of distribution $N(\omega)$ by the expression

$$\frac{|\overline{N}(\omega)-N(\omega)|}{N(\omega)} = O(\beta_1,\beta_2,\ldots) \;. \tag{4.49}$$

Analogously to (4.47), the asymptotic density of natural frequencies is introduced as

$$\overline{\nu}(\omega) = \frac{d\overline{N}(\omega)}{d\omega} \;. \tag{4.50}$$

Since differentiable or piecewise-differentiable functions are used as an asymptotic function of distribution, the asymptotic density [4.50) exists in the general sense everywhere, except possibly for a countable set of points.

Asymptotic distributions are not only of interest because they are analytically more convenient than exact distributions. In many problems in the theory of elastic vibrations, there are parameters which have little influence on the distribution of natural frequencies. For example, under certain conditions, the location of natural frequencies of elastic shells varies only slightly with the change of boundary conditions. For sufficiently high frequencies, in elastic plates, the average number of frequencies per unit frequency range does not depend much on the shape of the plate, etc. If the asymptotic frequency distribution contains only the essential parameters of the problem, it has the additional advantage of being applicable to a certain class of elastic systems, whereas the exact distribution is the characteristic of the given system only.

The asymptotic distribution, unlike the exact distribution, is not introduced in a unique way. Firstly, various asymptotic approximations can be used for one and the same class of systems. Secondly, adding to the asymptotic function of distribution terms of the order of an asymptotic error (4.49), we obtain a new distribution which is asymptotically equal to the initial one. And, lastly, narrowing the class of elastic systems and, hence, increasing the number of the parameters β_1, β_2, ..., we obtain more exactly defined asymptotic distributions.

Now, let us introduce the concept of the empirical distribution of natural frequencies [12]. Assume that the set $\underset{\sim}{\Theta}$ was determined by calculation or experimentally. Grouping frequencies over some sufficiently broad ranges, we shall estimate the mean density of the spectrum. We shall refer to this density as empirical and denote it by $\tilde{\nu}(\omega)$. Its relation to the exact distribution is given by the formula

$$\tilde{\nu}(\omega) = \frac{N(\Omega_k)-N(\Omega_{k-1})}{\Omega_k-\Omega_{k-1}} \quad (\Omega_{k-1} \leq \omega < \Omega_k) , \tag{4.51}$$

where Ω_1, Ω_2, ..., are the points of spectrum dissection. The empirical distribution function $\tilde{N}(\omega)$ will, analogously to (4.47) and (4.50), be introduced as the integral of the empirical density $\tilde{\nu}($ Thus, the empirical function of distribution is piecewise-linear, and the empirical density is a piecewise-constant function. It is evident that the empirical distribution for a given system may vary, i.e., varying the location of the points Ω_1, Ω_2, ..., we shall arrive at new distributions.

There is a useful analogy between the concepts of the theory of distribution of natural frequencies and some basic concepts of probability theory and mathematical statistics. The function of the distribution of natural frequencies corresponds to the function of probability distribution, and the density of natural frequencies to the probability density. The analogy will be closer if the spectrum is bounded by some limiting frequency ω_c, and the functions of the distribution of frequencies are normalized to unity, so that $N(\omega_c) = \overline{N}(\omega_c) = \tilde{N}(\omega_c) = 1$. Then, the function of the distribution of frequencies can be interpreted as the probability of a random event, that is, a frequency taken at random from the range $[0, \omega_c]$ will be smaller than ω. This interpretation becomes more valuable because the empirical distribution plays a role analogous to the role of a statistical estimate with respect to the exact distribution.

General Formulae for Calculating Asymptotic Functions of Distribution

Let the set θ be ordered by means of n parameters k_1, k_2, ..., k_n, which take on discrete positive values. These may be natural numbers, wave lengths characterizing the corresponding vibration modes, or values inversely proportional to wave lengths. Taking into consideration that the latter case is the most common in applications, we shall refer to the parameters k_1, k_2, ..., k_n as wave numbers (as we did in Section 4.1), and to their corresponding vector $\underset{\sim}{k}$ as the wave vector. In typical situations, the dimension of the space of wave numbers $\underline{K}$ coincides with the dimension of the physical space. The geometrical interpretation of the space $\underline{K}$ was given in Figure 22, where a lattice point corresponds to each frequency in the spectrum.

In order to obtain the asymptotic functions of distribution, it is necessary to have an exact or approximate analytical expression for natural frequencies as functions of the wave numbers. Assume that the natural frequency is a slowly varying function $\omega = \Omega(\underset{\sim}{k})$ of wave numbers. Then, in order to obtain approximate estimates, we can treat wave numbers as continuous arguments, while prescribing the spectrum of natural frequencies by means of the function $\Omega(\underset{\sim}{k})$ and the cell of volume $\Delta\underset{\sim}{k}$ falling within one frequency of the spectrum. The number of natural frequencies smaller than the prescribed value ω will be defined as the ratio of the volume of the region in the first octant $\underline{K}_+$ of the space limited by the surface $\Omega(\underset{\sim}{k}) = \omega$ versus the volume of one cell. Thus, we obtain the approximate formula for the function of distribution of natural frequencies*

$$\overline{N}(\omega) = \frac{1}{\Delta\underset{\sim}{k}} \int_{\Omega(\underset{\sim}{k})<\omega} d\underset{\sim}{k} \ . \qquad (4.52)$$

This formula is asymptotic in the sense that the greater the number of frequencies in the considered region is and the slower the function $\Omega(\underset{\sim}{k})$ varies, the more exact the formula is.

The asymptotic density of natural frequencies is obtained from (4.52) by differentiating with respect to ω. Transforming the volume integral (4.52) into the surface integral, it is not difficult to obtain the explicit formula for asymptotic density. Let $\Omega(\underset{\sim}{k})$ be a differentiable function. Then,

---

*This approach was suggested by Courant in: Courant, R., Hilbert, D., *Methods of Mathematical Physics*, Vol. 2, Moscow, Gostekhizdat, 1951.

$$\bar{\nu}(\omega) = \frac{1}{\Delta \underset{\sim}{k}} \int_{\Sigma(\omega)} \frac{d\Sigma}{|\text{grad } \Omega(\underset{\sim}{k})|} , \tag{4.53}$$

where $\Sigma(\omega) = \{\underset{\sim}{k}:\Omega(\underset{\sim}{k}) = \omega\}$, i.e., the integration in (4.53) is performed over the surface of equal values of the natural frequency in the space of wave numbers.

In the case when the function $\Omega(\underset{\sim}{k})$ is not single-valued, the single-valued branches $\Omega_j(\underset{\sim}{k})$ $(j = 1,2,\ldots,m)$ should be considered, and the formulae (4.52) and (4.53) should be applied to each branch separately. Instead of (4.52), we obtain the asymptotic estimate

$$\bar{N}(\omega) = \sum_{j=1}^{m} \int_{\Omega_j(\underset{\sim}{k})<\omega} \frac{d\underset{\sim}{k}}{\Delta_j \underset{\sim}{k}} , \tag{4.54}$$

where allowance is made for the cell volume not being constant within the scope of the considered space region $\underline{K}$. The formula (4.53) is generalized in the following way:

$$\bar{\nu}(\omega) = \sum_{j=1}^{m} \int_{\Sigma_j(\omega)} \frac{d\Sigma}{|\text{grad } \Omega_j(\omega)|\Delta_j \underset{\sim}{k}} . \tag{4.55}$$

Here, $\Sigma_j(\omega) = \{\underset{\sim}{k}:\Omega_j(\underset{\sim}{k}) = \omega\}$. The asymptotic formulae admit of further refinement at the expense of a correction of the integration region in the space $\underline{K}$. For example, the influence of boundary conditions can be refined by excluding the layers adjacent to the coordinate planes.

Asymptotic Points of Condensation of Natural Frequencies

The asymptotic density $\bar{\nu}(\omega)$ may have singularities which follow from formulae (4.53) and (4.55). Actually, if the gradient of the function $\Omega(\underset{\sim}{k})$ becomes zero at some point of the surface $\Sigma(\omega)$, the integral (4.53) may diverge for this frequency ω. In the latter case, we can say that there is an asymptotic point of condensation of natural frequencies. At this point, the tangent to the curve $\overline{N} = \overline{N}(\omega)$ becomes vertical. It should be noted that asymptotic condensation points are characteristics of the asymptotic (approximate) distributions, not of the exact distributions. The mean density of frequencies will increase near the condensation points, and the closer the asymptotic estimate is to the exact distribution, the sharper the increase will be.

The relation grad $\Omega(\underset{\sim}{k}) = 0$ is a necessary condition for the existence of an asymptotic condensation point. If the space $\underline{K}$ is one-dimensional, this condition will also be sufficient. In the two-dimensional case, the behaviour of the function $\Omega(\underset{\sim}{k})$ in the neighbourhood of the critical point grad $\Omega(\underset{\sim}{k}) = 0$ will determine the existence of a singularity. For example, if the Hessian of the function $\Omega(\underset{\sim}{k})$ at this point is different from zero, there will be singularity only in the case that the critical point is a saddle point.

Some Elementary and Classical Results

Consider a one-dimensional system. Let $\Omega(k)$ be a single-valued nondecreasing function, $k = K(\omega)$ the inverse function, and assume the interval on the k axis corresponding to one frequency to be of length $\Delta k = \text{const}$. Formulae (4.52) and (4.53) then take the form

$$\overline{N}(\omega) = \frac{1}{\Delta k} K(\omega) , \qquad \overline{\nu}(\omega) = \frac{1}{\Delta k} \frac{dK(\omega)}{d\omega} . \tag{4.56}$$

As the simplest example, let us take the longitudinal vibrations of a uniform prismatic rod of length ℓ. These vibrations are described by a one-dimensional wave equation. It is not difficult to show that the frequencies of natural vibrations, independently of boundary conditions, satisfy the asymptotic relation

$$\omega \sim k(E/\rho)^{1/2} , \qquad k\ell = m\pi + O(1) , \tag{4.57}$$

where E is the elasticity modulus and ρ is the density of the material. Noting that the size of the cell is $\Delta k = \pi/\ell$, we find from the formulae (4.56) the relations

$$\bar{N}(\omega) = \omega/\omega_0 , \quad \bar{\nu}(\omega) = 1/\omega_0 , \quad \omega_0 = \pi/\ell(E/\rho)^{1/2} . \tag{4.58}$$

Thus, the asymptotic function of the distribution of natural frequencies is linear, and the asymptotic density is constant. We obtain an analogous distribution for other problems that can be reduced to a one-dimensional wave equation.

Consider flexural vibrations of a uniform prismatic rod of length ℓ. Unlike (4.57), we have as the asymptotic value for the natural frequencies the expression

$$\omega \sim k^2(EI/\rho F)^{1/2} , \qquad k\ell = m\pi + O(1) , \tag{4.59}$$

which is independent of boundary conditions. Here, I is the moment of inertia and F is the cross-sectional area. It follows from formulae (4.56) that

$$\bar{N}(\omega) = \left(\frac{\omega}{\omega_0}\right)^{1/2} , \quad \bar{\nu}(\omega) = \frac{1}{2(\omega\omega_0)^{1/2}} , \quad \omega_0 = \frac{\pi^2}{\ell^2}\left(\frac{EI}{\rho F}\right)^{1/2} . \tag{4.60}$$

The asymptotic density of frequencies has a singularity for $\omega = 0$. It implies that the mean density of frequencies is maximal in the region adjacent to the origin of the coordinates. For example, for a simply-supported rod, the frequencies relate as 1:4:9... This simple example illustrates the meaning of the asymptotic condensation points.

The density of the natural frequencies of flexural vibrations for thin elastic plates of constant thickness was first estimated by Courant.* If the plate is rectangular, with sides a_1 and

---

*This approach was suggested by Courant in: Courant, R., Hilbert, D. *Methods of Mathematical Physics*, Vol. 2, Moscow, Gostekhizdat, 1951.

a_2, then, independently of boundary conditions, the asymptotic expression for natural frequencies has the form (4.27). The size of the cell on the surface $\underline{K}$ corresponding to one frequency will be $\Delta\underset{\sim}{k} = \pi^2/a_1a_2$. The application of formula (4.52) yields

$$\bar{N}(\omega) = \frac{\omega}{\omega_0}, \quad \bar{\nu}(\omega) = \frac{1}{\omega_0}, \quad \omega_0 = \frac{4\pi}{a_1a_2}\left(\frac{D}{\rho h}\right)^{1/2}. \tag{4.61}$$

Thus, the asymptotic distribution function is linear, and the asymptotic density of the frequencies is constant:

$$\bar{\nu}(\omega) = \nu_0 = \frac{a_1a_2}{4\pi}\left(\frac{\rho h}{D}\right)^{1/2}. \tag{4.62}$$

If the plate is different from a rectangle in shape, and the wave numbers are sufficiently large, the median surface of the plate can be approximately represented as a set of subregions, rectangular in plan. Non-degeneration of the dynamic boundary effect for plates (see Section 4.2) allows an approximate matching of solutions for adjacent subregions. As a result, we arrive at formula (4.62) in which the product of a_1a_2 is substituted by the median surface area of the plate A. An analogous result was obtained by Courant from energy conditions. He also considered the related problem of the distribution of natural frequencies of a homogeneous membrane.

If the typical wave lengths are comparable with the plate's thickness, then, either the refined theories of plates or complete equations of the theory of elasticity are to be used. The correction for the density of natural frequencies (4.62) based on the refined Reissner equations was given in [114]. The graph for the relation $\bar{\nu}(\omega)/\nu_0$ is given in Figure 25; ω_h denotes the typical frequency

$$\omega_h = \frac{12\gamma(1-\mu)}{2+\gamma(1-\mu)}\,\frac{1}{h^2}\left(\frac{D}{\rho h}\right)^{1/2}. \tag{4.63}$$

(γ is a nondimensional coefficient, generally assumed to be equal to 5/6, μ is Poisson's ratio). The results are valid for the principal branch of frequencies beyond the domain of degeneration of the dynamic boundary effect.

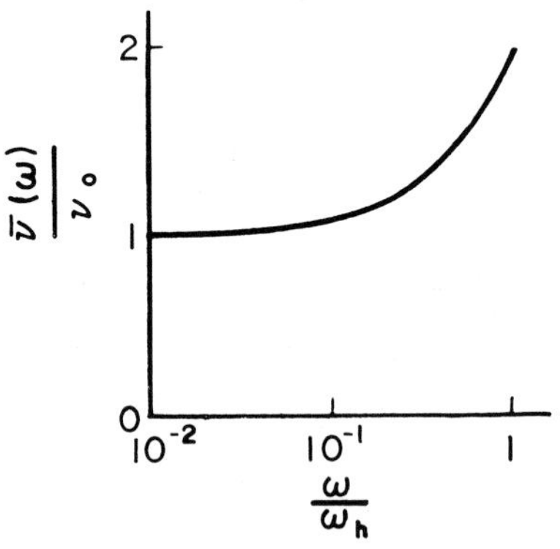

Figure 25

The density of the frequencies for a rectangular orthotropic plate whose elasticity axes are parallel to its sides was found by Wilkinson [172]:

$$\bar{\nu}(\omega) = \frac{a_1 a_2}{2\pi^2}\left(\frac{\rho^2 h^2}{D_{11} D_{22}}\right)^{1/4} K(c) , \qquad 2c^2 = 1 - \frac{D_{12}}{(D_{11} D_{22})^{1/2}} . \tag{4.64}$$

Here, $K(c)$ is Legendre's complete elliptic integral of the first kind, D_{11} and D_{22} are cylindrical stiffnesses, D_{12} is the mixed stiffness of the plate. As can be seen from formula (4.64), the asymptotic density is independent of the frequency. Wilkinson [172] also considered the case of a plate loaded on the median surface.

Let us consider natural vibtations of a rectangular parallelepiped with boundary conditions analogous to the support conditions assumed in the theory of rods, plates and shells. The formulae for natural frequencies of the parallelepiped with the sides a_1, a_2 and a_3 have the form

$$\omega = (k_1^2+k_2^2+k_3^2)^{1/2} c_\ell , \qquad \omega = (k_1^2+k_2^2+k_3^2)^{1/2} c_t ,$$

where c_ℓ is the velocity of the propagation of expansion waves, and c_t is the velocity of the propagation of shear waves. Wave numbers are introduced as

$$k_1 = \frac{m_1\pi}{a_1} , \qquad k_2 = \frac{m_2\pi}{a_2} , \qquad k_3 = \frac{m_3\pi}{a_3} \quad (m_1, m_2, m_3 = 1, 2, \ldots) ,$$

i.e., the size of one cell in the space $\underline{K}$ is $\pi^3/a_1a_2a_3$. Taking into account that frequencies corresponding to shear deformations have a double multiplicity, we find from formulae (4.50) and (4.52)

$$\bar{\nu}(\omega) = \frac{a_1a_2a_3\omega^2}{2\pi^2} \left(\frac{1}{c_\ell^3} + \frac{2}{c_t^3} \right) . \qquad (4.65)$$

The theory of heat capacity of solids is based on formula (4.65). The question whether the formula (4.65) holds true for a broader class of boundary conditions, and especially for elastic bodies of arbitrary form, calls for further investigations.

Let us consider a formula for the asymptotic density of the medium whose vibrations are described by the wave equation in the space of n dimensions. Let the medium be a rectangular parallelepiped with the sides $a_1, a_2, \ldots, a_n$. Then,

$$\bar{\nu}(\omega) = \frac{n a_1 a_2 \ldots a_n}{2^n \pi^{n/2} \Gamma(1+n/2)} \; \frac{\omega^{n-1}}{c^n} . \qquad (4.66)$$

Here, c is the velocity of wave propagation, and $\Gamma(x)$ is the gamma-function.

Effectiveness of Asymptotic Estimates

To make the estimate of frequency density sufficiently effective, it is necessary for the frequency range $\Delta\omega$, for which the estimate is used, to have a sufficiently large number of frequencies. Hence,

$$\bar{\nu}(\omega)\Delta\omega >> 1 . \tag{4.67}$$

For example, for a thin plate with $\Delta\omega >> \omega$, condition (4.67) with formula (4.62) taken into account has the form

$$A/(\lambda^2(\omega)) >> 1 . \tag{4.68}$$

Here, $\lambda(\omega)$ has order $\max\{k_1^{-1},k_2^{-1}\}$, i.e., has the meaning of a typical wave length. So, the wave length should be small as compared to the size of the median surface. The rigor of (4.68) increases if the range $\Delta\omega$ is sufficiently narrow (for example, if it is equal to the width of a resonance peak, or to the width of the passage of the spectrum analyser).

Criterion (4.67) also remains valid for empirical estimates. There is an additional requirement for the selection of the intervals $\Delta\omega_k = \Omega_k - \Omega_{k-1}$ in formula (4.51). Let us again take the example of a thin plate. Figure 22 showed the plane $\underline{K}$ of wave numbers. The natural frequencies (4.27) correspond to the lattice points on this plane. The lattice deforms with the change of boundary conditions. If we consider the empirical distribution of frequencies irrespective of the type of boundary conditions, the number of (lattice) points within the band $\omega < \Omega(\underset{\sim}{k}) < \omega+\Delta\omega$ can be interpreted as a random value. If $\Delta\omega$ = const., then, the higher the frequency ω, the greater is the fluctuation of the number of points in this narrowing band; and, consequently, the higher the level is of the fluctuations of the empirical density $\tilde{\nu}(\omega)$ in relation to the asymptotic density ν_0. It is not difficult to estimate the order of fluctuations. The greatest deviation of the value $\tilde{\nu}(\omega)\Delta\omega$ from the mean value $\nu_0\Delta\omega$ has the order of the number of cells $\Delta k_1 \Delta k_2$ in the area covering the line $\Omega(\underset{\sim}{k}) = \omega$. This number has the order $\nu_0^{1/2}\omega^{1/2}$. Hence,

$$\max\left|\frac{\tilde{\nu}(\omega)-\nu_0}{\nu_0}\right| \sim \frac{(\omega_0\omega)^{1/2}}{\Delta\omega}, \tag{4.69}$$

i.e., the level of fluctuations is inversely proportional to the width of the frequency interval and proportional to the square root of the frequency. To keep fluctuations on one level, it is necessary to increase the size of intervals as the frequency increases. This is taken into account in Figures 30 and 31 given below, where starting from a certain frequency, the double interval $\Delta\omega$ is introduced.

4.4 DENSITY OF NATURAL FREQUENCIES OF THIN ELASTIC SHELLS

Deriving the Basic Formulae

Apply the general relations (4.50) and (4.52) to thin elastic shells. Let all the conditions be satisfied under which the equations (4.41) and the asymptotic formula (4.43) for natural frequencies are valid and the mean density of frequencies is sufficiently large. Let us number the coordinates x_1, x_2 so that $\gamma = \kappa_2/\kappa_1 \leq 1$. Let us introduce the notations

$$\omega_1 = |\kappa_2|(E/\rho)^{1/2}, \qquad \omega_2 = |\kappa_1|(E/\rho)^{1/2}. \tag{4.70}$$

Here, $\omega_1 \leq \omega_2$. Substituting the asymptotic expression (4.43) into the formula (4.52), we shall turn to polar coordinates $k_1 = k\cos\Theta$, $k_2 = \sin\Theta$. After integration with respect to the polar radius k, we obtain [12]

$$\overline{N}(\omega) = \frac{a_1 a_2}{2\pi^2}\left(\frac{\rho h}{D}\right)^{1/2}\int_{\Theta_1}^{\Theta_2}[\omega^2-\omega_2^2(\gamma\cos^2\theta+\sin^2\theta)^2]^{1/2}d\Theta. \tag{4.71}$$

Here, $[\Theta_1, \Theta_2]$ is the domain of values Θ from the interval $[0, 1/2\pi]$ in which the expression in parenthesis under the integral sign is non-negative. Differentiating (4.71), we find the asymptotic density of natural frequencies [12]

$$\bar{\nu}(\omega) = \frac{a_1 a_2 \omega}{2\pi^2}\left(\frac{\rho h}{D}\right)^{1/2} \int_{\Theta_1}^{\Theta_2} [\omega^2 - \omega_2^2(\gamma \cos^2\Theta + \sin^2\Theta)^2]^{-1/2} d\Theta . \tag{4.72}$$

The integrals in (4.71) and (4.72) are calculated easily for $\gamma = 1$, and for $\gamma \neq 1$ they are expressed in terms of complete elliptic integrals. Here are the formulae for $\bar{\nu}(\omega)$:

$$\frac{\bar{\nu}(\omega)}{\nu_0} = \begin{cases} 0 , & \text{if } 1-\alpha\gamma < 0, \\ \dfrac{2}{\pi\sqrt{(1+\alpha)(1-\alpha\gamma)}} K\left[\sqrt{\dfrac{2\alpha(1-\gamma)}{(1+\alpha)(1-\alpha\gamma)}}\right], & \text{if } 1-\alpha\gamma > 0,\ \alpha < 1, \\ \dfrac{2}{\pi\sqrt{2\alpha(1-\gamma)}} K\left[\sqrt{\dfrac{(1+\alpha)(1-\alpha\gamma)}{2\alpha(1-\gamma)}}\right], & \text{if } 1-\alpha\gamma > 0,\ \alpha > 1, \\ \dfrac{1}{\sqrt{1-\alpha^2}} , & \text{if } 1-\alpha\gamma > 0,\ \gamma = 1. \end{cases} \tag{4.73}$$

In (4.73), $\alpha = \omega_2/\omega$, ν_0 is the asymptotic density for the plate.

Asymptotic Frequency Condensation Points and Their Interpretation

It follows from formulae (4.73) that the asymptotic density of frequencies has singularities [12]. For $\gamma = 1$, i.e., for a spherical shell, the density tends to infinity like $(\omega-\omega_2)^{-1/2}$ for the frequency $\omega_1 = \omega_2 = \omega_R$, where ω_R denotes the natural frequency of centrally-symmetrical (momentless) vibrations:

$$\omega_R = \frac{1}{R}\left(\frac{E}{\rho}\right)^{1/2} . \tag{4.74}$$

The graph for $\bar{\nu}(\omega)$ is shown in Figure 26(a). At $\gamma \neq 0$, the function $\bar{\nu}(\omega)$ can have a singularity of the same type as in the complete elliptic integral (logarithmic singularity). For shells of positive Gaussian curvature ($\gamma > 0$), the asymptotic density is equal to zero in the interval $0 \leq \omega \leq \omega_1$, has a singularity at $\omega = \omega_2$, and tends to the density ν_0 for $\omega \to \infty$ (Figure 26(b)). For a shell of zero Gaussian curvature, the asymptotic density is different from zero everywhere and has a singularity at $\omega = \omega_2$. In particular, for a circular cylindrical shell, $\omega_2 = \omega_R$, where ω_R is determined from (4.74), and R is the radius of the cross-section of the median surface. The frequency ω_R in this case is equal to the natural frequency of axisymmetrical vibrations of a ring cut out of the shell by two cross-sections. The graph for the density $\bar{\nu}(\omega)$ is shown in Figure 26(c). And, lastly, for shells of negative Gaussian curvature ($\gamma < 0$), there may be two singularities corresponding to ω_1 and ω_2 (Figure 26(d)).

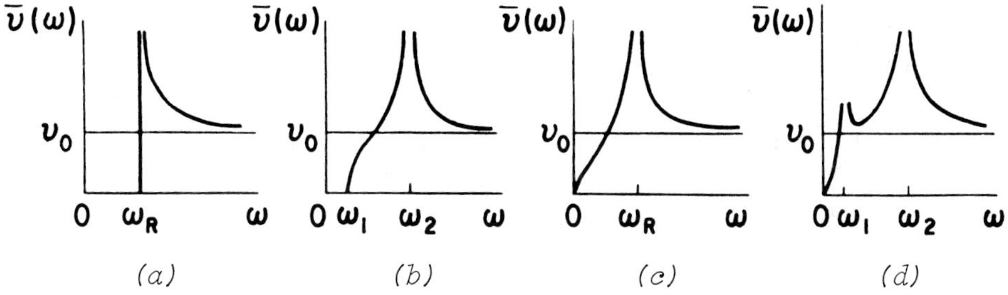

Figure 26

As it has already been mentioned, the considered condensation points relate only to the asymptotic estimates of the distribution of natural frequencies. If these estimates are sufficiently effective, the maximum of the empirical density will correspond to these condensation points. The condensation of frequencies in the vicinity of ω_1 and ω_2 can be physically substantiated in the following way: The frequencies ω_1 and ω_2 are certain typical natural frequencies of a momentless shell. A thin shell has the greatest stiffness in relation to momentless strains.

So, superposition of bending strains does not always cause an essential change of natural frequencies. Thus, a very large number of natural frequencies is concentrated in the vicinity of ω_1 and ω_2. Figures 27 and 28 show the results of a numerical experiment given in [114]. Figure 27 refers to a spherical panel with the parameters $a_1 = a_2 = 1\text{m}$, $R = 2\text{m}$, $h = 2 \cdot 10^{-3}\text{m}$, $E = 1.4 \cdot 10^{11}\text{N} \cdot \text{m}^{-2}$, $\mu = 0.3$, $\rho = 2.7 \cdot 10^3\text{kg} \cdot \text{m}^{-3}$. For Figure 28, the length of the cylindrical panel is $\ell = 2\text{m}$; the remaining data are unchanged. The edges of the panels are assumed to be simply supported.

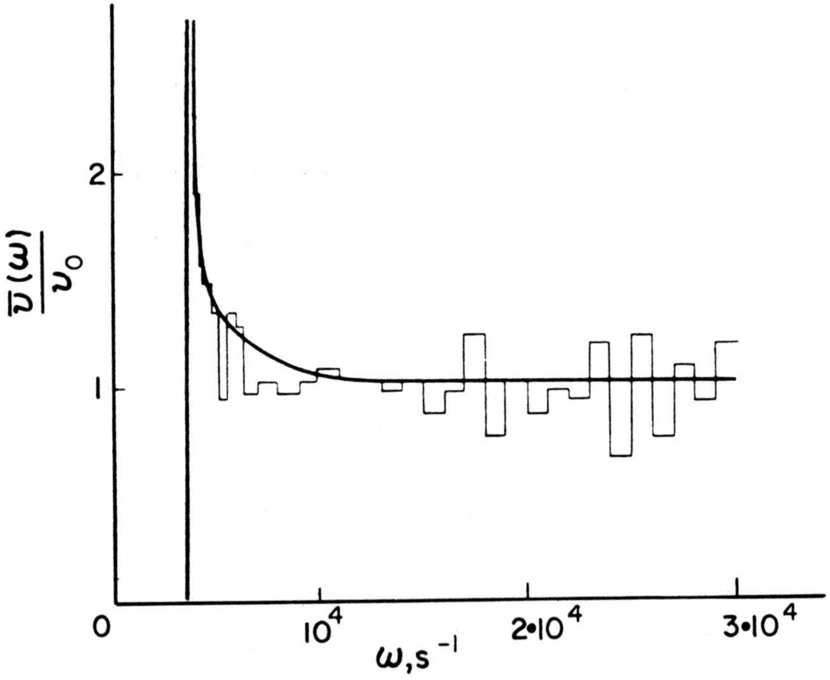

Figure 27

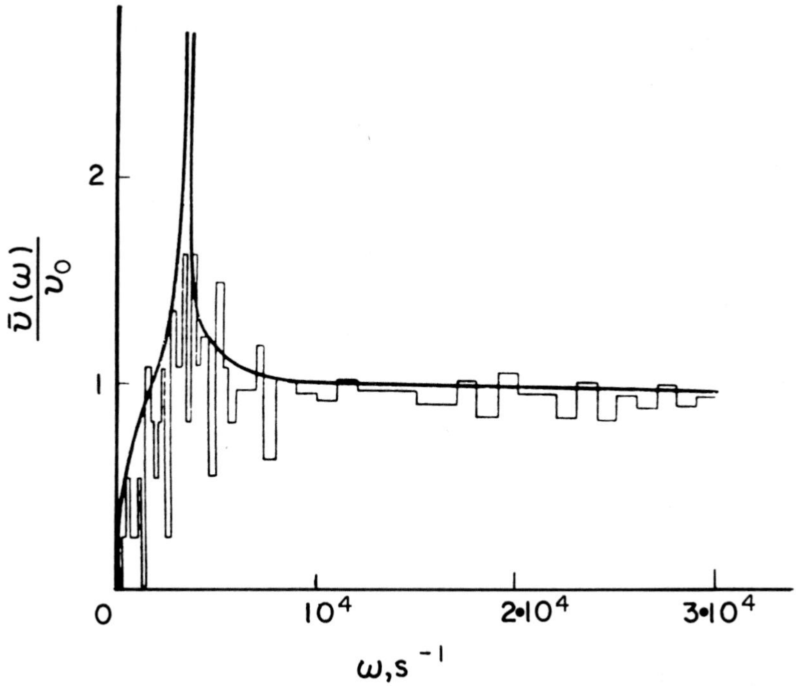

Figure 28

Closed Circular Cylindrical Shell

In this case, all natural frequencies except for those which correspond to axisymmetrical vibrations will have double multiplicity. Taking account of all the frequencies once, it is to be assumed that $a_1 = \pi R$, $a_2 = \ell$ in the formula (4.62) for ν_0. Here, ℓ is the length of the shell. As a result we obtain

$$\bar{\nu}(\omega) = \begin{cases} \dfrac{R\ell}{2\pi}\left(\dfrac{\rho h}{D}\right)^{1/2}\left(\dfrac{\omega}{2\omega_R}\right)^{1/2} K\left[\left(\dfrac{\omega+\omega_R}{2\omega_R}\right)^{1/2}\right], & \text{if } \omega < \omega_R , \\ \dfrac{R\ell}{2\pi}\left(\dfrac{\rho h}{D}\right)^{1/2}\left(\dfrac{\omega}{\omega+\omega_R}\right)^{1/2} K\left[\left(\dfrac{2\omega_R}{\omega+\omega_R}\right)^{1/2}\right], & \text{if } \omega > \omega_R . \end{cases} \tag{4.75}$$

The formula for the asymptotic density $\bar{\nu}_1(\omega)$, which takes account of the multiplicity of frequencies, has the form

$$\bar{\nu}_1(\omega) = 2\bar{\nu}(\omega) - \bar{\nu}_s(\omega) , \tag{4.76}$$

where $\bar{\nu}_s(\omega)$ is the density of the frequencies that correspond to axisymmetric modes. The density $\bar{\nu}_s(\omega)$ is determined by the formula

$$\bar{\nu}_s(\omega) = \begin{cases} 0\,, & \text{if } \omega < \omega_R\,, \\ \dfrac{\ell}{2\pi}\left(\dfrac{\rho h}{D}\right)^{1/4} \dfrac{\omega}{(\omega^2-\omega_R^2)^{3/4}}\,, & \text{if } \omega > \omega_R\,. \end{cases} \tag{4.77}$$

This formula follows from (4.56) if it is assumed that $\gamma = 0$, $k_1 = 0$, $k_2 = k$ in (4.43) and hence, determines the function $K(\omega)$. In Figure 29, the graphs of functions (4.76) and (4.77) with $h = 0.001R$, $\ell = 2R$, are plotted in thin lines. As can be seen from Figure 29, the contribution of axisymmetric modes to the asymptotic density of frequencies is not very large. So, the singularity of function $\bar{\nu}(\omega)$ at $\omega = \omega_R$ should not be ascribed only to the axisymmetric vibrations.

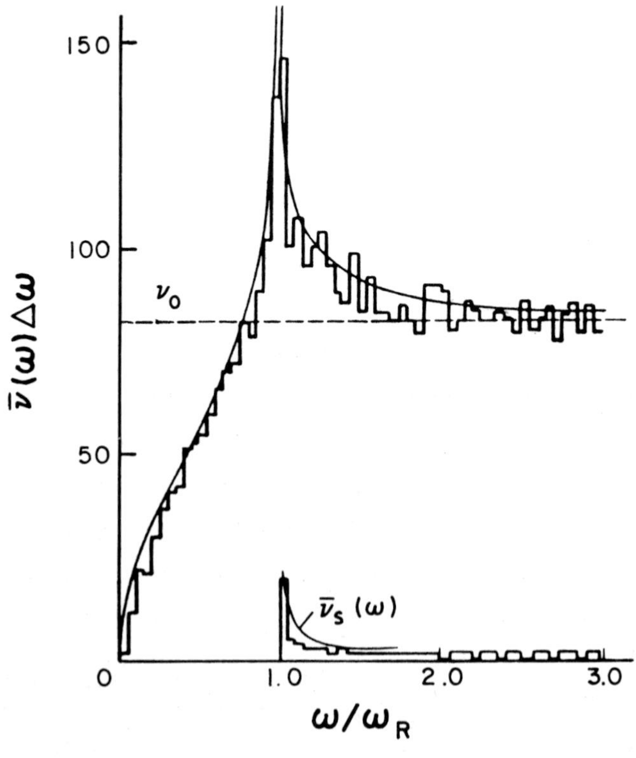

Figure 29

In Figure 29, the empirical distribution of frequencies for a closed circular cylindrical shell with clamped ends [46] is given. The parameters of the shell are $R = 1m$, $\ell = 2m$, $h = 10^{-3}m$, $E = 1.4 \cdot 10^{11} N \cdot m^{-2}$, $\rho = 2.7 \cdot 10^{3} kg \cdot m^{-3}$, $\mu = 0.3$. The modes of vibrations were approximated by the sum of products of beam functions for the axis coordinate, and by trigonometric functions for the circular coordinate. Calculations were performed with over 60 beam functions retained (the equations separate with respect to the circular coordinate). The total number of frequencies taken into account exceeded 3000. The results of the calculations by the asymptotic formula are shown by a thick unbroken line. Multiple frequencies were taken into account only once.

The effectiveness of asymptotic estimates for the frequency density is determined by criterion (4.67). For a cylindrical shell or a spherical one with $\Delta\omega = \omega_R$, this criterion yields

$$A/(Rh) \gg 1 ,$$

where A is the area of the median surface of the shell. Moreover, it is assumed that the mean frequency density is of the order ν_0 within the considered interval.

Equations (4.41) as well as formulae (4.43), (4.73), (4.75) and (4.77) are constructed without taking account of the tangential forces of inertia. Therefore, the natural frequencies related primarily to the tangential motion of the shell are not considered. The theory can be extended to the case when the tangential motion is taken into account, the asymptotic distributions of natural frequencies being determined by formulae (4.54), (4.55), where $j = 1,2,3$. The density of natural frequencies related to tangential motions is of the order $\nu_0 h/R$.

Figure 30 shows the data of a physical experiment reported in [155]. Steel shells ($R = 56mm$, $h = 1.6mm$, $\ell = 900mm$) were tested within the frequency range of 200 - 20000Hz. The frequency corresponding to the condensation point was 13,800 Hz. The points indicate the number of resonance peaks in the standard one third of the octave ranges. The elliptical integral in formula (4.72) was determined by numerical methods without taking account of the singularity at $\omega = \omega_R$. The scatter of the experimental data at small frequencies is due to the fact that for these the frequency density is small. The maximum of the empirical density shows up sufficiently clearly.

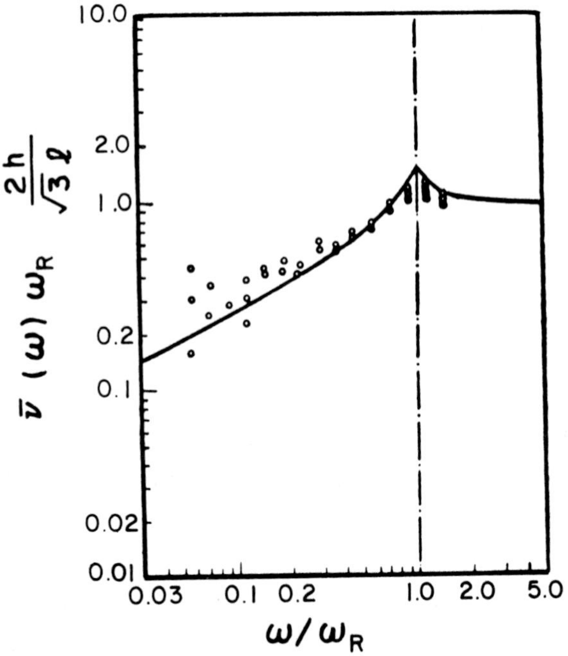

Figure 30

Some Other Results

Wilkinson [172] estimated the asymptotic density of natural frequencies for a three-layer spherical panel. Condensation points analogous to those in Figure 26(a) were obtained. Different from the homogeneous panel, here, the asymptotic density $\overline{\nu}(\omega)$ does not tend to a constant value with the increase of ω. In the same paper, a flat panel on an elastic base of Winkler's type is considered. In that case, additional asymptotic condensation points are found for the frequencies corresponding to the displacements of a rigid panel on an elastic base.

Vibrations of an orthotropic circular cylindrical shell were considered by N. Zhinzher and V. Khromatov [46]. The asymptotic density $\bar{\nu}(\omega)$ in this case is given by the relation

$$\bar{\nu}(\omega) = \frac{R\ell\omega}{2\pi}\left(\frac{\rho^2 h^2}{D_{11}D_{22}}\right)^{1/4} \times$$

$$\int_{\Theta_1(\omega)}^{\Theta_2(\omega)} \frac{d\Theta}{(1-c^2\sin^2 2\Theta)^{1/2}\left[\omega^2 - \dfrac{\omega_R^2\cos^4\Theta}{\cos^4\Theta+1/2 b_{12}d_{11}\sin^2 2\Theta+d_{11}^2 b_{11}\sin^4\Theta}\right]^{1/2}} . \tag{4.78}$$

Here, the notations

$$d_{11} = \left(\frac{D_{11}}{D_{22}}\right)^{1/2}, \quad b_{11} = \frac{B_{11}}{B_{22}}, \quad b_{12} = \frac{B_{12}}{B_{22}}, \quad \omega_R = \frac{1}{R}(\rho h B_{22})^{-1/2},$$

are used. In these relations, D_{jk} is the tensor of cylindrical stiffnesses, B_{jk} is the tensor of compliances for momentless deformation, and c^2 is determined from the second formula (4.64). The coordinate x_1 is measured along the generatrix, so the frequency ω_R, as in the case of an isotropic shell, has the meaning of the natural frequency for a corresponding ring. The asymptotic condensation point, as before, will be at $\omega = \omega_R$. If $R \to \infty$, the formula (4.78) transforms to (4.64). In the general case, formula (4.78) contains a hyperelliptical integral, which has been determined by a numerical method. In [46], the empiric distributions obtained by computer calculations of frequencies are given.

Khromatov [102] considered vibrations of a circular cylindrical shell, loading on the median surface by axial forces N_1 = const., N_2 = const. The integrals in the formulae of the type (4.72) are reduced to tabulated elliptic integrals. Let us consider the case when $N_2 = 0$, i.e., the shell is loaded along the generatrices. N_* will denote the critical axial compression force determined according to the classical theory, i.e.,

$$N_* = -\frac{1}{\sqrt{3(1-\mu^2)}} \frac{Eh^2}{R} .$$

If $0 < N_1 < |N_*|$, then, the singularity of asymptotic density for $\omega = \omega_R$ changes to an isolated maximum for the same frequency. For $N_1 \geq |N_*|$, the isolated maximum disappears. If the force N_1 is compressive and subcritical, i.e., $N_* < N_1 < 0$, the singularity stays but shifts in the direction of the frequency ω_*, where

$$\omega_* = \omega_R(1-N_1^2/N_*^2)^{1/2} .$$

Figure 31 shows an empirical distribution of the frequencies for a shell with the parameters $h/R = 10^{-3}$, $\ell/R = 2$, and with the tensile forces $N_1 = |N_*|$. The asymptotic distribution is shown by a thin unbroken line. Figure 32 gives an analogous graph for compressive forces $N_1 = 0.5N_*$.

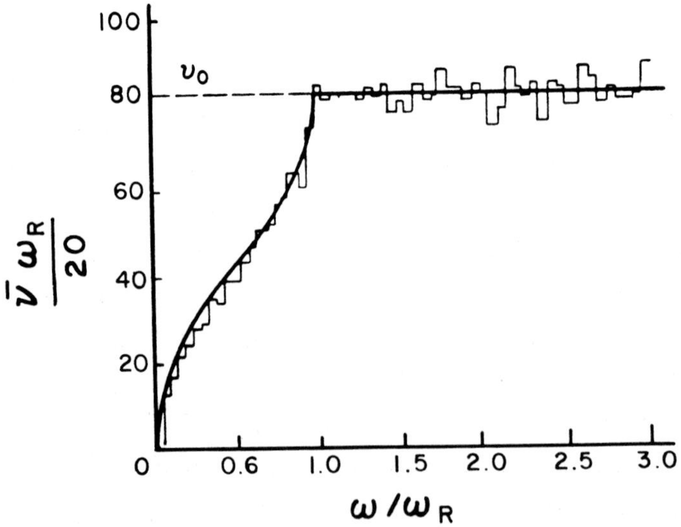

Figure 31

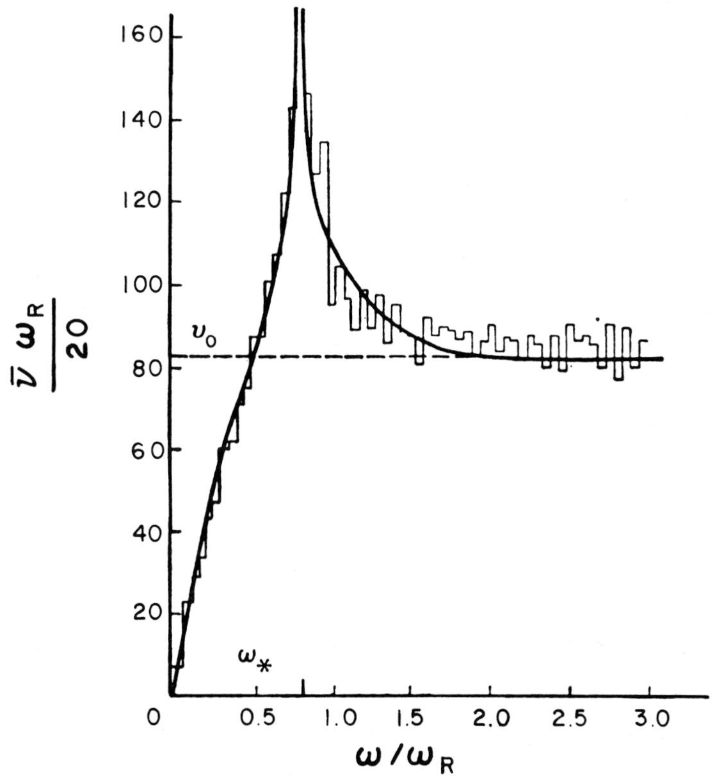

Figure 32

Shells of Variable Curvatures

The above mentioned books and papers considered shells of constant curvature or shells approximately interpreted as such. The asymptotic method, in principle, is applicable to equations with variable coefficients. The difficulty lies in constructing inner and corrective solutions. To avoid this difficulty, approximate methods, such as asymptotic ones, are usually applied, i.e., methods which make use of the slow variation of coefficients or of the existence of a small parameter for higher order derivatives. This approach to shells of revolution was introduced by Moskalenko [70] and Tovstick [97]. Initial equations have the form (4.41), where, instead of (4.42), it is necessary to take

$$\Delta = \frac{1}{r}\frac{\partial}{\partial x}\left(r\frac{\partial}{\partial x}\right) + \frac{1}{r^2}\frac{\partial^2}{\partial\Theta^2}\,, \qquad \Delta_k = \frac{1}{r}\frac{\partial}{\partial x}\left(\frac{r}{R_2}\frac{\partial}{\partial x}\right) + \frac{1}{r^2 R_1}\frac{\partial^2}{\partial\Theta^2}\,.$$

Here, x is the length of the meridianal arc, Θ is the polar angle in the cross-section of the shell, r(x) is the radius of this cross-section, $R_1(x)$ and $R_2(x)$ are principle radii of curvature. These radii are considered to be slowly varying functions as compared to the modes of shell vibrations. Distinguishing the rapidly varying factors in the solution of the equation (4.41), we assume

$$w(x,\theta) = W(x)e^{if(x)} \cos n\Theta\,,$$

$$\psi(x,\theta) = \Psi(x)e^{if(x)} \cos n\Theta\,, \qquad (n = 0,1,2,\ldots)\,.$$

Here, W(x), $\Psi(x)$ and f(x) are the slowly varying functions. The relation between the natural frequency ω and f(x) is given by the formula

$$\rho h\omega^2 = D\left(f' + \frac{n^2}{r^2}\right) + Eh\left(\frac{f'^2}{R_2} + \frac{n^2}{r^2 R_1}\right)^2 \left(f'^2 + \frac{n^2}{r^2}\right)^{-2}\,. \qquad (4.79)$$

Unlike (4.43), the analogs of the wave numbers f' and n/r depend on the coordinate x. With ω fixed, formula (4.79) can be treated as a differential equation for f(x). Natural frequencies are approximately determined from the condition that one of the solutions of equation (4.79) gives an increment of a multiple of π in the meridian's length. Thus, the asymptotic formula for natural frequencies takes the form $\omega = \Omega(m,n)$, where m = 1,2,... is the number of half-waves along the meridian, and n = 0,1,2,... is the number of half-waves in the circular direction. The number of natural frequencies smaller than the prescribed value ω is determined from a formula of the (4.52) type, i.e.,

$$\overline{N}(\omega) = \iint_{\Omega(m,n)<\omega} dm\,dn\,. \qquad (4.80)$$

Applying formulae (4.79) and (4.80) to a circular cylindrical shell and to a closed spherical shell, we arrive at the already known results (4.73).

Another approach to shells of varying curvature is based on an idea of Courant. If the median surface of a shell can be divided into subdomains, within each of which metrics and curvature can be assumed as constant, then, formulae of the type (4.71) and (4.72) are applicable in each domain. The function of distribution $\overline{N}(\omega)$ and the density $\overline{\nu}(\omega)$ will depend on the subdomain characteristics. Then, the question arises under which conditions can the function of distribution and the density of frequencies for the entire shell be approximately determined by the integration of formulae of the type (4.71) and (4.72) with respect to the entire median surface G, i.e., by the formulae

$$\overline{N}(\omega) = \frac{1}{2\pi^2} \iint_G \left(\frac{\rho h}{D}\right)^{1/2} \int_{\Theta_1}^{\Theta_2} [\omega^2 - \omega_2^2(\gamma \cos^2\theta + \sin^2\theta)^2]^{1/2} d\Theta dG , \qquad (4.81)$$

$$\overline{\nu}(\omega) = \frac{\omega}{2\pi^2} \iint_G \left(\frac{\rho h}{D}\right)^{1/2} \int_{\Theta_1}^{\Theta_2} [\omega^2 - \omega_2^2(\gamma \cos^2\Theta + \sin^2\Theta)^2]^{-1/2} d\Theta dG . \qquad (4.82)$$

Here, ω_2, γ, etc., are generally functions of the positions of points on the median surface x_1 and x_2.

It is easy to notice that in the case when the dynamic boundary effect on the boundaries of the median surface subdomains does not degenerate at any values of wave numbers, formulae (4.81) and (4.82) are asymptotically correct in the same sense as formulae (4.71) and (4.72). Recently, V. Lidksy and his co-workers showed [3] that the formulae (4.81) and (4.82) can probably be applied more extensively.

4.5 METHOD OF INTEGRAL ESTIMATES FOR THE ANALYSIS OF WIDE-BAND RANDOM VIBRATIONS

The Idea of the Method

As was already mentioned in Section 4.1, the high density of natural frequencies can be used for making relatively simple and effective estimates in problems of random vibrations under the action of broad spectrum forces. The method of integral estimates [11,113] is based on the asymptotic method in the theory of vibrations of elastic systems and on the theory of distribution of natural frequencies

Consider forced vibrations of a linear elastic system subjected to space-time loading. Although, in principle, the method can be formulated as applied to nonstationary vibrations, we shall consider here only systems undergoing stationary vibrations. The external loading is assumed to be wide-band in the sense that its spectral density is substantially different from zero within an interval containing a sufficiently large number of natural frequencies, and varies within the interval sufficiently slowly.

Let the following conditions be satisfied: the density of natural frequencies should be sufficiently high; the asymptotic expressions (Section 4.1) can be used for the frequencies ω_γ and for the vibration modes $\phi_\gamma(\underset{\sim}{x})$, the dependence on the wave vector $\underset{\sim}{k}$ substituting the dependence on the index γ; the damping coefficients ε_γ, the functions $r(\underset{\sim}{x})$ in formulae (3.19), (3.20), and the spectral densities of the generalized forces $S_{Q_j Q_k}(\omega)$ can be represented as functions of the wave vector $\underset{\sim}{k}$; the functions $\omega(\underset{\sim}{k})$, $\varepsilon(\underset{\sim}{k})$, $r(\underset{\sim}{x},\underset{\sim}{k})$, $\phi(\underset{\sim}{x},\underset{\sim}{k})$, and $S_Q(\underset{\sim}{k})$ vary slightly when passing to adjacent cells in the space of wave numbers $\underline{K}$. Under these conditions, sums entering in the formulae of the type (3.20) can be approximately estimated by integrals with respect to the non-negative octant $\underline{K}_+$ of the space $\underline{K}$.

Approximate Estimates for Variances

Consider, for example, the formula for variances (3.20). Let all the joint central moments of the generalized coordinates $u_j(t)$ be either identically equal to zero or their contribution of the variance of the parameter v be negligible. Also, let the generalized coordinates satisfy the equations (3.6), and the damping coefficients be sufficiently small. Then, from the formulae (3.20) and (3.32), we shall obtain an approximate expression for the variance:

$$D[v(\underset{\sim}{x},t)] \approx \sum_{j=1}^{\infty} \frac{\pi |r_j(\underset{\sim}{x})|^2 S_{Q_j Q_j}(\omega_j)}{2\varepsilon_j \omega_j^2} . \tag{4.83}$$

All the terms of the series (4.83) depend only on the number j. On the other hand, the functions of j involved are in the asymptotic sense fully determined by the values of the wave vector $\underset{\sim}{k}$. Hence,

$$D[v(\underset{\sim}{x},t)] \approx \sum_{j=1}^{\infty} \frac{\pi |r(\underset{\sim}{x},\underset{\sim}{k})|^2 S_Q(\underset{\sim}{k})}{2\varepsilon(\underset{\sim}{k})\omega^2(\underset{\sim}{k})} , \tag{4.84}$$

where $S_Q(\underset{\sim}{k})$ are diagonal elements of the matrix of spectral densities of the generalized forces. If all the terms of the series (4.84) are slowly varying functions of $\underset{\sim}{k}$, the sum on the right side can be approximately substituted by an integral. Hence, [11]

$$D[v(\underset{\sim}{x},t)] \sim \frac{\pi}{2\Delta \underset{\sim}{k}_1} \int_{\underline{K}_+} \frac{|r(\underset{\sim}{x},\underset{\sim}{k})|^2 S_Q(\underset{\sim}{k}) d\underset{\sim}{k}}{\varepsilon(\underset{\sim}{k})\omega^2(\underset{\sim}{k})} . \tag{4.85}$$

Here, integration is performed with respect to the non-negative octant $\underline{K}_+$ of the space $\underline{K}$; $\Delta \underset{\sim}{k}_1$ is the size of one cell in the space $\underline{K}$, corresponding to one term in the expansion (4.84). Generally, $\Delta \underset{\sim}{k}_1$ does not coincide with the $\Delta \underset{\sim}{k}$ from Section 4.2. For instance, for systems with symmetry, the summation in the series (4.83) and (4.84) can be performed over a part of the spectrum only.

The dimensionality of the integral (4.85) coincides with that of the space $\underline{K}$. If we take into account joint correlations of the generalized coordinates, double sums will enter in the formulae (4.83) and (4.84), and an approximate estimate for the variance will be expressed in terms of integrals whose dimensionality will be twice as large as that of the space $\underline{K}$.

The error of integral estimates can be determined analogously to the case of numerical determination of integrals by summation formulae. Other things being equal, the denser the spectrum of natural frequencies and the slower the functions under the integration sign vary, the smaller are the errors.

Approximate Estimates for Spectral Densities

In order to obtain the integral estimate for the spectral density $S_v = (\underset{\sim}{x},\omega)$, we shall do the following. First, we shall calculate the variance of parameter v when the spectrum of the input has the frequency ω_c as an upper bound. For this variance, we have a formula of the type (4.85), i.e.,

$$\underset{\sim}{D}[v_c(\underset{\sim}{x},t)] \sim \frac{\pi}{2\Delta \underset{\sim}{k}_1} \int_{\Omega(\underset{\sim}{k})<\omega_c} \frac{|r(\underset{\sim}{x},\underset{\sim}{k})|^2 S_Q(\underset{\sim}{k})\, d\underset{\sim}{k}}{\varepsilon(\underset{\sim}{k})\omega^2(\underset{\sim}{k})} .$$

Integration is performed over the region $\{k\colon\ \Omega(\underset{\sim}{k}) < \omega_c\}$, where $\omega = \Omega(\underset{\sim}{k})$ is an asymptotic expression for the natural frequency as a function of the wave vector. Differentiating this expression over the frequency ω_c, and substituting ω_c by ω, we obtain [11]

$$S_v(\underset{\sim}{x},\omega) \sim \frac{\pi}{4\Delta \underset{\sim}{k}_1} \frac{\partial}{\partial\omega} \int_{\Omega(\underset{\sim}{k})<\omega} \frac{|r(\underset{\sim}{x},\underset{\sim}{k})|^2 S_Q(\underset{\sim}{k})\, d\underset{\sim}{k}}{\varepsilon(\underset{\sim}{k})\omega^2(\underset{\sim}{k})} . \tag{4.86}$$

By its structure, this formula is analogous to the asymptotic estimate

$$\bar{\nu}(\omega) \sim \frac{1}{\Delta \underset{\sim}{k}} \frac{\partial}{\partial \omega} \int_{\Omega(\underset{\sim}{k})<\omega} d\underset{\sim}{k} \tag{4.87}$$

for the density of natural frequencies, which follows from the formulae (4.50) and (4.52).

To a certain degree, there is an explicit relation between the asymptotic estimate for the spectral density (4.86) and the asymptotic density of natural frequencies (4.87). Let the functions $\varepsilon(\underset{\sim}{k})$, $r(\underset{\sim}{x},\underset{\sim}{k})$ and $S_Q(\underset{\sim}{k})$ in formula (4.86) be transformed into functions of a single frequency ω by substituting $\Omega(\underset{\sim}{k}) = \omega$. Then, let these functions vary slowly as compared to the asymptotic function of distribution of natural frequencies $\bar{N}(\omega)$. Then, applying the mean value theorem and taking account of the formula (4.87), we shall represent (4.86) as

$$S_V(\underset{\sim}{x},\omega) \sim \pi \frac{\Delta \underset{\sim}{k}}{4\Delta \underset{\sim}{k}_1} \frac{|r(\underset{\sim}{x},\omega)|^2 S_Q(\omega)\bar{\nu}(\omega)}{\varepsilon(\omega)\omega^2} . \tag{4.88}$$

Thus, the spectral density $S_V(\underset{\sim}{x},\omega)$ proves to be proportional to the asymptotic density of the frequencies $\bar{\nu}(\omega)$. This result is understandable if we take into consideration the fact that the function $\bar{\nu}(\omega)$ characterizes the number of elementary oscillators per unit frequency range in the vicinity of the given frequency. If the function $\bar{\nu}(\omega)$ has a singularity at a certain frequency, this singularity is preserved in the spectral density. If the asymptotic density has a maximum at a certain frequency, the maximum of spectral density is to be expected in the vicinity of this frequency.

Taking Account of the Joint Correlation of Generalized Coordinates

The high density of the spectrum of natural frequencies is widely used in the integral method. As was shown in Section 3.1, the contribution of the joint correlation of generalized coordinates to the variance value at the output is comparable with the contribution of variances of generalized coordinates, in the case of closely spaced natural frequencies. Therefore, conditions (3.37) and (3.52) may prove to be not satisfied when the density of natural frequencies is high. Strictly speaking, the asymptotic estimates (4.85) and (4.86) correspond to the case when the secondary elements of the matrix $S_{Q_jQ_k}(\omega)$ are either identically equal to zero, or are negligible as compared to diagonal elements. If this condition is not satisfied, then, instead of formula (4.83) for the variance at the output, an analogous representation in terms of double sums is to be taken. The asymptotic estimate for variance will be expressed here in terms of integrals whose dimensionality is twice as large as that of the space.

Let the external loading be such that all $S_{Q_jQ_k}(\omega)$ are real functions of ω. Then, taking into account formulae (3.20), (3.28) and (3.31), we obtain

$$D[v(\underset{\sim}{x},t)] \approx \sum_j \frac{\pi|r_j(\underset{\sim}{x})|^2 S_{Q_jQ_j}(\omega_j)}{2\varepsilon_j\omega_j^2} +$$

$$\sum_m \sum_n \frac{4\pi r_m^*(\underset{\sim}{x}) r_n(\underset{\sim}{x})(\varepsilon_m+\varepsilon_n) S_{Q_mQ_n}(\omega_{mn})}{(\omega_m^2-\omega_n^2)^2+4(\varepsilon_m+\varepsilon_n)(\varepsilon_m\omega_n^2+\varepsilon_n\omega_m^2)} .$$

Here, $\omega_{mn} = \frac{1}{2}(\omega_m+\omega_n)$; the double sum does not contain terms for $m = n$. As before, we shall represent functions of the indices j, m and n in terms of functions of the wave vectors $\underset{\sim}{k}$ and $\underset{\sim}{k}'$ which take on discrete values in the region $\underline{K}_+$, and, then, by means of slowly varying continuous functions defined everywhere in $\underline{K}_+$, we shall perform their approximation. We shall denote by $S_Q(\underset{\sim}{k})$ the continuous function approximating diagonal elements of the matrix of spectral densities, as before. The remaining elements of the matrix will be approximated by means of the continuous functions $T_Q(\underset{\sim}{k},\underset{\sim}{k}')$. Generally, $T_Q(\underset{\sim}{k},\underset{\sim}{k}) \neq S_Q(\underset{\sim}{k})$. Turning from summation over indices to integration in the space of wave numbers, we come to the estimate

$$D[v(\underset{\sim}{x},t)] \sim \frac{\pi}{2\Delta \underset{\sim}{k}_1} \int_{\underline{K}_+} \frac{|r(\underset{\sim}{x},\underset{\sim}{k})|^2 S_Q(\underset{\sim}{k})\,d\underset{\sim}{k}}{\varepsilon(\underset{\sim}{k})\omega^2(\underset{\sim}{k})} +$$

$$\frac{4\pi}{\Delta \underset{\sim}{k}_1^2} \int_{\underline{K}_+} \int_{\Gamma(k)} \frac{r^*(\underset{\sim}{x},\underset{\sim}{k})r(\underset{\sim}{x},\underset{\sim}{k}')[\varepsilon(\underset{\sim}{k})+\varepsilon(\underset{\sim}{k}')]T_Q(\underset{\sim}{k},\underset{\sim}{k}')\,d\underset{\sim}{k}'d\underset{\sim}{k}}{[\omega^2(\underset{\sim}{k}')-\omega^2(\underset{\sim}{k})]^2+4[\varepsilon(\underset{\sim}{k})+\varepsilon(\underset{\sim}{k}')][\varepsilon(\underset{\sim}{k})\omega^2(\underset{\sim}{k}')+\varepsilon(\underset{\sim}{k}')\omega^2(\underset{\sim}{k})]} . \tag{4.89}$$

The first term on the right side coincides with the right side of formula (4.85), the second term takes account of the joint correlation of the generalized coordinates. Integration is performed with respect to the entire octant $\underline{K}_+$, except for the inner integral in the second term, where the region of integration is reduced to the band $\Gamma(\underset{\sim}{k})$ containing sufficiently closely spaced frequencies. Actually, the formula (3.31), which was applied for obtaining the estimate (4.89), was derived by using condition (3.26). On the other hand, the contribution of widely spaced frequencies $\omega(\underset{\sim}{k})$ and $\omega(\underset{\sim}{k}')$ to the value of the considered integral will be sufficiently small. Further, the integration region with respect to $\underset{\sim}{k}'$ will be taken as

$$\Gamma(\underset{\sim}{k}) = \{\underset{\sim}{k}' : |\omega^2(\underset{\sim}{k}')-\omega^2(\underset{\sim}{k})| < 4M\varepsilon(\underset{\sim}{k})\omega(\underset{\sim}{k}) , \tag{4.90}$$

where $M \geq 1$ is a certain positive number.

Let us give an approximate estimate for the inner integral value in formula (4.89). Applying the mean value theorem, we shall write out the approximate equation as

$$\int_{\Gamma(k)} \frac{r(\underset{\sim}{x},\underset{\sim}{k}')[\varepsilon(\underset{\sim}{k})+\varepsilon(\underset{\sim}{k}')]T_Q(\underset{\sim}{k},\underset{\sim}{k}')\,d\underset{\sim}{k}'}{[\omega^2(\underset{\sim}{k}')-\omega^2(\underset{\sim}{k})]^2+4[\varepsilon(\underset{\sim}{k})+\varepsilon(\underset{\sim}{k}')][\varepsilon(\underset{\sim}{k})\omega^2(\underset{\sim}{k}')+\varepsilon(\underset{\sim}{k}')\omega^2(\underset{\sim}{k})]} \approx$$

$$2r(\underset{\sim}{x},\underset{\sim}{k})\varepsilon(\underset{\sim}{k})T_Q(\underset{\sim}{k},\underset{\sim}{k})J(\underset{\sim}{k}) , \tag{4.91}$$

where the notation

$$J(\underset{\sim}{k}) = \int_{\Gamma(\underset{\sim}{k})} \frac{d\underset{\sim}{k}'}{[\omega^2(\underset{\sim}{k}')-\omega^2(\underset{\sim}{k})]^2+16\varepsilon^2(\underset{\sim}{k})\omega^2(\underset{\sim}{k})} \tag{4.92}$$

is used. Then, we expand the square of the frequency $\omega^2(\underset{\sim}{k}')$ into a power series in the vicinity of the point $\underset{\sim}{k}' = \underset{\sim}{k}$:

$$\omega^2(\underset{\sim}{k}') = \omega^2(\underset{\sim}{k})+\text{grad}\,\omega^2(\underset{\sim}{k})(\underset{\sim}{k}'-\underset{\sim}{k}) + 1/2(\underset{\sim}{k}'-\underset{\sim}{k})^T H(\underset{\sim}{k}'-\underset{\sim}{k}) + \cdots . \tag{4.93}$$

Here, H is the matrix of second derivatives of the function $\omega^2 = \Omega^2(\underset{\sim}{k})$ with respect to the components of the vector $\underset{\sim}{k}$. For instance, for the two-dimensional space $\underline{K}$,

$$H = \begin{bmatrix} \dfrac{\partial^2\Omega^2(\underset{\sim}{k})}{\partial k_1^2} & \dfrac{\partial^2\Omega^2(\underset{\sim}{k})}{\partial k_1 \partial k_2} \\[2ex] \dfrac{\partial^2\Omega^2(\underset{\sim}{k})}{\partial k_1 \partial k_2} & \dfrac{\partial^2\Omega^2(\underset{\sim}{k})}{\partial k_2^2} \end{bmatrix} .$$

If grad $\omega^2(\underset{\sim}{k}) \neq 0$ everywhere in $\underline{K}_+$, and if the band $\Gamma(\underset{\sim}{k})$ is sufficiently narrow, it is sufficient to retain the linear terms of the expansion in (4.93). Then, the integral (4.92) takes the form

$$J(\underset{\sim}{k}) = \int_{\Gamma_1(\underset{\sim}{k})} \frac{d\underset{\sim}{k}''}{|\text{grad}\,\omega^2(\underset{\sim}{k})\underset{\sim}{k}''|^2+16\varepsilon^2(\underset{\sim}{k})\omega^2(\underset{\sim}{k})} , \tag{4.94}$$

and the region of integration is defined as

$$\Gamma_1(\underset{\sim}{k}) = \{\underset{\sim}{k}'': |\text{grad}\,\omega^2(\underset{\sim}{k})\underset{\sim}{k}''| < 4M\varepsilon(\underset{\sim}{k})\omega(\underset{\sim}{k})\} .$$

The integral (4.94) is easily calculated. For instance, for a one-dimensional system with M = 1, we have

$$J(k) = \frac{\pi}{8\varepsilon(k)\omega(k)} \left|\frac{\partial\omega^2(k)}{\partial k}\right|^{-1} .$$

Generally, for $M \sim 1$ and an arbitrary dimensionality m of the space $\underline{K}$, an order of magnitude estimate is found as

$$J(\underset{\sim}{k}) \sim \frac{\varepsilon^{m-2}(\underset{\sim}{k})\omega^{m-2}(\underset{\sim}{k})}{|\text{grad } \omega^2(\underset{\sim}{k})|^m} \tag{4.95}$$

We substitute the estimate (4.95) into formula (4.89) to obtain

$$D[v(\underset{\sim}{x},t)] \sim \frac{\pi}{2\Delta \underset{\sim}{k}_1} \int_{\underline{K}_+} \frac{|r(\underset{\sim}{x},\underset{\sim}{k})|^2 S_Q(\underset{\sim}{k})d\underset{\sim}{k}}{\varepsilon(\underset{\sim}{k})\omega^2(\underset{\sim}{k})} +$$

$$\frac{4\pi}{\Delta \underset{\sim}{k}_1^2} \int_{\underline{K}_+} \frac{|r(\underset{\sim}{x},\underset{\sim}{k})|^2 \varepsilon^{m-1}(\underset{\sim}{k})\omega^{m-2}(\underset{\sim}{k})T_Q(\underset{\sim}{k},\underset{\sim}{k})d\underset{\sim}{k}}{|\text{grad } \omega^2(\underset{\sim}{k})|^m} . \tag{4.96}$$

If $T_Q(\underset{\sim}{k},\underset{\sim}{k})$ is of the order $S_Q(\underset{\sim}{k})$, the contribution of the second integral to the variance will be small as compared to the contribution of the first integral only in the case when

$$|\text{grad } \omega^2(\underset{\sim}{k})|^m \Delta \underset{\sim}{k}_1 \gg \varepsilon^m(\underset{\sim}{k})\omega^m(\underset{\sim}{k}) . \tag{4.97}$$

Thus, we have arrived at the analog of condition (3.37). The fact that grad $\omega^2(\underset{\sim}{k})$ enters in the asymptotic representation for the variance shows the substantial role of the density of natural frequencies. For $m = 1$, the second integral in formula (4.96) does not contain damping parameters. The contribution of each pair of close frequencies increases with the decrease of damping. But, the number of frequencies in the band $\Gamma(\underset{\sim}{k})$ decreases, i.e., the number of vibration modes strongly interacting with each other grows smaller. In one-dimensional systems, these two effects compensate each other.

If the equality grad $\omega^2(\underset{\sim}{k}) = 0$ holds at some point of region $\underline{K}$, the quadratic terms should be retained in the expansion

(4.93). As was shown above, this may take place if there are asymptotic points of condensation in the spectrum of natural frequencies.

4.6 APPLICATION OF THE METHOD OF INTEGRAL ESTIMATES

Some Examples

The general relations of the method were given in Section 4.5. Here, we shall consider applications of the method to some problems of random vibrations of rods, plates and shells. A thin plate of a linear viscoelastic material under broad-band space-time random loading, stationary in time, will serve as a basic example. First, we shall consider a rectangular plate (though, as will be seen later, most results are valid for a more general case).

Let the conditions of application of the method of integral estimates be satisfied (Section 4.5). For the natural frequencies of the plate, we shall take the asymptotic expression according to formula (4.27)

$$\omega \sim k^2 (D_r/\rho h)^{1/2} . \tag{4.98}$$

Here, D_r is the real part of the cylindrical stiffness, $k^2 = k_1^2 + k_2^2$, where k_1 and k_2 are wave numbers. For the asymptotic value of the damping coefficients in formulae (4.85), and subsequently, we shall take the relation

$$2\varepsilon \sim \chi(k)\omega(k) , \tag{4.99}$$

where the loss tangent χ depends on the modulus of the wave number. Due to the asymptotic formula (4.98), any dependence of the damping coefficient on the number of a natural mode can be approximated by means of formula (4.99). For natural modes in an inner region, we shall take the asymptotic expression (4.26), i.e.,

$$\phi(x_1,x_2) = \sin k_1(x_1-\xi_1)\sin k_2(x_2-\xi_2) , \tag{4.100}$$

where ξ_1 and ξ_2 are phase constants depending on boundary conditions. In the zone of the boundary effect, the asymptotic formulae taking account of boundary conditions at the nearest part of the contour are to be used instead of (4.100). For example, close to the clamped edge $x_1 = 0$, the mode of natural vibrations is given by the asymptotic expression which follows from (4.33), i.e.,

$$\phi(x_1,x_2) = \frac{k_1}{2^{1/2}(k_1^2+k_2^2)^{1/2}} \left\{ \frac{(k_1^2+2k_2^2)^{1/2}}{k_1} \sin k_1x_1 - \right.$$

$$\left. \cos k_1x_1+\exp[-x_1(k_1^2+2k_2^2)^{1/2}] \right\} \sin k_2(x_2-\xi_2) . \tag{4.101}$$

For the time being, we shall consider the intensity of an external loading $q(\underset{\sim}{x},t)$ as a delta-correlated function of spatial coordinates and an arbitrary stationary random function of time. The time spectral density of loading is given here by the formula (1.33). Substituting the latter into formula (3.17) and taking into account that

$$\int_G \int_G \phi(\underset{\sim}{x},\underset{\sim}{k})\phi(\underset{\sim}{x}',\underset{\sim}{k}')\delta(\underset{\sim}{x}-\underset{\sim}{x}')d\underset{\sim}{x}d\underset{\sim}{x}' = 1/4\ a_1a_2 ,$$

(a_1 and a_2 are the lengths of plate's sides), we shall find the joint spectral densities of the generalized forces as

$$S_{Q_\alpha Q_\beta}(\omega) = \frac{4\Psi(\omega)\delta_{\alpha\beta}}{\rho^2h^2a_1a_2} . \tag{4.102}$$

Thus, in the considered case, the matrix of joint spectral densities of the generalized forces turns out to be diagonal. There is no cross correlation between generalized forces and, hence, between generalized coordinates. So, formulae (4.85), (4.86) etc., by not taking account of this correlation, are in this case exact (in the asymptotic sense). Assuming that $\Psi(\omega)$ is a continuous slowly varying function, we obtain for a continuous analog of the spectral densities (4.102) the expression

$$S_Q(\underset{\sim}{k}) = \frac{4\Psi(k)}{\rho^2 h^2 a_1 a_2} \,. \tag{4.103}$$

Here, when substituting the argument according to (4.98), the notation for the function $\Psi(\omega)$ is the same as before.

Calculation of Variances in the Plate's Inner Region

For the variance of the field $v(\underset{\sim}{x},t)$, we have the formula (4.85) which, by taking account of the relation (4.99), takes the form

$$\underset{\sim}{D}[v(\underset{\sim}{x},t)] \sim \frac{\pi}{\Delta \underset{\sim}{k}_1} \int_{\underline{K}_+} \frac{|r(\underset{\sim}{x},\underset{\sim}{k})|^2 S_Q(\underset{\sim}{k})\, d\underset{\sim}{k}}{\chi(\underset{\sim}{k})\,\omega^3(\underset{\sim}{k})} \,. \tag{4.104}$$

Substituting here the expression (4.103), and turning to polar coordinates $k_1 = k\cos\theta$, $k_2 = k\sin\theta$, we obtain

$$\underset{\sim}{D}[v(\underset{\sim}{x},t)] \sim \frac{4}{\pi n D_r^2}\left(\frac{D_r}{\rho h}\right)^{1/2} \int_0^{\pi/2}\int_0^{\infty} \frac{|r(\underset{\sim}{x},k,\theta)|^2 \Psi(k)\, dk\, d\theta}{\chi(k)\, k^5} \,. \tag{4.105}$$

Here, n denotes the ratio of the areas of two cells in the plane of wave numbers

$$n = \Delta \underset{\sim}{k}_1 / \Delta \underset{\sim}{k} \,. \tag{4.106}$$

Here, $\Delta \underset{\sim}{k}$ corresponds to one mode of natural vibrations, i.e., $\Delta \underset{\sim}{k} = \pi^2/a_1 a_2$. The cell $\Delta \underset{\sim}{k}_1$ corresponds to one term of the series (4.84). To illustrate the meaning of the parameters n, we shall consider a plate simply supported on the contour. Then,

$\xi_1 = \xi_2 = 0$. Let the variance of the deflection w in the centre of the plate be calculated. The series for this deflection will include only those generalized coordinates to which modes of vibration with an odd number of waves on each coordinate will correspond. Therefore, four cells $\Delta\underset{\sim}{k}$ will correspond to each considered mode of vibration. Thus, for the deflection in the centre of the plate we have $r = \pm 1$, $n = 4$. Formula (4.105), after integration with respect to θ, takes the form [11,113]:

$$\underset{\sim}{D}[w] \sim \frac{1}{2D_r^2}\left(\frac{D_r}{\rho h}\right)^{1/2} \int_0^\infty \frac{\Psi(k)\,dk}{\chi(k)\,k^5} \,. \qquad (4.107)$$

The result does not depend on the plate's dimensions a_1, a_2. Therefore, it can be expected to remain valid (in some asymptotic sense) for a plate of an arbitrary shape and with arbitrary conditions on the contour. Actually, formula (4.107) can be obtained by means of the following considerations: Let the modes of natural vibrations in the inner region of the plate have the form (4.100), where the phases $k_1\xi_1$ and $k_2\xi_2$ are random and equally probable in the region $0 \leq k_1\xi_1 \leq 2\pi$, $0 \leq k_2\xi_2 \leq 2\pi$. In formula (4.105) we assume that

$$r(\underset{\sim}{x},\underset{\sim}{k}) = \sin k_1(x_1-\xi_1)\sin k_2(x_2-\xi_2)$$

and perform an additional averaging with respect to the phases. All the modes of vibration are to be taken into account in the corresponding series, i.e., $n = 1$. Averaging with respect to the phases, denoted by a bar above the formula, yields

$$\overline{r^2(\underset{\sim}{x},\underset{\sim}{k})} = \overline{\sin^2 k_1(x_1-\xi_1)\sin^2 k_2(x_2-\xi_2)} = 1/4 \,, \qquad (4.108)$$

from which we again obtain formula (4.107). Averaging the deflection variance over a certain region of the plate containing a sufficiently large number of half-waves in the modes of vibration gives the same result. Indeed, for example

$$\frac{1}{a_1 a_2}\int_0^{a_1}\int_0^{a_2} \sin^2 k_1(x_1-\xi_1)\sin^2 k_2(x_2-\xi_2)dx_1dx_2 = 1/4 .$$

The variance of other parameters characterizing the vibration field in the inner region of the plate is calculated analogously. Consider for instance the normal stresses σ_{11} at the points with ordinates $z = \pm h/2$. Henceforth, we shall denote this stress by σ. Taking into account that at these points

$$\sigma = \pm \frac{6D}{h^2}\left(\frac{\partial^2 w}{\partial x_1^2} + \mu \frac{\partial^2 w}{\partial x_2^2}\right), \tag{4.109}$$

we find

$$r(\underset{\sim}{x},\underset{\sim}{k}) = \pm \frac{6D_r(1+i\eta)}{h^2} (k_1^2+\mu k_2^2)\sin k_1(x_1-\xi_1)\sin k_2(x_2-\xi_2) .$$

Calculations analogous to the previous ones give the stress variance in the inner region $\underset{\sim}{x} \in G$ as

$$\underset{\sim}{D}[\sigma] \sim \frac{9}{4h^4}\left(\frac{D_r}{\rho h}\right)^{1/2} (3+2\mu+3\mu^2) \int_0^\infty \frac{\Psi(k)dk}{\chi(k)k} . \tag{4.110}$$

It is taken into consideration here that within the scope of the assumptions made, $\chi^2 \ll 1$.

Comparison with Results Obtained by the Method of Canonical Integral Representation

As can be seen from the preceding material, the variances of displacements and stresses in the inner region are not dependent (in the asymptotic approximation) on a plate's dimensions in plan. The corresponding formulae remain valid for the case of an unbounded plate. These formulae can be obtained from somewhat different considerations, for instance, by the method of integral canonical expansions [74].

As an example, we shall calculate the variance of a deflection for a plate under a normal loading whose time spectral density is specified by (1.33). The general formula for the variance of deflection has the form (3.137). It includes the space-time spectral density $S_q(\underset{\sim}{k},\omega)$. Taking into account the relations (1.29), we find

$$S_q(\underset{\sim}{k},\omega) = \frac{1}{4\pi^2}\int_{-\infty}^{\infty} S_q(\underset{\sim}{\rho},\omega)e^{-i\underset{\sim}{k}\underset{\sim}{\rho}}d\underset{\sim}{\rho} = \frac{\Psi(\omega)}{4\pi^2}. \tag{4.111}$$

Let the damping in a system be sufficiently small ($\chi^2 \ll 1$), and $\Psi(\omega)$ be a sufficiently slowly varying function of ω. Substituting the expression (4.111) into formula (3.137), we shall place the slowly varying part outside the integration sign with respect to ω in order to obtain

$$\int_{-\infty}^{\infty} \frac{\Psi(\omega)\,d\omega}{(k^4D_r-\rho h\omega^2)^2+(\chi k^4 D_r)^2} \approx \frac{\Psi(\omega_0)}{\rho^2h^2}\int_{-\infty}^{\infty}\frac{d\omega}{(\omega_0^2-\omega^2)^2+(\chi\omega_0^2)^2}.$$

Here, $\omega_0(k)$ is the natural frequency corresponding to the wave number k. Furthermore,

$$\int_{-\infty}^{\infty}\frac{d\omega}{(\omega_0^2-\omega^2)^2+(\chi\omega_0^2)^2} \approx \frac{\pi}{\chi_0\omega_0^3},$$

where $\chi_0 = \chi(\omega_0)$ is the value of the loss tangent at this frequency. The formula for the variance of the deflection, after turning to polar coordinates and integrating with respect to the polar angle, takes the form

$$\underset{\sim}{D}[w] \sim \frac{1}{2\rho^2h^2}\int_0^{\infty}\frac{\Psi[\omega_0(k)]k\,dk}{\chi_0(k)\omega_0^3(k)}.$$

Substituting here the expression (4.98), setting $\Psi[\omega_0(k)] \equiv \Psi(k)$, omitting the indices for $\chi_0(k)$ and $\omega_0(k)$, we again obtain formula

(4.107). It is possible to analogously obtain an asymptotic expression for the variance of stresses (4.110), as well as for a number of other formulae, which due to the asymptotic nature of the method remain valid for both finite and infinite (or semi-infinite) systems.

Influence of Some Parameters of the Problem on the Variance of the Displacements and Stresses

For the sake of definiteness, we assume that the intensity of the loading is truncated white noise. Let ω_* and ω_{**} be the lower and the upper frequencies of the spectrum, respectively. Then, for $\Psi_0 > 0$, the spectral density $\Psi(\omega)$ is specified by the relation

$$\Psi(\omega) = \begin{cases} \Psi_0 = \text{const.}, & |\omega| \in [\omega_*, \omega_{**}] , \\ 0 , & |\omega| \notin [\omega_*, \omega_{**}] . \end{cases} \tag{4.112}$$

Then, following [11], we shall consider three types of energy dissipation for vibrations:

$$\chi = \chi_* \omega_* / \omega , \tag{4.113a}$$

$$\chi = \chi_* , \tag{4.113b}$$

$$\chi = \chi_* \omega / \omega_* . \tag{4.113c}$$

Here, χ_* is the value of the loss tangent at the frequency ω_*. The case (a) corresponds to "external" friction; case (b) to a loss tangent that does not depend on frequency; and case (c) to Voigt's friction. In order to represent the results visually, we introduce the characteristic deflection w_0 and the characteristic stress σ_0:

$$w_0^2 = \frac{\Psi_0}{4D_r^2 k_*^4 \chi_*} \left(\frac{D_r}{\rho h}\right)^{1/2} , \quad \sigma_0^2 = \frac{9\Psi_0 (3+2\mu+3\mu^2)}{8h^4 \chi_*} \left(\frac{D_r}{\rho h}\right)^{1/2} , \tag{4.114}$$

where k_* is the wave number corresponding to the frequency ω_*. Substituting the expressions (4.112), (4.113), and (4.114) into the formula (4.107), we obtain

$$\frac{D[\underset{\sim}{w}]}{w_0^2} \sim \begin{cases} 1 - \omega_*/\omega_{**} \,, & (4.115a) \\ \frac{1}{2}\,[1-(\omega_*/\omega_{**})^2] \,, & (4.115b) \\ \frac{1}{3}\,[1-(\omega_*/\omega_{**})^3] \,. & (4.115c) \end{cases}$$

For the variance of stresses, taking formula (4.110) into account, we obtain

$$\frac{D[\underset{\sim}{\sigma}]}{\sigma_0^2} \sim \begin{cases} \omega_{**}/\omega_* - 1 \,, & (4.116a) \\ \ell n(\omega_{**}/\omega_*) \,, & (4.116b) \\ 1-\omega_*/\omega_{**} \,. & (4.116c) \end{cases}$$

Let $\omega_*/\omega_{**} \to 0$, which corresponds approximately to the limit transition towards white noise. Then, the deflection variances calculated by the formulae (4.115) are in the ratio 1:2:3, and for the first two types of friction, the variance of stresses is infinite. Formulae (4.115) and (4.116) show explicitly and quite simply that the response of an elastic system to a random loading of this kind depends on the character of the damping forces. This example illustrates the advantage of integral estimates over direct calculations by the method of eigen elements.

In Figure 33, numerical results are compared. For the numerical calculations, the values $a_1 = a_2 = 1\text{m}$, $h = 2\cdot 10^{-3}\text{m}$, $E = 1.4\cdot 10^{11}\text{N}\cdot\text{m}^{-2}$, $\mu = 0.3$, $\rho = 2.7\cdot 10^3\text{kg}\cdot\text{m}^{-3}$, $\omega_* = 3.7\cdot 10^4\text{s}^{-1}$, were taken; the loss tangent was assumed as independent of the frequency. The results of the calculations according to the formulae (4.115b) and (4.116b) are plotted as smooth lines, the results of the series summation as stepped lines. The discrepancy for variance of the deflection $D[\underset{\sim}{w}]$ is quite large. It accumulates mainly at lower frequencies. The contribution of lower frequencies to the variance of stresses, $D[\sigma]$, is comparatively small. Therefore, there is a better agreement here between asymptotic estimates and the results of empirical calculations.

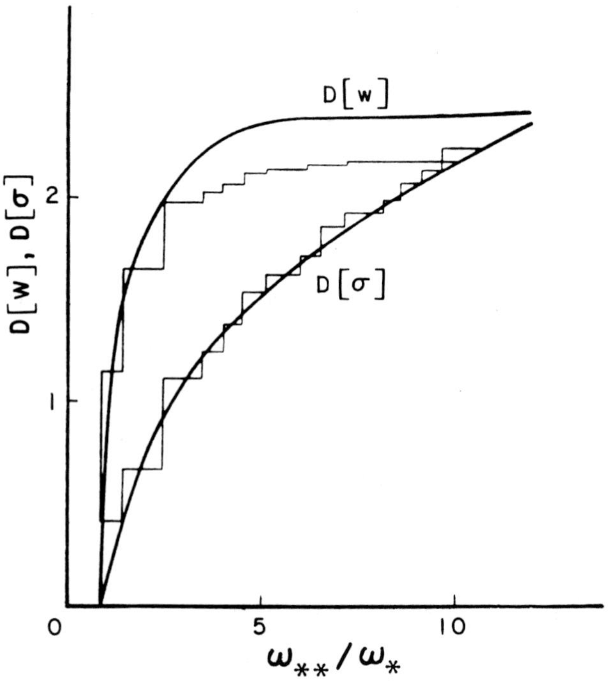

Figure 33

Calculation of Variances in Zones of Boundary Effects

As an example, we shall consider the boundary effect at the clamped edge of a plate. Let it be the side $x_1 = 0$. Modes of vibration in this zone are given by the expression (4.101). We calculate the normal stress in the clamped edge by means of formula (4.109). Using the expression (4.101), we obtain

$$r(0,x_2,\underset{\sim}{k}) = \pm\, 6\sqrt{2}D_r(1+i\eta)k_1kh^{-2} \sin k_2(x_2-\xi_2) \ . \qquad (4.117)$$

Let us consider the problem of determining the coefficient n in formula (4.106). If the sides of the plate perpendicular to the side $\xi_1 = 0$ are simply supported, then $\xi_2 = 0$, $k_2 = \pi/a_2$. Then, for the middle of the clamped side, only those modes of vibration to which an odd number of half-waves with respect to the coordinate x_2 corresponds should be taken into account. Here, $n = 2$, and

$$r(0, a_2/2, \underset{\sim}{k}) = \pm\, 6\sqrt{2} D_r (1 + i\eta) k_1 k h^{-2} \, . \tag{4.118}$$

Averaging the expression (4.117) with respect to the uniformly distributed phase $k_2\xi_2$ or with respect to a sufficiently large segment of the clamped side, we arrive at the same result. Substituting into this formula the expression (4.105) yields $(\chi^2 \ll 1)$

$$\underset{\sim}{D}[\sigma]_{x_1=0} \sim \frac{36}{h^4}\left(\frac{D_r}{\rho h}\right)^{1/2} \int_0^\infty \frac{\Psi(k)\,dk}{\chi(k)\,k} \, . \tag{4.119}$$

The obtained formula is analogous to formula (4.110). The relation between variances of stresses for the clamped side and those for the inner region does not depend on the type of spectral density and the type of damping. This relation characterizes the concentration of stresses at the clamped edge and is equal [11] to

$$\frac{\underset{\sim}{D}[\sigma]_{x_1=0}}{\underset{\sim}{D}[\sigma]} \sim \frac{16}{3 + 2\mu + 3\mu^2} \, . \tag{4.120}$$

For $\mu = 0.3$, the square root of this relation (the coefficient of concentration with respect to mean-square stresses) turns out to be approximately equal to two.

Some Problems of Random Vibrations of Thin Shells

We shall proceed from the engineering theory equations of the type (4.41). Extending these equations to shells of linear viscoelastic material, we shall write them for the case of forced vibrations as

$$\underline{D}\Delta\Delta w - \left(\frac{1}{R_2}\frac{\partial^2\psi}{\partial x_1^2} + \frac{1}{R_1}\frac{\partial^2\psi}{\partial x_2^2}\right) + \rho h \frac{\partial^2 w}{\partial t^2} = q \ ,$$

$$\Delta\Delta\psi + \underline{E}h\left(\frac{1}{R_2}\frac{\partial^2 w}{\partial x_1^2} + \frac{1}{R_1}\frac{\partial^2 w}{\partial x_2^2}\right) = 0 \ . \tag{4.121}$$

Here, $\underline{D}$ and $\underline{E}$ are linear viscoelastic operators and $q(\underset{\sim}{x},t)$ is the intensity of a normal loading. To simplify further calculations, we shall assume that Poisson's ratio μ is a number (and not an operator). Then,

$$\frac{D_i(\omega)}{D_r(\omega)} = \frac{E_i(\omega)}{E_r(\omega)} = \eta(\omega) \ , \tag{4.122}$$

where $\eta(\omega)$ is the loss tangent, extended oddly for $\omega < 0$. The deflection $w(\underset{\sim}{x},t)$ for the inner region will, as before, have the form (4.100). For frequencies of natural vibrations, we have the asymptotic formula (4.43), i.e.,

$$\omega^2 \sim \left[(k_1^2+k_2^2)^2 + \frac{k_0^4(k_1^2\gamma+k_2^2)^2}{(k_1^2+k_2^2)^2}\right]\frac{D_r}{\rho h} \ , \tag{4.123}$$

where the notations (4.44) are used.

The normal loading $q(\underset{\sim}{x},t)$ is taken in the same form as in the previous example. And here, the formula (4.102) for the spectral densities of the generalized forces as well as the expression derived from it, (4.103), retain their sense. From formula (4.85), we calculate the deflection variance in the inner region. Substituting the expressions (4.103) and (4.123) into the general relation (4.85) and turning to polar coordinates, we obtain [115]

$$D[\underset{\sim}{w}] \sim \frac{4}{\pi n D_r^2}\left(\frac{D_r}{\rho h}\right)^{1/2} \int_0^{\pi/2}\int_0^{\infty} \frac{|r(\underset{\sim}{x},k,\Theta)|^2 \Psi(k,\Theta)\,k\,dk\,d\Theta}{\chi(\underset{\sim}{x},\Theta)\,[k^4+k_0^4(\gamma\cos^2\Theta+\sin^2\theta)^2]^{3/2}} \,. \quad (4.124)$$

The notations are the same as in formula (4.105).

Calculations are somewhat simpler in the case of a spherical shell. Let $R_1 = R_2 = R$. Then, the natural frequency (4.123) depends only on the modulus k of the wave vector $\underset{\sim}{k}$. Due to this, the loss tangent χ and the function Ψ will depend only on k. For example, we shall determine an asymptotic estimate for the variance of a normal displacement. Here $r(\underset{\sim}{x},k,\theta) = 1$, $n = 4$. Formula (4.124) yields

$$D[\underset{\sim}{w}] \sim \frac{1}{2D_r^2}\left(\frac{D_r}{\rho h}\right)^{1/2} \int_0^{\infty} \frac{\Psi(k)\,k\,dk}{\chi(k)\,(k^4+k_0^4)^{3/2}} \,. \quad (4.125)$$

If the spectral density $\Psi(\omega)$ is specified in the form (4.112) and damping as (4.113), we obtain from the formula (4.125) the expressions

$$\frac{D[\underset{\sim}{w}]}{w_0^2} \sim \begin{cases} \dfrac{\omega_R}{\omega_*}\left[\tan^{-1}\left(\dfrac{\omega_{**}^2}{\omega_R^2}-1\right)^{1/2} - \tan^{-1}\left(\dfrac{\omega_*^2}{\omega_R^2}-1\right)^{1/2}\right], & (4.126a)\\[2ex] \left(1-\dfrac{\omega_R^2}{\omega_{**}^2}\right)^{1/2} - \left(1-\dfrac{\omega_R^2}{\omega_*^2}\right)^{1/2}, & (4.126b)\\[2ex] \dfrac{1}{2}\dfrac{\omega^*}{\omega_R}\left[\dfrac{\omega_R^2}{\omega_{**}^2}\left(\dfrac{\omega_{**}^2}{\omega_R^2}-1\right)^{1/2} - \dfrac{\omega_R^2}{\omega_*^2}\left(\dfrac{\omega_*^2}{\omega_R^2}-1\right)^{1/2} + \right. & \\ \left. \tan^{-1}\left(\dfrac{\omega_{**}^2}{\omega_R^2}-1\right)^{1/2} - \tan^{-1}\left(\dfrac{\omega_*^2}{\omega_R^2}-1\right)^{1/2}\right]. & (4.126c) \end{cases}$$

Here, the term w_0 is used according to the first formula (4.114) with k_0 instead of k_*.

The formulae (4.126) are given for the case when $\omega_* > \omega_R$, where ω_R is the frequency (4.74) of radial vibrations of a closed shell. If $\omega_* < \omega_R$, then it should be assumed that $\omega_* = \omega_R$ in formulae (4.126). Thus, components of external loading whose frequencies are smaller than ω_R make no contribution to the deflection variance. The reason for this is that the asymptotic density of the spectrum in the interval $[0,\omega_R]$ is equal to zero.

As another example, we shall consider the calculation of the normal stress $\sigma = 6M_{11}/h^2+N_{11}/h$ in the inner region. Here,

$$r(\underset{\sim}{x},\underset{\sim}{k}) = \pm \frac{6D_r(1+i\eta)}{h^2}\left[k_1^2 + \mu k_2^2 + k_0^2 k_2^2\left(\frac{1-\mu^2}{2}\right)^{1/2} \frac{\gamma k_1^2+k_2^2}{(k_1^2+k_2^2)^2}\right] \times$$

$$\sin k_1(x_1-\xi_1)\sin k_2(x_2-\xi_2) \;, \tag{4.127}$$

where the notations (4.44) are used. Hence, after appropriate transformations of formula (4.124), we obtain ($\chi^2 \ll 1$):

$$\underset{\sim}{D}[\sigma] \sim \frac{36}{\pi h^4}\left(\frac{D_r}{\rho h}\right)^{1/2} \times$$

$$\int_0^{\pi/2}\int_0^{\infty} \frac{\{k^2 f_1(\Theta)+k_0^2[(1-\mu^2)/3\,]^{1/2} f_2(\Theta)\sin^2\Theta\}^2}{\chi(k,\Theta)\,[k^4+k_0^4 f_2(\Theta)\,]^{3/2}}\,\Psi(k,\Theta)\,k\,dk\,d\Theta \;, \tag{4.128}$$

where $f_1(\Theta) = \cos^2\Theta+\mu\sin^2\Theta$ and $f_2(\Theta) = \gamma\cos^2\Theta + \sin^2\Theta$.

In the case of a spherical shell with the spectral density of loading (4.112) and the loss tangent defined by (4.113a), formula (4.128) yields

$$\frac{\underset{\sim}{D}[\sigma]}{\sigma_0^2} \sim \frac{\omega_R}{\omega_*}\left\{\left(\frac{\omega_{**}^2}{\omega_R^2} - 1\right)^{1/2} - \left(\frac{\omega_*^2}{\omega_R^2} - 1\right)^{1/2} + \left(\frac{1-\mu^2}{3}\right)^{1/2} \frac{1+3\mu}{3+2\mu+3\mu^2}\,\ln\frac{\omega_{**}^2}{\omega_*^2} - \right.$$

$$\left. \frac{2(1+\mu+2\mu^2)}{3+2\mu+3\mu^2}\left[\tan^{-1}\left(\frac{\omega_{**}^2}{\omega_R^2} - 1\right)^{1/2} - \tan^{-1}\left(\frac{\omega_*^2}{\omega_R^2} - 1\right)^{1/2}\right]\right\} . \tag{4.129}$$

Here, σ_0 is determined from the second formula (4.114). Some other examples can be found in references [11,20,115].

Asymptotic Estimate for Spectral Characteristics

A general asymptotic formula for the spectral density of the field v(x,t) has the form (4.86). We shall illustrate the application of this formula by a simple example. We shall show how the spectral density of stresses close to a clamped edge of a thin elastic plate is determined. Consider a loading specified by (1.33). Analogously to formula (4.119) we write

$$\underset{\sim}{D}[\sigma(\omega_c)] \sim \frac{36}{h^4}\left(\frac{D_r}{\rho h}\right)^{1/2} \int_0^{k_c} \frac{\Psi(k)\,dk}{\chi(k)k}, \qquad k_c = \omega_c^{1/2}(\rho h/D_r)^{1/4},$$

where ω_c is the cut-off frequency of the input process. Differentiating the right side with respect to this frequency, we obtain the estimate for the spectral density [11]

$$S_\sigma(\omega) \sim \frac{9a_1a_2}{h^4}\left(\frac{D_r}{\rho h}\right)^{1/2} \frac{\Psi(\omega)}{\chi(\omega)\omega}. \tag{4.130}$$

It should be noted that this estimate is of an asymptotic character. The spectral density calculated by it is flattened relative to peaks which correspond to an individual contribution of every natural mode. It is evident that an error in determining spectral densities may substantially exceed errors in determining mean squares. Nevertheless, formulae of the type (4.130) can be very useful for drawing general conclusions about the influence of various factors on the character of the spectral density. Figure 34 shows the graphs for the flattened spectral density $S_w(\omega)$ of the deflection for the case when the loading is spatial coordinate white noise and time "quasi-white" noise, i.e., the spectral density $\Psi(\omega)$ is given by the expression (4.112). It can be seen in the graph how the character of the spectral density varies when passing from external friction (case 1) to Voigt's friction (case 3). An analogous graph for the spectral density $S_\sigma(\omega)$ of the stresses is given in Figure 34(b). In the case of external friction, the spectral density of the stress is similar to the time spectral density of the loading. In the case of damping independent of the frequency, as well as in the case of Voigt's friction, higher modes of vibrations are more damped than the lower ones.

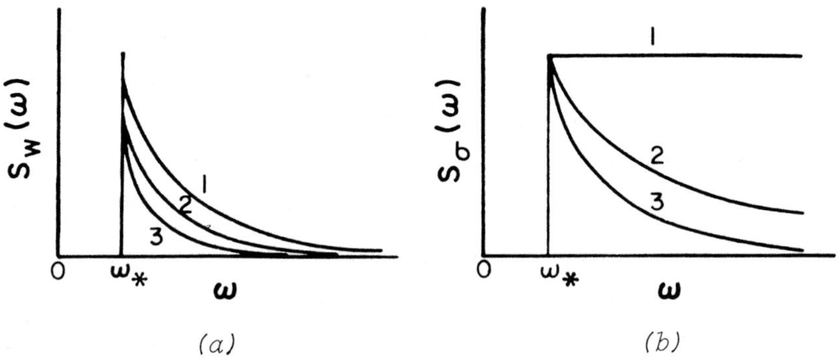

Figure 34

It is possible to obtain asymptotic estimates for some other spectral characteristics. For instance, when determining the extent of damage and the average longevity of a structure (see Chapter 7), it is necessary to know the effective frequency ω_e of the variation of stresses

$$\omega_e^2 = \frac{\int_0^\infty \omega^2 S_\sigma(\omega)\,d\omega}{\int_0^\infty S_\sigma(\omega)\,d\omega} .$$

Using the asymptotic estimate (4.130) for the spectral density $S_\sigma(\omega)$, we obtain the following simple formulae for ω_e^2:

$$\omega_e^2 = \begin{cases} 1/3(\omega_{**}^2+\omega_{**}\omega_*+\omega_*^2) \ , & (4.131a) \\ (\omega_{**}^2-\omega_*^2)/\ln(\omega_{**}/\omega_*)^2 \ , & (4.131b) \\ \omega_*\omega_{**} \ . & (4.131c) \end{cases}$$

These formulae make it possible to qualitatively estimate the influence of the width of the excitation time spectrum on the expected longevity of a structure.

Joint Correlation of Generalized Coordinates

The above given examples related to the case when the time spectral density of a loading was specified by (1.33). And, for these examples, mutual spectral densities of the generalized forces and, consequently, joint spectral densities of the generalized coordinates turned out to be equal to zero. Therefore, an approximation of the integral method in the form (4.85) is exact in the sense that additional terms of joint correlation are in this case identically equal to zero. Now, we shall consider problems in which these additional terms are to be taken into account. To simplify calculations, we shall consider a one-dimensional system - a rod of length a, or a plate vibrating under a cylindrical deflection. As before, we shall calculate the variance of the function v(x,t) linearly related to the deflection w(x,t). If the latter is represented in the form of a series with respect to the natural modes $\phi_j(x)$, the variance of the function v(x,t) is defined as

$$\underset{\sim}{D}[v(x,t)] \approx \sum_j \frac{\pi |r_j(x)|^2 S_{Q_jQ_j}(\omega_j)}{\chi_j \omega_j^3} +$$

$$\sum_m \sum_n \frac{2\pi r_m^*(x) r_n(x)(\chi_m\omega_m+\chi_n\omega_n) S_{Q_mQ_n}(\omega_{mn})}{(\omega_m^2-\omega_n^2)^2+(\chi_m\omega_m+\chi_n\omega_n)(\chi_m\omega_n+\chi_n\omega_m)\omega_m\omega_n} \quad . \qquad (4.132)$$

We obtain the corresponding asymptotic estimate taking account of the relations (4.89), (4.91) and (4.92):

$$\underset{\sim}{D}[v(x,t)] \sim \frac{\pi}{\Delta k_1} \int_0^\infty \frac{|r(x,k)|^2 S_Q(k)\,dk}{\chi(k)\omega^3(k)} +$$

$$\frac{4\pi}{n_1 \Delta k_1^2} \int_0^\infty \int_{\Gamma(k)} \frac{|r(x,k)|^2 \chi(k)\omega(k) T_Q(k,k')\,dk'\,dk}{[\omega^2(k)-\omega^2(k')]^2 + 4\chi^2(k)\omega^4(k)} . \qquad (4.133)$$

The first term on the right side of (4.133) is analogous to the right side in (4.104). The second term is equivalent to the double sum in (4.132) and takes account of the joint correlation of the generalized coordinates. Here $T_Q\ (k,k')$ is a continuous analog of the joint spectral densities $S_{Q_m Q_n}(\omega)$, $\Gamma(k)$ being the domain of the wave number k, and n_1 a positive integer that takes into account the possibility of omissions in the double sum of formula (4.132). The frequency ω and the size of the linear cell Δk are in this case given by the relations

$$\omega = k^2 (D_r/(\rho h))^{1/2} , \qquad \Delta k = \pi/a . \qquad (4.134)$$

Let the function $T_Q(k,k')$ satisfy the conditions for which the estimate (4.95) is applicable. Then, instead of (4.133), it is possible to write

$$\underset{\sim}{D}[v(x,t)] \sim \frac{\pi}{\Delta k_1} \int_0^\infty \frac{|r(x,k)|^2 S_Q(k)\,dk}{\chi(k)\omega^3(k)} +$$

$$\frac{\pi^2}{n_1 \Delta k_1^2} \int_0^\infty \frac{|r(x,k)|^2 \chi(k) T_Q(k,k)\,dk}{\omega(k)\,d\omega^2(k)/dk} . \qquad (4.135)$$

For further calculations, the time spectral density of the loading $S_q(\xi,\omega)$ will be taken in the form

$$S_q(\xi,\omega) = \Psi(\omega) e^{-\alpha|\xi|} . \qquad (4.136)$$

Here, as before, $\xi = x'-x$; also, the function $\Psi(\omega)$ characterizes the time correlation and the constant α^{-1} has the sense of a correlation scale with respect to the coordinate x. For the estimate of variances for the inner region, we shall calculate spectral densities of the generalized forces as for a simply supported rod. All the necessary calculations were performed in Section 3.3. The final result has the form

$$S_{Q_mQ_n}(\omega) = \frac{4\Psi(\omega)}{\rho^2h^2} J_{mn} ,$$

$$J_{mn} = \frac{1}{n^2\pi^2(1+\alpha_n^2)} \begin{cases} \alpha a + \dfrac{2}{1+\alpha_m^2}[1-e^{-\alpha a}(-1)^m] & (m = n), \\ \dfrac{2n}{m(1+\alpha_m^2)}[1-e^{-\alpha a}(-1)^m] , & \\ & (|m \pm n| - \text{even number}) , \\ 0 & (|m \pm n| - \text{odd number}) . \end{cases} \qquad (4.137)$$

Diagonal and secondary elements of the matrix of joint spectral density of the generalized forces are different in form. So, they should be approximated by means of different functions, for which provisions were made in the general theory (see Section 4.5). Besides, spectral densities take on different values depending on whether the numbers m and $|m \pm n|$ are odd or even. All this should be taken into consideration when choosing the numbers n_1, the functions $S_Q(k)$ and $T_Q(k,k)$, this choice being mutually stipulated. Let us illustrate this by an example.

We calculate the variance of the deflection in the middle of a rod's span. With $|m \pm n|$ being odd, we have $S_{Q_mQ_n}(\omega) = 0$. In other cases $S_{Q_mQ_n}(\omega) \neq 0$. The series in formula (4.132) contain only those summands to which the odd numbers of half-waves along the axis x correspond. In formulae (4.133) and (4.135), we must assume $r^2 = 1$, $n_1 = 2$. Continuous analogs of the spectral density of the generalized forces, based on (4.137), take the form

$$S_Q(k) = \frac{4\Psi(k)}{\rho^2h^2a^2}\left[\frac{\alpha a}{k^2+\alpha^2} + \frac{2k^2(1+e^{-\alpha a})}{(k^2+\alpha^2)^2}\right],$$

$$T_Q(k,k) = \frac{8\Psi(k)}{\rho^2h^2a^2}\,\frac{k^2(1+e^{-\alpha a})}{(k^2+\alpha^2)^2}\,. \qquad (4.138)$$

Substituting expressions (4.138) and the above-mentioned values of r^2 and n_1 into the formulae (4.133) and (4.135) gives an approximate estimate for the variance of the deflection in the middle of the span. Here, the following question arises - under what conditions can this estimate be extended to a rod with arbitrary conditions at the ends $x = 0$ and $x = a$. Evidently, the scale of spatial coordinate correlation of the loading should be sufficiently small as compared to the span a. Assuming $\alpha \gg 1$ in the formula (4.138), we obtain

$$S_Q(k) = \frac{4\Psi(k)}{\rho^2h^2a^2}\,\frac{\alpha a}{k^2+\alpha^2}\,,$$

$$T_Q(k,k) = \frac{8\Psi(k)}{\rho^2h^2a^2}\,\frac{k^2}{(k^2+\alpha^2)^2}\,.$$

Substitution of these expressions into formula (4.135), with (4.134) taken into account, yields

$$\underset{\sim}{D}[w(\,a/2\,)] \sim \frac{2\alpha}{D_r^2}\left(\frac{D_r}{\rho h}\right)^{1/2}\int_0^\infty \frac{\Psi(k)}{\chi(k)k^6(k^2+\alpha^2)}\left[1 + \frac{\chi^2(k)k^3}{4\alpha(k^2+\alpha^2)}\right]dk\,.$$

The length of the span a does not appear in the right side of this formula, which may indicate that the result holds true for a rod with arbitrary support at the ends.

Some Results of Direct Calculations

Direct calculations of spectral densities, accelerations and stresses in circular cylindrical shells and spherical shells were performed by V. Khromatov [103]. Some results of the calculations are shown in Figure 35. The calculation was performed for a square spherical panel under the action of a normal pressure q(x,t) assumed as a homogeneous function of the spatial coordinates and a stationary function of time, with the correlation function

$$K_q(\underset{\sim}{x},\underset{\sim}{x}';t,t') = \langle q^2\rangle a_1 a_2 \exp(-\omega_0|t-t'|)\delta(\underset{\sim}{x}-\underset{\sim}{x}') \ .$$

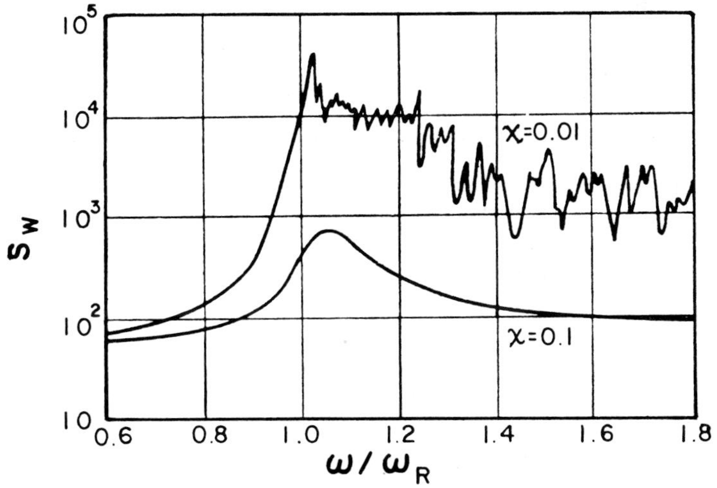

Figure 35

Along the horizontal axis, a dimensionless frequency $\bar{\omega} = \omega/\omega_R$ is laid off. Here, ω_R is the natural frequency (4.127) corresponding to the condensation point. Along the vertical axis, a dimensionless spectral density of the deflection in the centre of the panel is laid off:

$$s_W = S_W(1/2\ a_1,\ 1/2\ a_2;\omega)\omega_0(\rho h\omega_R^2)^2\langle q^2\rangle^{-1} \ .$$

The contributions of specific natural modes at small damping ($\chi = 0.01$) can be seen in the graph. These contributions become indiscernable with the increase of damping. In both cases, the general course of the graphs of the spectral density follows that of the graph for the asymptotic density of natural frequencies.

Crandall [125] performed direct calculations of the mean squares of displacements, velocities and accelerations in plates subjected to the action of concentrated forces delta-correlated in time. He found that if a plate has elements of symmetry, the corresponding elements of symmetry make their appearance as lines of increased intensity of a vibration field. This interesting phenomenon can be accounted for by selective properties of natural modes in relation to concentrated forces. In the spectrum of generalized forces those will be predominant to which natural modes correspond having an antinode in the neighbourhood of the point of application of a concentrated force. In symmetrical plates this may cause the appearance of extended antinodes with the corresponding elements of symmetry (see Figure 36 where the points of force application are shown by circles, and extended antinodes by double lines).

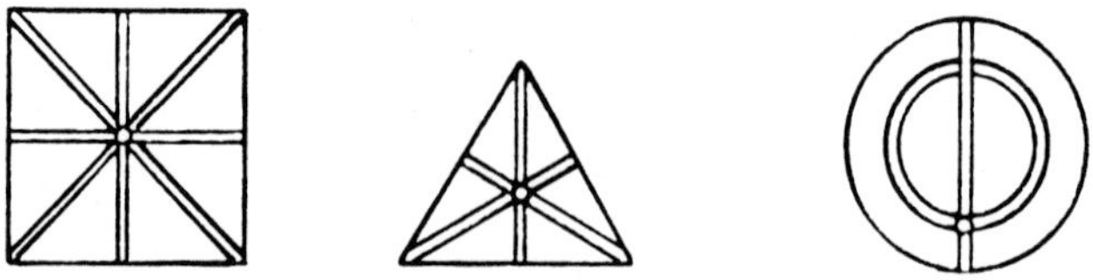

Figure 36

Chapter 5
Parametrically Excited Random Vibrations

5.1 INTRODUCTION

Equations of Parametric Vibrations

Vibrations that are caused by a variation in the parameters of a system are called parametrically excited, or simply parametric vibrations. A classical example is a pendulum with a suspension point oscillating in the direction of the gravity field. The natural frequency of the pendulum in the moving system of coordinates is a function of time, which, for certain relations of parameters, may build up oscillations of the pendulum. As an example of parametric vibrations of continuous systems, we can take the vibrations of a straight elastic rod loaded by a longitudinal force that varies periodically in time. For certain relations between the frequency of the external force and the natural frequencies of the rod's flexural vibrations, the rod's rectilinear shape becomes unstable, i.e., a small initial perturbation causes the development of flexural vibrations of a large amplitude.

From the mathematical point of view, the feature common to all parametric vibrations is that they are described by differential equations with coefficients depending explicitly on time. For example, the equation of small oscillations of a pendulum with a moving suspension point is written as

$$\frac{d^2\phi}{dt^2} + 2\varepsilon \frac{d\phi}{dt} + \omega_0^2 \left(1 + \frac{1}{g}\frac{d^2u}{dt^2}\right)\phi = 0 , \tag{5.1}$$

where $\phi(t)$ is a small deflection angle of the pendulum, $\omega_0 = (g/\ell)^{1/2}$ is its natural frequency, g is the acceleration of gravity, ℓ is the length of the pendulum, ε is the damping coefficient, and u(t) is the displacement of the suspension point. The equation of the perturbed motion for the problem of parametric vibrations of an elastic rod has under certain assumptions has the form

$$EI \frac{\partial^4 w}{\partial x^4} + 2m\varepsilon \frac{\partial w}{\partial t} + m \frac{\partial^2 w}{\partial t^2} - N(t) \frac{\partial^2 w}{\partial x^2} = 0 . \tag{5.2}$$

Here, w(x,t) is the deflection of the rod, EI is the flexural stiffness, m is the mass per unit length, and ε is the damping coefficient (all these parameters are assumed to be constant), N(t) denotes the longitudinal forces in the rod, and is a function of time.

In the general case, linearized equations of parametric vibrations of an elastic system can be represented in operator form by

$$A \frac{\partial^2 \underset{\sim}{u}}{\partial t^2} + \gamma B \frac{\partial \underset{\sim}{u}}{\partial t} + [C+\mu F(t)]\underset{\sim}{u} = \underset{\sim}{0} . \tag{5.3}$$

The notations in this equation are the same as in equation (3.1). In addition, the damping parameter γ and the parametric term $\mu F(t)$ are introduced. Here, F is a linear operator depending on time as a parameter, μ is a certain number (modulation parameter). An effective method of analysis, as for problems of forced random vibrations, is based on an expansion of the function $\underset{\sim}{u}(\underset{\sim}{x},t)$ into the series (3.5) of eigenfunctions of the related problem (3.2). Sometimes, this method leads to separable equations for the generalized coordinates $u_k(t)$. For example, if the equation (5.2) is solved with boundary conditions being equal to

$$w = \partial^2 w/\partial x^2 = 0 \,, \qquad (x = 0,\ x = \ell) \,,$$

which corresponds to a rod of length ℓ, simply supported at its ends, the substitution

$$w(x,t) = \sum_{k=1}^{\infty} u_k(t) \sin \frac{k\pi x}{\ell} \,,$$

leads to the equations

$$\frac{d^2 u_k}{dt^2} + 2\varepsilon \frac{du_k}{dt} + \omega_k^2 \left[1 + \frac{N(t)}{N_k}\right] u_k = 0 \,, \qquad (k = 1,2,\ldots) \,. \tag{5.4}$$

Here, ω_k are the natural frequencies of the rod, and N_k is its Euler force, i.e.,

$$\omega_k = \frac{k^2\pi^2}{\ell^2}\left(\frac{EI}{m}\right)^{1/2} \,, \qquad N_k = \frac{k^2\pi^2 EI}{\ell^2} \,.$$

In the general case, separation of the generalized coordinates it not achieved. Equation (5.3) is then approximately solved by the method of reduction, i.e., retention of a finite number of terms in the series (3.5). Let this number be equal to n. Introduce for the vector-function of the generalized coordinates the notation $\underset{\sim}{u}(t) = \{u_1(t), u_2(t), \ldots, u_n(t)\}$. For the components of this vector, we obtain a system of ordinary differential equations that in matrix form is

$$A \frac{d^2 \underset{\sim}{u}}{dt^2} + \gamma B \frac{d\underset{\sim}{u}}{dt} + [C + \mu F(t)]\underset{\sim}{u} = \underset{\sim}{0} \,. \tag{5.5}$$

Here, A, B, C and F(t) are the matrix analogs of the corresponding operators of the original equation (5.3).

Parametric Resonances

A distinctive feature of equations (5.4) and (5.5) is that for a certain class of their coefficients (functions of time) a trivial solution of these equations may prove to be unstable in the sense of Liapunov. For example, if we assume in the equations (5.4) that $N(t) = N_0 \cos \Theta t$, where N_0 and Θ are constants, each of them becomes a Mathieu equation with a dissipative term:

$$\frac{d^2u}{dt^2} + 2\varepsilon \frac{du}{dt} + \omega_0^2(1+\mu \cos \Theta t)u = 0 . \qquad (5.6)$$

Here, index k is omitted, and the notation $\mu = N_0/N_k$ is introduced. The regions of instability of a trivial solution of equation (5.6) in the plane (Θ,μ) have the shape of wedges whose points are located for small ε close to the values

$$\Theta = 2\omega_0/p , \qquad (p = 1,2,\ldots) . \qquad (5.7)$$

These relations correspond to the so-called parametric resonances.

Now let us consider the system

$$A \frac{d^2\underset{\sim}{u}}{dt^2} + \gamma B \frac{d\underset{\sim}{u}}{dt} + (C+\mu G \cos \Theta t)\underset{\sim}{u} = \underset{\sim}{0} , \qquad (5.8)$$

where G is a matrix with constant coefficients. Let the damping parameter γ be sufficiently small. In this case, we distinguish simple parametric resonances

$$\Theta = 2\omega_k/p , \qquad (k = 1,2,\ldots,n;\ p = 1,2,\ldots) , \qquad (5.9)$$

combination sum resonances

$$\Theta = (\omega_j+\omega_k)/p \ , \quad (j,k = 1,2,\ldots,n;\ p = 1,2,\ldots) \ , \tag{5.10}$$

and combination difference resonances

$$\Theta = (|\omega_j-\omega_k|)/p \ , \quad (j,k = 1,2,\ldots,n;\ p = 1,2,\ldots) \ . \tag{5.11}$$

Here, ω_k are the natural frequencies of the corresponding conservative system, defined as the roots of the characteristic equation

$$\det(C-\omega^2 A) = 0 \ . \tag{5.12}$$

If A, B and C are symmetric matrices, the type of the combination resonance is determined by the structure of the matrix G. For example, for a symmetric matrix G only the combination sum resonances occur. If the matrix G is not symmetric, resonances of the difference type occur.

Parametric vibrations in deterministic systems have been investigated in great detail.* Extension of the theory to stochastic systems whose behaviour is described by means of differential equations with coefficients varying stochastically in time, is of interest. Again, there arises the problem of stability of the trivial solutions of these equations. Stability is understood in the stochastic sense, i.e., as stability with respect to probability, to mathematical expectations, to a set of moment functions of the second order, etc. The stochastic instability of a trivial solution is analogous to the parametric resonances of the corresponding deterministic system.

Parametric resonances arise when relations of the type (5.9) - (5.11) between the frequency of parametric action and the

---

*See, for example: Bolotin, V.V., *Dynamic Stability of Elastic Systems*, Gostekhizdat, Moscow, 1956; Schmidt, G., *Parametric Vibrations*, Mir, Moscow, 1978.

natural frequencies of a system are satisfied. If the parametric action is a random process with a latent periodicity, analogous resonance phenomena can be expected to occur in the stochastic system. Parametric actions in the form of narrow-band stationary random processes are of special interest. Such processes are a suitable model for real vibration actions, whose amplitudes and phases, as a rule, vary slowly and stochastically in time.

Stochastic Stability

The theory of stochastic stability first came into existence mainly in connection with problems of control theory. It was here that extending the classical theory of the stability of motion to stochastic systems became necessary. The purpose of the theory is to supply methods and criteria for solving the following problem: will the deviation of a system's trajectory from the unperturbed trajectory be sufficiently small in some stochastic sense in the case when initial data are perturbed or when perturbations occur in motion?

The mathematical aspects of the theory are treated in [101,147]. To make further references easier, we shall dwell here on some basic definitions. Let the system be stochastic in the sense that at each moment of time t its state is described by the random vector $\underset{\sim}{x}(t)$ in some phase space $\underline{U}$. As applied to equation (5.5), we have $\underset{\sim}{x} = \{\underset{\sim}{u},\dot{\underset{\sim}{u}}\}$, i.e., the dimensionality of the space $\underline{U}$ is equal to 2n. The stochastic character of the behaviour of the system can be determined by a random specification of the initial conditions as well as by the action of random factors in the process of the system's motion. We shall consider the second case. The evolution of the system is described by the differential equation for realizations $\underset{\sim}{x}(t)$ of the random process:

$$\frac{d\underset{\sim}{x}}{dt} = \underset{\sim}{f}(\underset{\sim}{x},t) \; . \tag{5.13}$$

Let $\underset{\sim}{x} \equiv 0$ be a solution of this equation. We consider the initial condition $\underset{\sim}{x}(t_0) = \underset{\sim}{x}_0$ as deterministic. We introduce the norm $||\underset{\sim}{x}||$ into the space $\underline{U}$. By analogy with the classical definition of stability according to Liapunov, we introduce the following definitions of stability with respect to probability.

The solution $\underset{\sim}{x}(t) \equiv \underset{\sim}{0}$ of equation (5.13) is called stable with respect to probability if for any $\varepsilon > 0$, $\rho > 0$ it is possible to find a $\delta(\varepsilon,\rho) > 0$ such that $||\underset{\sim}{x}_0|| < \delta$ implies

$$\underset{\sim}{P}\{ \sup_{t_0<t<\infty} ||\underset{\sim}{x}(t)|| < \varepsilon\} > 1 - \rho \; . \tag{5.14}$$

The meaning of relation (5.14) is that for the stable solution $\underset{\sim}{x}(t) \equiv \underset{\sim}{0}$ initial perturbations can always be selected in such a way that the probability of an a priori specified small deviation of the system from the origin of coordinates at $t > t_0$ will be smaller than any a priori specified value of ρ.

The solution $\underset{\sim}{x}(t) \equiv \underset{\sim}{0}$ is called asymptotically stable with respect to probability if it is stable with respect to probability and if, besides, for any $\varepsilon > 0$ the condition

$$\lim_{t\to\infty} \underset{\sim}{P}\{||\underset{\sim}{x}(t)|| < \varepsilon\} = 1 \; , \tag{5.15}$$

is satisfied.

The solution $\underset{\sim}{x}(t) \equiv 0$ is called stable with respect to the mathematical expectation of the norm $||\underset{\sim}{x}||$ in the space $\underline{U}$ if for any $\varepsilon > 0$ a $\delta(\varepsilon) > 0$ can be found such that $||\underset{\sim}{x}_0|| < \delta$ implies

$$\sup_{t_0<t<\infty} \langle||x(t)||\rangle < \varepsilon \; . \tag{5.16}$$

If the solution $\underset{\sim}{x}(t) \equiv \underset{\sim}{0}$ is stable with respect to the mathematical expectation of the norm and, in addition, if the condition

$$\lim_{t\to\infty} \langle || \underset{\sim}{x}(t) || \rangle = 0 , \tag{5.17}$$

is satisfied, it is called asymptotically stable with respect to the mathematical expectation of the norm. If the norm $||\underset{\sim}{x}||$ is introduced by means of the relation

$$||\underset{\sim}{x}|| = \left(\sum_{k=1}^{n} |x_k|^p \right)^{1/p} ,$$

p-stability is implied. For $p = 2$ this corresponds to stability in the mean-square sense.

Stability with Respect to a Set of Moment Functions

In applied analysis, definitions of stability are widely used which are based on moments of phase variables of second and higher orders. Consider the moment functions of the process $\underset{\sim}{x}(t)$ that are equal to the mathematical expectations of the process components and their products at coinciding moments of time. We distinguish moment functions of the first order, i.e., mathematical expectations of components, moment functions of the second order, i.e., mathematical expectations of squares and pairwise component products, etc. Henceforth the notation

$$m_{\underbrace{jk\ell\ldots}_{r}}(t) = \langle \underbrace{x_j(t)x_k(t)x_\ell(t)\ldots}_{r} \rangle , \tag{5.18}$$

will be used, where the number of indices r is equal to a moment's order. We shall form a vector

$$\underset{\sim}{m}_r(t) = \{m_{\underbrace{11\ldots1}_{r}}(t), m_{\underbrace{11\ldots2}_{r}}(t), \ldots\} , \tag{5.19}$$

of moment functions of r-th order. Taking symmetry of moment functions into account, the dimensionality of this vector is reduced.

The corresponding vector space will be denoted by $\underline{M}_r$, and the norm in this space by $||\underset{\sim}{m}_r||$. Let us introduce the following definition [116].

The solution $\underset{\sim}{x}(t) \equiv \underset{\sim}{0}$ is called stable with respect to a set of moment functions of r-th order if for each $\varepsilon > 0$ a $\delta(\varepsilon) > 0$ can be found such that $||\underset{\sim}{m}_r(t_0)|| < \delta$ implies

$$\sup_{t_0<t<\infty} ||\underset{\sim}{m}_r(t)|| < \varepsilon \ . \tag{5.20}$$

If the solution $\underset{\sim}{x}(t) \equiv \underset{\sim}{0}$ is stable in the above mentioned sense, and if the condition

$$\lim_{t\to\infty} ||\underset{\sim}{m}_r(t)|| = 0 \tag{5.21}$$

is also satisfied, the solution is called asymptotically stable with respect to the set of moment functions of r-th order.

In a number of problems, moment functions of various orders are connected in such a way that it is impossible to analyze their behaviour in time separately. In this case, it is expedient to modify the definition of stability. We introduce the vector

$$\underset{\sim}{m}_1^r(t) = \{\underset{\sim}{m}_1(t), \underset{\sim}{m}_2(t), \ldots, \underset{\sim}{m}_r(t)\} \ , \tag{5.22}$$

whose components are moment functions of the process from the first to the r-th order inclusive. The dimensionality of this vector is reduced if the symmetry of the moment functions is taken into account. We shall denote the corresponding vector space by $\underline{M}_1^r$, and the norm in this space by $||\underset{\sim}{m}_1^r||$. For example, the Euclidean norm is written in the form

$$||\underset{\sim}{m}_1^r|| = \left(\sum_{j=1}^{n} m_j^2 + \sum_{j=1}^{k} \sum_{k=1}^{n} m_{jk}^2 + \cdots + \sum_{j=1}^{k} \sum_{k=1}^{\ell} \cdots m^2_{\underbrace{jk\ell\ldots}_{r}} \right)^{1/2} . \tag{5.23}$$

The solution $\underset{\sim}{x}(t) \equiv \underset{\sim}{0}$ is called stable with respect to moment functions up to r-th order inclusive if for each $\varepsilon > 0$ a $\delta(\varepsilon) > 0$ can be found such that $||\underset{\sim}{m}_1^r(t_0)|| < \delta$ implies

$$\sup_{t_0<t<\infty} ||\underset{\sim}{m}_1^r(t)|| < \varepsilon \ . \tag{5.24}$$

If, in addition, the condition

$$\lim_{t\to\infty} ||\underset{\sim}{m}_1^r(t)|| = 0 \tag{5.25}$$

is satisfied, the solution $\underset{\sim}{x}(t) \equiv \underset{\sim}{0}$ is called asymptotically stable with respect to the set of moment functions up to order r inclusive. Henceforth, for brevity, we shall speak of stability in the spaces $\underline{M}_r$ and $\underline{M}_1^r$, respectively.

5.2 THE METHOD OF STOCHASTIC LIAPUNOV FUNCTIONS

Fundamentals of the Method

In order to estimate the stability of an unperturbed motion, it is necessary to analyze the behaviour of a set of contiguous realizations bringing this analysis into agreement with one of the definitions of stochastic stability. When applied to systems whose evolution is a diffusion Markov process, the method of Liapunov's functions takes a form very similar to the classical one. As in the classical method, Liapunov's functions play the role of test functions. One will remember that the function $v(\underset{\sim}{x},t)$, specified in some neighbourhood of the half-line $\underset{\sim}{x} = \underset{\sim}{0}$, $t \geq t_0$, is called positive definite according to Liapunov if the conditions $v(\underset{\sim}{0},t) = w(\underset{\sim}{0}) = 0$, $v(\underset{\sim}{x},t) > w(\underset{\sim}{x}) > 0$ for $\underset{\sim}{x} \neq \underset{\sim}{0}$ are satisfied. The function $v(x,t)$ is considered to have an infinitesimal upper limit (uniformly small with respect to x), if for any $h > 0$ a $\delta(h) > 0$ can be found such that for any $t \geq t_0$ condition $||\underset{\sim}{x}(t)|| < \delta$ implies $|v(\underset{\sim}{x},t)| < h$.

In the method of stochastic Liapunov functions, the generating differential operator of the Markov type [101] is used instead of the operator by means of which the value of the derivative of Liapunov's function along phase trajectories is calculated. This differential operator is introduced on the set of differentiable functions $v(\underset{\sim}{x},t)$ as

$$L = \frac{\partial}{\partial t} + \sum_{\alpha=1}^{n} \chi_\alpha \frac{\partial}{\partial x_\alpha} + \frac{1}{2} \sum_{\alpha=1}^{n} \sum_{\beta=1}^{n} \chi_{\alpha\beta} \frac{\partial^2}{\partial x_\alpha \partial x_\beta} , \tag{5.26}$$

where $\chi_\alpha(\underset{\sim}{x},t)$ and $\chi_{\alpha\beta}(\underset{\sim}{x},t)$ are the intensities of the Markov process (2.87). The generating operator can be represented as

$$L = \frac{\partial}{\partial t} + \Lambda^* , \tag{5.27}$$

where the operator Λ is given by formula (2.93), and the adjoint operator Λ^* coincides with the operator of the inverse Kolmogorov equation (2.92). In the literature dealing with the theory of Markov processes, the operator Λ^* is also called the generating operator.

The value Lv has the meaning of a derivative of the conditional mathematical expectation of the function $v(\underset{\sim}{x},t)$ with respect to time (the function being calculated on the condition that the process has the value $\underset{\sim}{x}$ at the moment of time t). The task is to construct functions $v(\underset{\sim}{x},t)$ such that it would be possible to estimate stochastic stability (instability) by means of the sign of Lv in some region of the phase space. The solution to this problem is given by theorems of stochastic stability whose formulation and techniques of application are analogous to the corresponding results of the classical theory of the stability of motion. We shall consider three basic theorems formulated for an investigation of the stability of the trivial solution $\underset{\sim}{x}(t) \equiv \underset{\sim}{0}$.

Let a continuous positive definite function $v(\underset{\sim}{x},t)$ relating to the region of definition of the operator L and satisfying the condition $Lv \leq 0$ for $\underset{\sim}{x} \neq \underset{\sim}{0}$, exist in the vicinity of $\underset{\sim}{x} = \underset{\sim}{0}$, $t \geq t_0$. Then, the solution $\underset{\sim}{x}(t) \equiv \underset{\sim}{0}$ is stable with respect to probability.

Let a continuous positive definite function $v(\underset{\sim}{x},t)$ having an infinitesimal limit relating to the region of definition of the operator L, and satisfying the condition $Lv < 0$ for $\underset{\sim}{x} \neq \underset{\sim}{0}$, exist in the vicinity of $\underset{\sim}{x} = \underset{\sim}{0}$, $t \geq t_0$. Then, the solution $\underset{\sim}{x}(t) \equiv \underset{\sim}{0}$ is asymptotically stable with respect to probability.

Suppose a quadratic positive definite form $w(\underset{\sim}{x},t)$ satisfying the condition

$$Lv = -w \tag{5.28}$$

is found for some positive definite quadratic form $v(\underset{\sim}{x},t)$ with coefficients that are continuous bounded functions of time. In this case, the solution $\underset{\sim}{x}(t) \equiv \underset{\sim}{0}$ is asymptotically stable in the mean square sense.

Examples

Consider a linear oscillating system with one degree of freedom whose damping coefficient is perturbed by white noise. This system was already considered in Section 2.4, where the intensities of a corresponding Markov process were calculated under two assumptions: (1) that the white noise $\xi(t)$ is Ito's white noise, and (2) that it is Stratonovich's white noise. To make further calculations simpler we shall introduce dimensionless variables:

$$t' = \omega_0 t \ , \quad s' = s/\omega_0 \ , \quad \gamma = \varepsilon/\omega_0 \ , \tag{5.29}$$

assuming further that $s' = 1$ and omitting the prime in t'. Then, equation (2.104) takes the form

$$\frac{d^2u}{dt^2} + [2\gamma+\mu\xi(t)]\frac{du}{dt} + u = 0 . \tag{5.30}$$

The intensities of the process $\underset{\sim}{x} = \{u,\dot{u}\}$ are determined from the formulae (2.106) and (2.107), if in them we assume that $\omega_0 = s = 1$, $\varepsilon = \gamma$.

For Ito's white noise the generating operator (5.26) takes the form

$$L = \frac{\partial}{\partial t} + x_2\frac{\partial}{\partial x_1} - (2\gamma x_2+x_1)\frac{\partial}{\partial x_2} + \frac{1}{2}\mu^2x_2^2\frac{\partial^2}{\partial x_2^2} . \tag{5.31}$$

As a Liapunov function, we take the quadratic form

$$v = c_{11}x_1^2 + 2c_{12}x_1x_2 + c_{22}x_2^2 , \tag{5.32}$$

whose coefficients c_{jk} are to be chosen in such a way that the condition of one of the theorems of stochastic stability is satisfied. In particular, in order to satisfy the conditions of the asymptotic stability theorem in the mean-square sense, we should require that

$$Lv = -(x_1^2+x_2^2) . \tag{5.33}$$

Substituting the quadratic form (5.32) into relation (5.33), we find the values of its coefficients to be

$$c_{11} = 2(4\gamma-\mu^2)^{-1} + \gamma , \quad c_{22} = 2(4\gamma-\mu^2)^{-1} , \quad c_{12} = 1/2 .$$

The quadratic form (5.32) will be positive definite if the conditions $c_{11} > 0$, $c_{22} > 0$ are satisfied. From this, we obtain a sufficient condition of asymptotic stability with respect to probability,

$$\mu^2 < 4\gamma . \tag{5.34}$$

If we interpret white noise $\xi(t)$ according to Stratonovich, the intensities of the Markov process should be calculated from the formulae (2.107). The generating operator will take the form:

$$L = \frac{\partial}{\partial t} + x_2 \frac{\partial}{\partial x_1} - \left(2\gamma x_2 + x_1 - \frac{1}{2}\mu^2 x_2 \right) \frac{\partial}{\partial x_2} + \frac{1}{2}\mu^2 x_2^2 \frac{\partial^2}{\partial x_2^2} ,$$

and the coefficients of the quadratic form (5.32) are determined as

$$c_{11} = (2\gamma - \mu^2)^{-1} + \frac{1}{2}(2\gamma - 1/2\,\mu^2) ,$$
$$c_{22} = (2\gamma - \mu^2)^{-1} , \quad c_{12} = \frac{1}{2} . \tag{5.35}$$

Assuming that $c_{11} > 0$, $c_{22} > 0$, we arrive at the condition of asymptotic stability in the mean square sense [101], namely,

$$\mu^2 < 2\gamma . \tag{5.36}$$

Comparing relations (5.34) and (5.36), we can see that the choice of modelling white noise substantially influences the results. In this case, though, sufficient conditions of stability are compared. Further below, it will be shown that a discrepancy remains even in the case when necessary and sufficient conditions are compared.

5.3 METHOD OF MOMENT FUNCTIONS

Three Typical Situations in Applying this Method

The general characteristics of the method of moment functions as applied to a wide class of statistical dynamics problems were given in Section 2.1. In problems of stability, the method of moment functions permits one to reduce the analysis of stability of solutions of stochastic differential equations to the analysis of the stability of deterministic differential equations describing the evolution of the moment functions. Therefore, one of the definitions of stability with respect to a set of moment functions is used for a definition of stochastic stability.

To obtain equations for moment functions we shall take the Kolmogorov equation (2.90), i.e.,

$$\frac{\partial p}{\partial t} = - \sum_{\alpha=1}^{n} \frac{\partial}{\partial x_\alpha} [\chi_\alpha(\underset{\sim}{x},t)p] + \frac{1}{2} \sum_{\alpha=1}^{n} \sum_{\beta=1}^{n} \frac{\partial^2}{\partial x_\alpha \partial x_\beta} [\chi_{\alpha\beta}(\underset{\sim}{x},t)p] \ . \tag{5.37}$$

Let the intensities χ_α and $\chi_{\alpha\beta}$ be polynomials of components of the vector $\underset{\sim}{x}$. Then, a system of ordinary differential equations generally relating various moment functions (5.18) to each other can be derived from equation (5.37). For instance, in order to obtain an equation on whose left side there is a time derivative with respect to the moment function $m_{jk\ell}\ldots(t)$, it is necessary to multiply each term of equation (5.37) by the product $x_j x_k x_\ell \ldots$ and integrate over the entire phase space $\underline{U}$. After the necessary transformations of integrals, we arrive at the desired equation.

We may encounter the following three situations: Firstly, it may turn out that all the equations for moment functions fall into groups, each of which contains only moment functions of one and the same order r, e.g.,

$$\frac{d\underset{\sim r}{m}}{dt} = \Phi(\underset{\sim r}{m},t) \ , \qquad (r = 1,2,\ldots) \ . \tag{5.38}$$

This will occur, for example, if the intensities of the Markov process are independent of $\underset{\sim}{x}$. The system of equations (5.38) admits an analysis of stability of the solution $\underset{\sim}{x}(t) \equiv \underset{\sim}{0}$ of a corresponding stochastic system in the space $\underline{M}_r$.

Secondly, equations for moment functions may have the form

$$\frac{d\underset{\sim}{m}_\rho}{dt} = \Phi_\rho\,(\underset{\sim}{m}_1,\ldots,\underset{\sim}{m}_\rho)\ , \qquad (\rho = 1,2,\ldots,r)\ , \tag{5.39}$$

where the right sides depend not only on moment functions of ρ-th order, but also on moment functions of a lower order. The system (5.39) makes it possible to analyze stability with respect to a set of moment functions up to the order r (inclusive), i.e., stability in the space $\underline{M}_1^r$. When stability in $\underline{M}_r$ is to be analyzed, it is necessary to consider the stability of the solution $\underset{\sim}{m}^r(t) \equiv \underset{\sim}{0}$ of the system (5.39) with respect to some variable.

And lastly, there are such cases when for the moment functions there will be an infinite system of coupled equations, say,

$$\frac{d\underset{\sim}{m}_\rho}{dt} = \Phi_\rho(\underset{\sim}{m}_1,\ldots,\underset{\sim}{m}_\rho,\ldots)\ , \qquad (\rho = 1,2,\ldots)\ , \tag{5.40}$$

whose right sides contain moment functions of higher order. Such a case is typical for nonlinear systems, and also for parametric systems excited by non-white noise. Here, along with considering stability in relation to a set of moment functions of various orders, we have to deal with the problem of reducing the infinite system of equations (5.40) to a system of a finite number of equations of the type (5.39). This reduction, which can only be made approximately, makes the analysis of stability not quite rigorous, and the error due to the reduction cannot always be estimated.

Example: Stability in the Mean

As an example we shall consider the system from Section 5.2. Equation (5.37) takes the form

$$\frac{\partial p}{\partial t} = -\frac{\partial}{\partial x_1}(\chi_1 p) - \frac{\partial}{\partial x_2}(\chi_2 p) + \frac{1}{2}\frac{\partial^2}{\partial x_2^2}(\chi_{22} p)\ , \tag{5.41}$$

where, depending on the method of interpretation of white noise, the intensities χ_1, χ_2, and χ_{22} are calculated either from the formulae (2.106) or from the formulae (2.107). In order to obtain the equations relating the moments of first order m_1 and m_2, we multiply equation (5.41) once by x_1, and another time by x_2, then we integrate over the phase space and make the necessary transformations. For the intensities (2.106), we arrive at the equations

$$\frac{dm_1}{dt} = m_2 \, , \qquad \frac{dm_2}{dt} = -(m_1 + 2\gamma m_2) \, .$$

The solution $m_1 = m_2 = 0$ of this system is asymptotically stable if the condition $\gamma > 0$ is satisfied. From this, we conclude that the solution $\underset{\sim}{x}(t) \equiv \underset{\sim}{0}$ of the stochastic system is asymptotically stable with respect to moments of the first order (in the mean) for white noise of any intensity only if damping is positive.

If white noise is interpreted according to Stratonovich, i.e., as the limiting case of a process with finite variance, the intensities calculated by formulae (2.107) should be substituted into the equation (5.41). For the moments of first order, we obtain the equations

$$\frac{dm_1}{dt} = m_2 \, , \qquad \frac{dm_2}{dt} = -m_1 - \left(2\gamma - \frac{1}{2}\mu^2\right) m_2 \, ,$$

whose trivial solution is asymptotically stable if $\mu^2 < 4\gamma$. Stability with respect to the moment functions m_{11}, m_{12}, m_{22} is analyzed analogously (evidently, it is equivalent to the stability in the mean square).

Stochastic Analog of the Mathieu-Hill Equation

Let us apply the method of moment functions to the analysis of conditions of parametric excitation in a system with an equation of the type (5.1), (5.4) or (5.6):

$$\frac{d^2u}{dt^2} + 2\varepsilon \frac{du}{dt} + \omega_0^2[1+\mu\xi(t)]u = 0 \ . \qquad (5.42)$$

Here, as distinct from the above mentioned equations, $\xi(t)$ is a stationary normal white noise. In relation to the phase variables $x_1 = u$, $x_2 = \dot{u}$, this system is of the Markov type. Introducing, as before, the dimensionless parameters (5.29), assuming $s' = 1$, and omitting the prime in t', we write the corresponding Ito stochastic differential equations as

$$dx_1 = x_2 dt \ , \qquad dx_2 = -(2\gamma x_2 + x_1)dt - \mu dw \ . \qquad (5.43)$$

The intensities of the Markov process $\underset{\sim}{x}(t) = \{x_1(t), x_2(t)\}$ calculated from the formulae (2.102) and (2.103) are

$$\chi_1 = x_2 \ , \quad \chi_2 = -(2\gamma x_2 + x_1) \ , \quad \chi_{11} = \chi_{12} = 0 \ , \quad \chi_{22} = \mu^2 x_1^2 \ . \qquad (5.44)$$

The Kolmogorov equation (5.37) takes the form

$$\frac{\partial p}{\partial t} = 2\gamma p - x_2 \frac{\partial p}{\partial x_1} + (2\gamma x_2 + x_1) \frac{\partial p}{\partial x_2} + \frac{1}{2} \mu^2 x_1^2 \frac{\partial^2 p}{\partial x_2^2} \ . \qquad (5.45)$$

Using equation (5.45), it is not difficult to obtain the equations for the moment functions in a standard fashion. These equations are of the type (5.38), i.e., an uncoupled system of equations for moment functions of a certain order r is obtained. This allows one to carry out the analysis of stability of the zero solution of equation (5.42) using the definitions of stability in relation to moment functions of various orders.

Specifically, for moment functions of the first order, the system

$$\frac{dm_1}{dt} = m_2 \ , \qquad \frac{dm_2}{dt} = -m_1 - 2\gamma m_2 \ ,$$

is obtained, from which it follows that the condition of asymptotic stability in the mean is $\gamma > 0$. However high the level of excitation is, the system proves to be stable in the mean. But, as in the previous example, this shows the inadequacy of the definition of stability in the mean rather than the impossibility of exciting parametric oscillations by means of a random action of the white noise type.

For moments of the second order, we analogously obtain the system of equations

$$\frac{dm_{11}}{dt} = 2m_{12}\,, \qquad \frac{dm_{22}}{dt} = \mu^2 m_{11} - 2m_{12} - 4\gamma m_{22}\,,$$

$$\frac{dm_{12}}{dt} = -m_{11} - 2\gamma m_{12} + m_{22}\,.$$

Its solution $m_{11} = m_{12} = m_{22} = 0$ will be asymptotically stable if

$$\mu^2 < 4\gamma\,. \tag{5.46}$$

This yields the necessary and sufficient condition for asymptotic stability in the mean square for the solution $\underset{\sim}{x}(t) \equiv \underset{\sim}{0}$ of the system (5.42).

Influence of the Moments' Order on the Conditions of Stability

It has already been shown from previous examples that conditions of stability substantially depend on the definition of stochastic stability that is used as the basis. Specifically, the results vary when passing to the moment functions of higher order. A systematic analysis of this phenomenon with the system (5.42) as an example is given in [71].

The equation of the type (5.38) for the vector $\underset{\sim}{m}_r$, whose components are equal to the moment functions of the r-th order, has the form

$$\frac{dm_r}{dt} = H\underset{\sim}{m}_r \, . \tag{5.47}$$

Here, with symmetry taken into consideration, dimensionality of the vector $\underset{\sim}{m}_r$ and of the matrix H equals r+1. The non-zero elements of the matrix H, when properly numbered, have the form

$$h_{j,j} = -2(j-1)\gamma \, , \qquad h_{j,j-1} = -(j-1) \, ,$$
$$h_{j,j+1} = r+1-j \, , \qquad h_{j,j-2} = \frac{1}{2}(j-1)(j-2)\mu^2 \, . \tag{5.48}$$

Analysis of the stability of the solution $\underset{\sim}{x}(t) \equiv \underset{\sim}{0}$ of system (5.42) with respect to a set of moment functions of r-th order is reduced to determining the region of stability of a trivial solution of the system (5.47) with constant coefficients determined from the formulae (5.48).

Determination of the boundaries of the stability region is facilitated by the fact that the points corresponding to small μ's on the plane (γ,μ) are known to pertain to the stability region. Therefore, moving away from the stability region, it is sufficient to verify two conditions: positiveness of the determinant of the matrix H and positiveness of Hurwitz' principal determinant. For example, the stability region with respect to moment functions of the third order is given by the inequalities

$$\mu^2 < 4\gamma^2(4+37\gamma^2+40\gamma^4) \, , \tag{5.49}$$

$$\mu^2 < (1+8\gamma^2)/(4\gamma) \, . \tag{5.50}$$

When the inequality (5.49) is not satisfied, the principal Hurwitz determinant changes its sign, and when the inequality (5.50) is not satisfied, the determinant of H changes its sign. If the damping coefficient γ is sufficiently small, the boundary of the stability region is determined from condition (5.49). If the damping coefficients are large (actually when $\gamma > 0.25$), condition (5.50) governs.

Figure 37 shows the boundaries of the stability regions that are found for various definitions of stochastic stability. The numbers on the curves are equal to the order of the moment functions which are used in the definition of stability. Comparing the conditions that correspond to even r values, we can see that with higher order of moments, the stability conditions become more restrictive. However, the boundaries calculated for r = 2 and r = 3 intersect. Analogously, the boundaries calculated for r = 4 and r = 5 also intersect. The higher the order of moment functions, the smaller the difference between the boundaries of the stability regions, which can be seen, for instance, when we compare the boundaries for r = 17 and r = 20 in Figure 37.

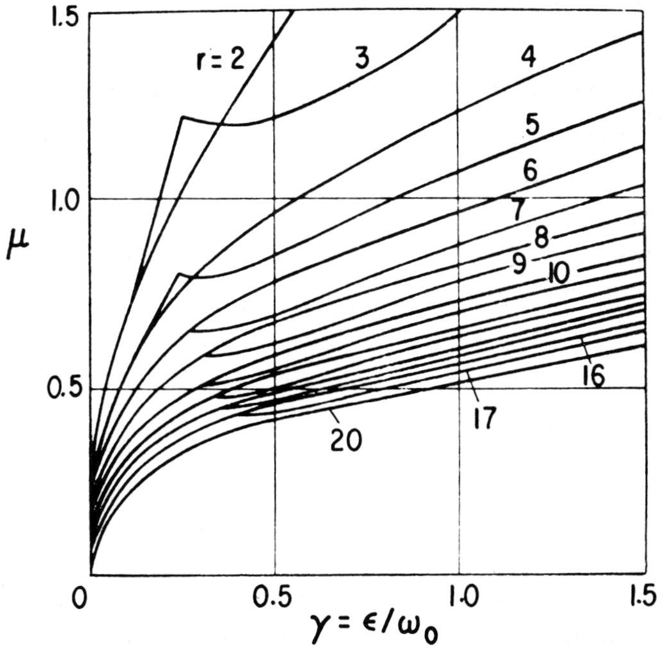

Figure 37

5.4 MODIFICATION OF THE METHOD OF MOMENT FUNCTIONS

Preliminary Remarks

Below, we shall consider parametric oscillations in systems excited by stationary processes with a finite variance. As an example we take an oscillatory system with one degree of freedom whose linearized equation of perturbed motion has the form

$$\frac{d^2u}{dt} = 2\varepsilon \frac{du}{dt} + \omega_0^2[1+\mu\Phi(t)]u = 0 . \tag{5.51}$$

Here, $\Phi(t)$ is a stochastic time function with a mathematical expectation equal to zero. The task is to find regions in the space of parameters ε, ω_0, μ, and also of additional parameters characterizing the properties of the function $\Phi(t)$, where the solution $u(t) \equiv 0$ of equation (5.51) is stable in the sense of one of the definitions of stochastic stability.

The analysis of stability is greatly compounded due to the fact that the function $\Phi(t)$ is not white noise. Approximate methods of statistical dynamics, based on the theory of nonlinear oscillations, are usually applied for the analysis. Let the system (5.51) be of a sufficiently high quality, i.e., $\varepsilon^2 \ll \omega_0^2$. Then, it is natural to assume that for a wide class of actions $\Phi(t)$ the process $u(t)$ will be sufficiently narrow-banded. We seek the solution of equation (5.51) in the form

$$u(t) = A(t) \sin[\omega_0 t+\psi(t)] , \tag{5.52}$$

where $A(t)$ and $\psi(t)$ are the slowly varying random amplitude and phase. Approximate equations for the newly introduced functions are obtained by using the Krylov-Bogolubov-Mitropolski method. Since the functions $A(t)$ and $\psi(t)$ vary slowly as compared to $\Phi(t)$, they should be considered as components of a two-dimensional Markov process. Then, subsequently, one of the above described methods is used.

The approach described above was used by some authors, for example, in [60,93,170]. In these references, approximate stability conditions were found. Omitting in these conditions the terms which have the error order of the used method, it is possible to reduce the conditions to a unified form:

$$\mu^2 S_\Phi(2\omega_0) < \frac{2\gamma}{\pi\omega_0} . \tag{5.53}$$

The spectral density value of the process $S_\Phi(\omega)$ for the frequency $\omega = 2\omega_0$, i.e., for the double natural frequency of the system, enters into the left side. If $\Phi(t)$ is a narrow-band random process with the carrying frequency Θ, the spectral density $S_\Phi(\omega)$ will have a sharp maximum close to $\omega = \Theta$. Therefore, if the relation $\Theta = 2\omega_0$ is satisfied, condition (5.53) may turn out to be violated for sufficiently small μ. Thus, we come to the analog of the parametric resonance (5.7) for $p = 1$. Incidentally, it can be mentioned that condition (5.53) differs from condition (5.46) in that the value of the spectral density $S_\Phi(\omega)$ for $\omega = 2\omega_0$ multiplied by 2π (instead of the dimensional intensity of white noise) enters into it. This result is "almost obvious": because the system (5.51) is lightly damped, it will respond primarily to that part of the spectrum of the parametric action $\Phi(t)$ whose frequencies are close to $2\omega_0$. Hence, substituting the action with the spectral density $S_\Phi(\omega)$ by white noise with the intensity $s = 2\pi S_\Phi(2\omega_0)$, and applying condition (5.46), we arrive at condition (5.53).

If $\Phi(t)$ is a narrow-band random process or a process with latent periodicity, by analogy with the deterministic Mathieu equation (5.6), the regions of instability in the plane of the parameters (μ,Θ) can be expected to have a fairly complex form. Specifically, if a process is sufficiently narrow-banded, and damping of the system is sufficiently small, a series of wedges can be expected to appear close to relations (5.7). However, the approximate methods described above permit one to detect only the region of instability close to the relation $\Theta = 2\omega_0$, which corresponds to the main parametric resonance. The task is to explore the possibility of analogs of secondary resonances appearing for $p > 1$.

A method of calculating the fine structure of instability regions for systems of stochastic equations of a very wide class was suggested in [116]. Description of this method and its results will be given further on in this chapter.

Extension of the Phase Space

Let the equation of a dynamic system have the form

$$\frac{d\underset{\sim}{x}}{dt} = f[\underset{\sim}{x}(t),\underset{\sim}{q}(t)] \ , \tag{5.54}$$

where the right side is dependent of the vector $\underset{\sim}{q}(t)$ of external actions. We shall assume that the components of this vector are stationary and jointly stationary normal random processes with finite variances and rational spectral densities.

The phase space of the system (5.54) will be denoted by $\underline{U}$. Evidently, the process $\underset{\sim}{x}(t)$ cannot be interpreted as a Markov process in this space. Analogously to Section 2.4, we shall consider, along with the prescribed system (5.54), the system describing processes in a certain filter, by means of which a multidimensional normal white noise is transformed into a normal process $\underset{\sim}{a}(t)$. The vector of phase variables for the filter will be denoted by $\underset{\sim}{y}(t)$, and the corresponding phase space by $\underline{V}$. The dimensionality of this space depends on the structure of the polynomials in the expressions for the spectral densities of the process $\underset{\sim}{q}(t)$, and will be denoted by ν. Considering the extended phase space $\underline{U} \oplus \underline{V}$ with the elements

$$z(t) = \{x_1(t),\ldots,x_n(t);y_1(t),\ldots,y_\nu(t)\} , \tag{5.55}$$

we find that the evolution of the vector $\underset{\sim}{z}(t)$ is a diffusive Markov process. The task is to draw a conclusion from the characteristics of this process about the stochastic stability of the solution $\underset{\sim}{x}(t) \equiv \underset{\sim}{0}$ of equation (5.54). This can be done by studying the behaviour of the moment functions of the process $\underset{\sim}{z}(t)$:

$$m_{jk\ell\ldots}(t) = \langle z_j(t)z_k(t)z_\ell(t)\ldots\rangle . \tag{5.56}$$

There are two difficulties in solving the considered problem by the method of moment functions. The first one is caused by the fact that, generally, the process $\underset{\sim}{z}(t)$ is described by non-linear equations. For instance, in the case of a linear parametric system, the components of the vector $\underset{\sim}{q}(t)$ enter into the equation (5.54) as coefficients of the phase variables $x_k(t)$, but when the phase space is extended, these components will be included among phase variables. Therefore, in forming differential equations for moment functions (5.56), we shall get infinite systems of coupled equations of the type (5.40). For the actual analysis, it is necessary to carry out a reduction to a finite system. In connection with this, there arises the problem of an expedient method of reduction (closure) of an infinite system.

Another difficulty may be caused by the fact that not only the moment functions of the analyzed process $\underset{\sim}{x}(t)$, but also joint moment functions of the component processes $\underset{\sim}{x}(t)$ and $\underset{\sim}{y}(t)$, will be among the moment functions (5.56). This may require that the definition of stability be modified.

Problem of Closure

The reduction of infinite systems of equations for probabilistic characteristics to finite systems is often required in problems of statistical dynamics. The best-known example is the problem of closure of equations of the statistical theory of turbulence [65]. In order to eliminate spectral characteristics of the third order from the field of homogeneous isotropic turbulence, they are assumed to be equal to zero. Another example is the closure of a chain of kinetic equations in statistical physics. The closure of equations is based either on qualitative considerations of the properties of the considered system or on some theoretical-probabilistic considerations. For instance, assuming the spectral characteristics of the third order in the equations of turbulence theory to be equal to zero, we may utilize the hypothesis that the properties of the field of turbulent pulsations are sufficiently close to a normal field.

Various methods of closure may be suggested for application to the systems of equations for moment functions. The simplest one is to accept the hypothesis of normality of the process $\underset{\sim}{z}(t)$. In this case, all the central moments of an odd order (starting from the third) turn out to be equal to zero, and moments of an even order are related by certain expressions to moments of the second order. Specifically, if moments of an order not higher than the third enter into equations for moments of the second order (which is typical for linear parametric systems), the condition of all the moments of the third order being equal to zero closes the system at moments of the second order.

The hypothesis of quasi-normality, first formulated in application to the equations of turbulence theory by Millionshchikov, proves to be more flexible.* According to this hypothesis, it is assumed that all the moments higher than a certain order (closure level) are related to one another by the same expressions as the corresponding moments of a normal process. No additional conditions are superimposed on moments of lower order, so they are

* Millionshchikov, M., "On the Theory of Homogeneous Isotropic Turbulence", *Dokladi Academy of Sciences, U.S.S.R.*, Vol. 32, 1941, No. 9.

determined from a system of equations obtained by reduction. If the closure level is equal to two, the hypothesis of quasi-normality becomes the hypothesis of normality. At higher closure levels, the hypothesis of quasi-normality may be expected to give better results than the hypothesis of normality at the same closure levels.

The hypothesis of quasi-normality is best formulated in terms of cumulant functions. By definition, the cumulants $\nu_{jk\ell\cdots}$ of the random vector $\underset{\sim}{z}(t)$ are introduced in terms of its characteristic function

$$\phi(\underset{\sim}{\Theta}) = \left\langle \exp\left(i \sum_{j=1}^{n_1} \Theta_j z_j \right) \right\rangle , \tag{5.57}$$

in the following way

$$\underbrace{\nu_{jk\ell\cdots}}_{\rho} = (-i)^{\rho} \left. \frac{\partial^{\rho} \ln\phi(\underset{\sim}{\Theta})}{\underbrace{\partial\Theta_j \partial\Theta_k \partial\Theta_\ell \cdots}_{\rho}} \right|_{\underset{\sim}{\Theta}=\underset{\sim}{0}} . \tag{5.58}$$

Here $\underset{\sim}{\Theta}$ is a numerical vector of dimensionality $n_1 = n+\nu$. As can be seen from formula (5.58), the cumulants are naturally expressed in terms of the coefficients of an expansion of the characteristic function $\Phi(\underset{\sim}{\Theta})$ into a Taylor series. For normal vectors, all cumulants from the third order on are equal to zero, so the expansion breaks off with the terms of the second order. By breaking off this expansion at the r-th term, we take all cumulants of order greater than r to be equal to zero. Thus, the hypothesis of quasi-normality as applied to the vector process $\underset{\sim}{z}(t)$ takes the form

$$\underbrace{\nu_{jk\ell\cdots}}_{\rho}(t) \equiv 0 , \qquad (\rho > r) . \tag{5.59}$$

The hypothesis of quasi-normality is widely used in the statistical theory of turbulence, in the classical and in quantum field theory [65]. All attempts made to prove convergence of the method based on the hypothesis have failed. Moreover, examples have been given refuting the notion of the generality of the method. The assumption that all cumulants of an order higher than r are equal to zero is equivalent to breaking off the Taylor series for a characteristic function, which is inadmissible. It may turn out that there are no distributions for which (with specified values of lower moments) all cumulants of an order higher than r are equal to zero. Therefore, approximation based on the hypothesis of quasi-normality are, evidently, of an asymptotic character in the sense that for each problem there is a closure level which gives the best approximation.

So far, a more effective method of closure has not been found. For instance, a method based on neglecting the central moments of higher orders (this method can be interpreted as a variant of the method of a small parameter) gives far worse results. Other, more particular methods, which use for example, considerations of quasi-independence of components of the analyzed process or considerations of quasi-stationarity, etc., are not very reliable.

The hypothesis of quasi-normality as applied to parametrically excited random oscillations was treated in [116]. There, results were obtained which were not found by other approximate methods. Specifically, the author succeeded in defining the fine structure of parametric resonances excited by narrow-band random processes.

Formulae for Moments of Higher Order

Let us formulate the hypothesis of quasi-normality in terms of central moments

$$\mu_{jk\ell\cdots} = \langle (z_j - m_j)(z_k - m_k)(z_\ell - m_\ell)\cdots \rangle \ . \qquad (5.60)$$

The condition of closure of the equations on the r-th level is represented as

$$\underbrace{\mu_{jk\ell\cdots}}_{\rho} = \begin{cases} 0 & (\rho = 2p+1 > r) , \\ \Sigma\, \mu_{\alpha_2\alpha_2}\mu_{\alpha_3\alpha_4}\cdots\mu_{\alpha_{2p-1}\alpha_{2p}} & (\rho = 2p > r) . \end{cases} \qquad (5.61)$$

The sum on the right side includes all the possible partitions of 2p indices $\alpha_1, \alpha_2, \ldots, \alpha_{2p}$ (including the repeated ones) into p pairs. The total number of summands is equal to (2p-1)!!.

Since it is not the central moments (5.60), but the initial ones (5.56) that enter the equations of the method of moment function, it is expedient to give here the relations following from (5.61), by which initial moments are related. If cumulants of the third order are equal to zero, we have the formula expressing the third order moments in terms of lower order moments as

$$m_{jk\ell} = \Sigma\, m_{\alpha_1\alpha_2} m_{\alpha_3} - 2m_j m_k m_\ell \, . \qquad (5.62)$$

Assuming that cumulants of the fourth order become zero we find that

$$m_{jk\ell m} = \Sigma\, m_{\alpha_1\alpha_2\alpha_3} m_{\alpha_4} + \Sigma\, m_{\alpha_1\alpha_2} m_{\alpha_3\alpha_4} - 2\, \Sigma\, m_{\alpha_1\alpha_2} m_{\alpha_3} m_{\alpha_4} + 6 m_j m_k m_\ell m_m \, . \qquad (5.63)$$

If the cumulants of the fifth order are equal to zero, we obtain the formula

$$m_{jk\ell mn} = \Sigma\, m_{\alpha_1\alpha_2\alpha_3\alpha_4} m_{\alpha_5} - \Sigma\, m_{\alpha_1\alpha_2\alpha_3} m_{\alpha_4} m_{\alpha_5} + \Sigma\, m_{\alpha_1\alpha_2} m_{\alpha_3} m_{\alpha_4} m_{\alpha_5} - 4 m_j m_k m_\ell m_m m_n \, . \qquad (5.64)$$

Permutations of indices $m_{\alpha_1\alpha_2}$, $m_{\alpha_1\alpha_2\alpha_3}$ and $m_{\alpha_1\alpha_2\alpha_3\alpha_4}$ in the formula (5.62)-(5.64) are not performed. For example (5.62) is written in detailed form as

$$m_{jk\ell} = m_{jk}m_{\ell} + m_{j\ell}m_{k} + m_{k\ell}m_{j} - 2m_{j}m_{k}m_{\ell} \ . \tag{5.65}$$

The right side of formula (5.63) in the detailed form will have 14 summands, and the right side of the formula (5.64) will have 26 summands.

Modification of the Definition of Stability

Using the moment functions (5.56) up to the order r inclusive, we form a vector

$$\underset{\sim}{m}_1^r = \{m_1(t),\ldots,m_n(t);m_{11}(t),\ldots,m_{nn_1}(t);\ldots\} \ , \tag{5.66}$$

eliminating from its components the known (prescribed) moments of the process $\underset{\sim}{y}(t)$ and reducing its dimensionality on account of symmetry. Unlike the vector (5.19), the vector (5.66) involves joint moment functions of the processes $\underset{\sim}{x}(t)$ and $\underset{\sim}{y}(t)$. If it is difficult or not expedient to eliminate these joint functions from the analysis of stability, then the definition of stability of the process $\underset{\sim}{x}(t)$ with respect to a set of moment functions (including the joint functions) should be modified. If the vector $\underset{\sim}{m}_1^r(t)$ is treated in the sense of (5.66), the definitions of stability in which the relations (5.24) and (5.25) are used are valid. Introducing the extended vector space $\underline{N}_1^r$ with the elements of (5.66), we shall speak, for brevity, of the stability in the space $\underline{N}_1^r$.

Since $\underline{M}_1^r \subset \underline{N}_1^r$, stability in $\underline{N}_1^r$ leads to stability in $\underline{M}_1^r$ (for the same level of closure r). A partial answer to the converse assumption can be obtained by using the Buniakovsky-Schwarz inequality. Let $\chi(\underset{\sim}{z})$ be a non-negative function of the variable

$\tilde{z}$ integrable in $W \subset \underline{R}^{n+\nu}$; $\phi(\tilde{z})$ and $\psi(\tilde{z})$ are functions of the same variable which are square integrable in $\underline{R}^{n+\nu}$ with a weight function $\chi(\tilde{z})$. Then,

$$\Big(\int_W \phi\psi\chi d\tilde{z}\Big)^2 \leq \int_W \phi^2\chi d\tilde{z} \cdot \int_W \psi^2\chi d\tilde{z} \ . \qquad (5.67)$$

The joint moments are calculated with the help of their joint probability density $p(\tilde{x},\tilde{y};t)$:

$$m_{\underbrace{jk\ell\ldots}_{\rho}\underbrace{\alpha\beta\gamma\ldots}_{\rho}}(t) = \int_{\underline{R}^n}\int_{\underline{R}^\nu} \underbrace{x_j x_k x_\ell \ldots}_{\rho}\underbrace{y_\alpha y_\beta y_\gamma \ldots}_{\rho_1} p(\tilde{x},\tilde{y}:t)d\tilde{y}d\tilde{x} \ . \qquad (5.68)$$

The Latin indices correspond to the output process, i.e., $j,k,\ell,\ldots = 1,2,\ldots,n$. The Greek indices correspond to the process $y(t)$, that is, $\alpha,\beta,\gamma,\ldots = n+1,n+2,\ldots,n+\nu$. Substituting the product $x_j x_k x_\ell \ldots$, instead of the function $\phi(\tilde{z})$, the product $y_\alpha y_\beta y_\gamma \ldots$ instead of the function $\psi(\tilde{z})$, and the probability density $p(\tilde{x},\tilde{y};t)$ instead of the function $\chi(\tilde{z})$ into the inequality (5.67), we obtain

$$m^2_{\underbrace{jk\ell\ldots}_{\rho}\underbrace{\alpha\beta\gamma\ldots}_{\rho_1}}(t) \leq \text{const}\cdot m_{\underbrace{jk\ell\ldots}_{\rho}\underbrace{jk\ell\ldots}_{\rho}}(t) \ . \qquad (5.69)$$

The constant on the right side is expressed in terms of the known moments of the input process. Let the system be stable in $\underline{M}_1^{r_1}$. Then all the moment functions of the process $\tilde{x}(t)$ for which $r < r_1$ will have proper characteristics resulting from the corresponding stability definition. All joint moments in which $\rho \leq r_1/2$ in the inequalities (5.69) will be bounded in absolute value by these inequalities, and thus will also have proper characteristics. Hence, it follows that stability in $\underline{M}_1^{r_1}$ leads to stability in $\underline{N}_1^r$, where $r \leq \min\{r_1, \frac{1}{2} r_1+1\}$. In the specific case when $r_1 = 2$, stability in $\underline{M}_1^2$ is a necessary and sufficient condition for stability in $\underline{N}_1^2$.

Techniques for Determining Regions of Stochastic Stability

Retaining the equations involving derivatives of the functions $m_{jk\ell\ldots}(t)$ of the order not higher than r in the system (5.40), and expressing on the right sides the functions $m_{jk\ell\ldots}(t)$ of higher order according to the hypothesis of quasi-normality, we arrive at the closed system of equations for the components of the vector $\underset{\sim}{m}_1^r(t)$. The problem reduces to finding the conditions of stability of the trivial solution $\underset{\sim}{m}_1^r(t) \equiv \underset{\sim}{0}$ of this system. This is, evidently, equivalent to stochastic stability of the solution of the original system with respect to a set of moment functions of the first, second, etc., order including the order r, at which closure of the system (5.40) takes place.

After closure, the system (5.40) remains nonlinear. We shall apply to it Liapunov's stability theorem in the first approximation. Linearizing the system in the vicinity of the solution $\underset{\sim}{m}_1^r(t) \equiv \underset{\sim}{0}$, we shall reduce it to the form

$$\frac{d\underset{\sim}{m}_1^r}{dt} = H\underset{\sim}{m}_1^r , \tag{5.70}$$

where H is a matrix. If the conditions of the theorem are satisfied, i.e., if all the characteristic indices (eigenvalues of the matrix H) have negative real parts, then the trivial solution of the nonlinear system is stable according to Liapunov.

Thus, the problem is reduced to that of Routh-Hurwitz for polynomials with real coefficients. For large n, ν, and r, the dimensionality of the matrix H may prove to be too large for an analytical solution.

Table 5.1 gives an idea of the extent of the calculations. It gives the number of unknown moment functions $m_{jk\ell\ldots}(t)$ (numerator) and dimensionality of the matrix H (denominator). The difference is due to the fact that not all moment functions enter the equation (5.70), which leads to reducing the dimensionality.

Table 5.1

Correlation Function $K_\Phi(\tau)$	Closure Levels		
	r = 2	r = 3	r = 4
$\sigma_\Phi^2 e^{-\alpha\|\tau\|}$	9/7	19/16	34/30
$\sigma_\Phi^2 e^{-\alpha\|\tau\|}\left(\cos\Theta_\alpha\tau + \frac{a}{\Theta_\alpha}\sin\Theta_\alpha\|\tau\|\right)$	14/9	34/25	69/55

The table is constructed for processes of two types at the input: for an exponentially-correlated process with the correlation function

$$K_\Phi(\tau) = \sigma_\Phi^2 e^{-\alpha|\tau|} , \tag{5.71}$$

and for a process with the correlation function

$$K_\Phi(\tau) = \sigma_\Phi^2 e^{-\alpha|\tau|}\left(\cos\Theta_\alpha\tau + \frac{a}{\Theta_\alpha}\sin\Theta_\alpha|\tau|\right) . \tag{5.72}$$

In the first case, the equation of the formative filter has the form

$$\frac{d\Phi}{dt} + \alpha\Phi = \xi(t) . \tag{5.73}$$

And, to obtain a correlation function in the form (5.71), we should assume that

$$s = 2\alpha\sigma_\Phi^2 . \tag{5.74}$$

In the second case, the filter equation has the form

$$\frac{d^2\Phi}{dt^2} + 2\alpha\frac{d\Phi}{dt} + \Theta^2\Phi = \xi(t) , \tag{5.75}$$

and, to obtain a correlation function in the form (5.72), we should assume that

$$s = 4\alpha\Theta^2\sigma_\Phi^2 , \qquad \Theta^2 = \Theta_\alpha^2 + \alpha^2 > 0 . \tag{5.76}$$

The dimensionality of the phase space for the formative filter is $\nu = 1$ and $\nu = 2$, respectively. The data listed in the table relates to a system with one degree of freedom.

To determine the boundaries of the stability regions numerically, the classical criterion of Routh-Hurwitz can be used, and proceeding from a known region of stability, it is sufficient to verify the sign of det H and the sign of Hurwitz's principal determinant. It is also possible to use Zubov's criterion of stability,* based on the mapping of the left half-plane of characteristic indices onto the inside of the unit circle. The criterion is realized by means of raising the matrix

$$H_1 = (H-I)^{-1}(H+I) \tag{5.77}$$

to powers of higher order (for example, to power H_1^{2p}, where $p \sim 10$) and comparing an easily calculated norm of these powers to a certain preassigned small positive number. The latter method is preferable in the case of high dimensionality of the matrix H.

A substantial use of the hypothesis of quasi-normality makes the modified method of moment functions heuristic. The method may lead to some precise solution of the stability problem for the original system when the order r of the moments in the system (5.70) increases. However, it should be noted that, with the increase of r, the definition of stability actually changes (since stability is considered with respect to a set of moment functions of higher order).

---

*Zubov, V.I., *Mathematical Methods of Analyzing Automatic Control Systems*, Mashinostroenie, Leningrad, 1974.

5.5 PARAMETRIC RESONANCES IN STOCHASTIC SYSTEMS

A Stochastic Analog of the Mathieu-Hill Equation

We shall apply the modified method of moment functions to parametrically excited systems with one or two degrees of freedom.* First, we shall consider the equation (5.51) where $\Phi(t)$ is a stationary normal process with the correlation function (5.72). Supplementing equation (5.51) by the filter equation (5.75), we shall introduce the phase vector $\underset{\sim}{z}(t)$ with the components $z_1 = u$, $z_2 = \dot{u}$, $z_3 = \Phi$, $z_4 = \dot{\Phi}$. Let us write Ito's stochastic differential equation

$$\begin{aligned} dz_1 &= z_2 dt \ , \\ dz_2 &= -(2\gamma z_2 + z_1 + \mu z_1 z_3)dt \ , \\ dz_3 &= z_4 dt \ , \\ dz_4 &= -(2\rho z_4 + \theta^2 z_3)dt + \sqrt{s}dw \ . \end{aligned} \tag{5.78}$$

Here, the dimensionless variables (5.29) are used, and we have the notations $\rho = \alpha/\omega_0$ and $\theta = \Theta/\omega_0$. If, as in the previous example, we assume that $\sigma_\Phi^2 = 1$, we then have $s = 4\rho\theta^2$ for the intensity of white noise s in accordance with the first formula (5.76).

Vector $\underset{\sim}{z}(t)$ describes a Markov diffusion process. The probability density $p(\underset{\sim}{z},t)$ of this process satisfies the Kolmogorov equation

$$\begin{aligned} \frac{\partial p}{\partial t} &= 2(\gamma+\rho)p - z_2 \frac{\partial p}{\partial z_1} + (2\gamma z_2 + z_1 + \mu z_1 z_3) \frac{\partial p}{\partial z_2} - \\ & z_4 \frac{\partial p}{\partial z_3} + (2\rho z_4 + \theta^2 z_3) \frac{\partial p}{\partial z_4} + \frac{1}{2} s \frac{\partial^2 p}{\partial z_4^2} \ . \end{aligned} \tag{5.79}$$

---

*Sections 5.5 and 5.6 are based mostly on the following papers: Bolotin, V.V., Moskvin, V.G., "On Parametric Resonances in Stochastic Systems", Izvestia of Academy of Sciences, U.S.S.R., *Mekhanika Tverdovo Tela,* 1972, No. 4; and Bolotin, V.V., Moskvin, V.G., "Excitation of Parametric Vibrations in Stochastic Systems with Two Degrees of Freedom", Izvestia of Academy of Sciences, *Mekhanika Tverdovo Tela,* 1973, No. 4.

The equations for the moment functions are obtained from equation (5.79) by multiplying the latter by the components of vector $\underset{\sim}{z}(t)$ or by their products, and by performing integration by parts. First, we shall consider stability in $\underline{N}_1^2$, restricting ourselves to equations for a set of moments of the first and second order. The total number of these moments (taking account of symmetry) equals 14. Those moments which only include components of the input process $\Phi(t)$ are independent of time and are not taken into account when stability is considered. Let us write their values:

$$m_3 = m_4 = m_{34} = 0 \ , \quad m_{33} = 1 \ , \quad m_{44} = \theta^2 \ . \tag{5.80}$$

As a result, there remains a system of nine equations for the components of the vector $\underset{\sim}{m}_1^2(t)$:

$$\frac{dm_1}{dt} = m_2 \ ,$$

$$\frac{dm_2}{dt} = -(m_1 + 2\gamma m_2 + \mu m_{13}) \ ,$$

$$\frac{dm_{11}}{dt} = 2m_{12} \ ,$$

$$\frac{dm_{12}}{dt} = m_{22} - (m_{11} + 2\gamma m_{12} + \mu m_{113}) \ ,$$

$$\frac{dm_{13}}{dt} = m_{14} + m_{23} \ , \tag{5.81}$$

$$\frac{dm_{14}}{dt} = m_{24} - (\theta^2 m_{13} + 2\rho m_{14}) \ ,$$

$$\frac{dm_{22}}{dt} = -2(m_{12} + 2\gamma m_{22} + \mu m_{123}) \ ,$$

$$\frac{dm_{23}}{dt} = m_{24} - (m_{13} + 2\gamma m_{23} + \mu m_{133}) \ ,$$

$$\frac{dm_{24}}{dt} = -[m_{14} + \theta^2 m_{23} + 2(\gamma + \rho)m_{24} + \mu m_{134}] \ .$$

The equations (5.81) contain the third moments. Taking into account formulae (5.65), we obtain

$$m_{113} = 2m_1m_{13}\ , \qquad m_{123} = m_1m_{23}+m_2m_{13}\ ,$$
$$m_{134} = m_1m_{34} = 0\ , \qquad m_{133} = m_1m_{33} = m_1\ , \tag{5.82}$$

where (5.80) is used. Having substituted these relations into (5.81), and linearizing, we obtain a system of the type (5.70) with the matrix

$$H = \begin{bmatrix} 0 & 1 & 0 & 0 & 0 & 0 & 0 & 0 & 0 \\ -1 & -2\gamma & 0 & 0 & 0 & -\mu & 0 & 0 & 0 \\ 0 & 0 & 0 & 0 & 2 & 0 & 0 & 0 & 0 \\ 0 & 0 & 0 & -4\gamma & -2 & 0 & 0 & 0 & 0 \\ 0 & 0 & -1 & 1 & -2\gamma & 0 & 0 & 0 & 0 \\ 0 & 0 & 0 & 0 & 0 & 0 & 1 & 1 & 0 \\ 0 & 0 & 0 & 0 & 0 & -\theta^2 & -2\rho & 0 & 1 \\ -\mu & 0 & 0 & 0 & 0 & -1 & 0 & -2\gamma & 1 \\ 0 & 0 & 0 & 0 & 0 & 0 & -1 & -\theta^2 & -2(\gamma+\rho) \end{bmatrix} . \tag{5.83}$$

Matrix (5.83) decomposes into two blocks, which follows from using the hypothesis of quasi-normality, and is not repeated in passing over to higher degrees of closure (see also [143]).

Discussion of Numerical Results

The boundary of the stability region for $\gamma = 0.025$, $\rho = 0.02$, is plotted in Figure 38 by a heavily drawn line (the stability region is on the left). Already in this approximation, parametric resonances can be found in stochastic systems. For instance, the instability region has the form of a clearly pronounced wedge at $\theta = 2$. This wedge corresponds to the main parametric resonance $\theta = 2\omega_0$, which enters into the relations (5.7) for $p = 1$. There is also a second wedge at $\theta = 1$. It corresponds to the ratio of the frequencies $\Theta = \omega_0$, i.e., to the parametric resonance (5.7) for $p = 2$.

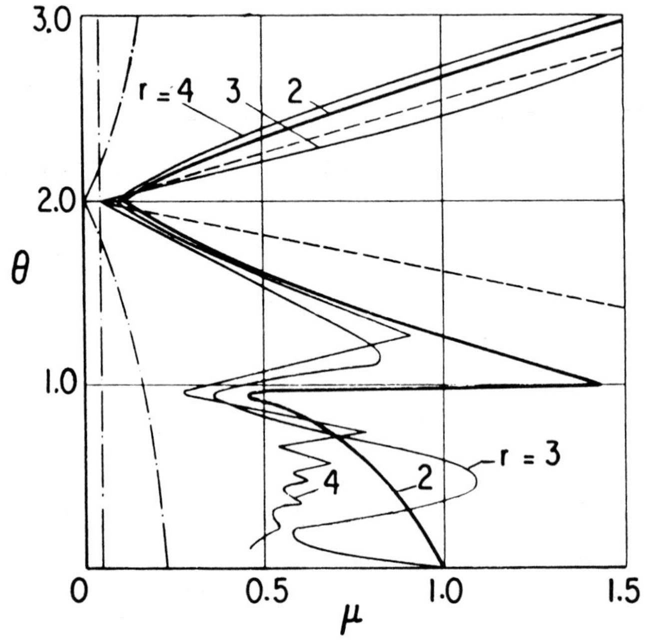

Figure 38

The nature of instability varies along the boundary of the region. For small values of θ, the transition from stability to instability is a consequence of the matrix H determinant becoming zero. Thus, for small values of θ, the variation of the vector $\underset{\sim}{m}_1^2(t)$ in time, close to the boundary of the stability region, is of a quasi-monotonic character. If the parameter θ is sufficiently large, the principal Hurwitz determinant becomes zero. The behaviour of the vector $\underset{\sim}{m}_1^2(t)$ close to the boundary is oscillatory in this case. For the lower branch of the boundary of the stability region in $\underline{N}_1^2$, the analytical expression $\mu = \mu_*(\theta,\gamma,\rho)$ can be obtained, where

$$\mu_*^2 = \frac{4(\gamma+\rho)(\gamma\theta^2+\rho)-(\theta^2-1)^2}{1+4\rho(\gamma+\rho)-\theta^2}\ , \quad \theta < 1+4\rho(\gamma+\rho)\ . \qquad (5.84)$$

The calculation of further approximations is very time-consuming. For instance, for the approximation of the fourth order moments, the matrix H has an order equal to 55. The results of the calculations for $r = 3$ and $r = 4$ are plotted by thin solid lines in Figure 38. A more complex structure of the instability region is disclosed which has analogs of secondary resonances (5.7) for $p = 2$, $p = 3$ or, possibly, for $p = 4$. Since the matrix is badly conditioned, the calculations of critical values are not exact. At any rate, the first two wedges (for $p = 1$ and for $p = 2$) are determined with sufficient accuracy.

Comparison with Other Methods

Sufficient conditions of almost reliable stability were considered in [119,136,140,146]. The vertical dash-and-dot line in Figure 38 is plotted according to Infante's criterion [140]

$$\mu \le 2\gamma\sqrt{1-\gamma^2}\ , \qquad (\gamma \le 1/\sqrt{2})\ . \qquad (5.85)$$

The line showing the dependence on the frequency is plotted according to Gray's criterion [136]:

$$\mu^2 \le \gamma(1-\gamma^2)\,|\theta - 2\sqrt{1-\gamma^2}|\ , \qquad (\gamma \le 1/\sqrt{2})\ . \qquad (5.86)$$

As can be seen from Figure 38, these criteria are not sufficiently effective to determine the boundary of the instability region.

A great number of papers and books [1,60,93,107,170] are explicitly or implicitly based on the assumption that the process $u(t)$ is a narrow-band one, with a carrying frequency close to ω_0. This assumption is admissible only for $\gamma << 1$. Final formulae relating the parameters on the boundary of the instability region may differ from one another. But, as shown above, (accurate to the error of the method), approximate stability conditions are reduced to the unified simple form (5.53). For example, according to Weidenhammer [170] and Graefe [135],

$$\mu^2 S_\Phi(2\omega_0\sqrt{1-\gamma^2}) < \frac{2\gamma}{\pi\omega_0}(1-\gamma^2)\ , \tag{5.87}$$

and according to Alekseev and Valeev [1],

$$\mu^2[S_\Phi(2\omega_0)-S_\Phi(0)] < \frac{2\gamma}{\pi\omega_0}\ , \tag{5.88}$$

etc. The stability region according to criterion (5.53) is plotted by a dashed line in Figure 38. This boundary agrees with our results only in the vicinity of the relation $\Theta = 2\omega_0$.

The influence of the damping parameter γ and the wide-band parameter ρ on the boundaries of the stability regions is shown in Figures 39 and 40.

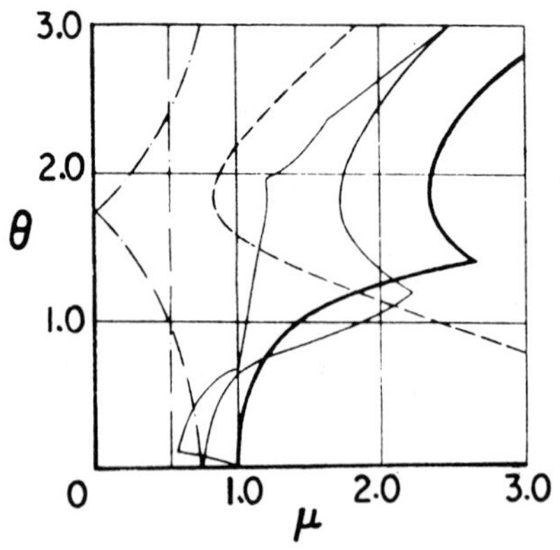

Figure 39

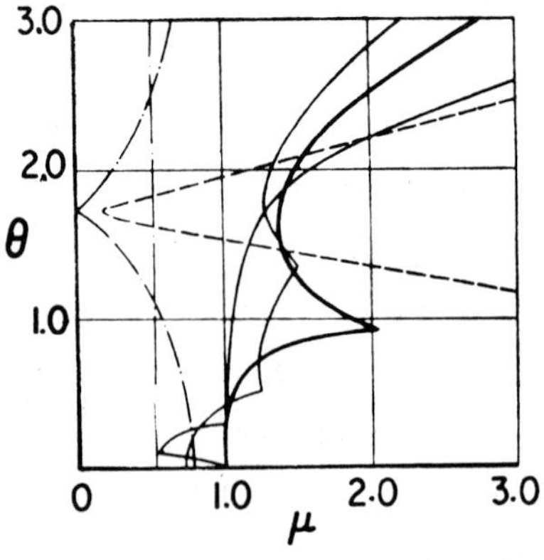

Figure 40

Figure 39 is constructed for the case when $\gamma = 0.5$, $\rho = 0.5$, i.e., for a system with a relatively large damping and a relatively wide-band input process. Figure 40 is constructed for $\gamma = 0.5$,

$\rho = 0.02$. The results of calculations according to the criteria of the above-mentioned works are also shown here. The notations are the same as in Figure 38. Stability regions are located on the left. It must be noted that the criterion (5.53) and the related criteria (5.87) and (5.88), as applied to the data of Figures 39 and 40, become invalid since the condition $\gamma \ll 1$ is violated. Both the increase of damping and the widening of the spectrum of the parametric action (for σ_Φ = const.) lead to a stabilization of the system. The selective nature of instability close to the relation of frequencies $\Theta = 2\omega_0$ smoothes over to a large extent.

System with Two Degrees of Freedom

Consider a system described by equations of the type

$$\frac{d^2u_1}{dt^2} + 2\varepsilon_1 \frac{du_1}{dt} + \omega_1^2[u_1+\mu\Phi(t)(g_{11}u_1+g_{12}u_2)] = 0 ,$$
$$\frac{d^2u_2}{dt^2} + 2\varepsilon_2 \frac{du_2}{dt} + \omega_2^2[u_2+\mu\Phi(t)(g_{21}u_1+g_{22}u_2)] = 0 . \qquad (5.89)$$

Here, ω_1 and ω_2 are the frequencies of natural vibrations of the corresponding conservative system, ε_1 and ε_2 are damping coefficients, μ is a numerical parameter (all these values are assumed to be positive), $\Phi(t)$ is a stationary normal process with a mathematical expectation equal to zero, and g_{jk} are nonstochastic numerical coefficients.

System (5.89) is an analog of the system of differential equations with periodic coefficients (5.5) for $n = 2$. According to the general theory, the wedges of instability regions of a deterministic system at small damping can be located close to the frequencies

$$\Theta = 2\omega_1/p , \qquad \Theta = 2\omega_2/p , \qquad (p = 1,2,\ldots) . \qquad (5.90)$$

and also close to the frequencies

$$\Theta = (\omega_1+\omega_2)/p \ , \qquad \Theta = |\omega_1-\omega_2|/p \ , \qquad (p = 1,2,\dots) \ . \tag{5.91}$$

Combination resonances bearing the plus sign are to be expected in systems which belong to the canonic type with $\varepsilon_1 = \varepsilon_2 = 0$. A combination resonance with the minus sign is typical for essentially noncanonic systems. Typical examples of the matrix G having the elements g_{jk} are given below:

$$G_1 = \begin{bmatrix} 1 & 0 \\ 0 & 1 \end{bmatrix} , \quad G_2 = \begin{bmatrix} 1 & 1 \\ 1 & 1 \end{bmatrix} , \quad G_3 = \begin{bmatrix} 0 & 1 \\ 1 & 0 \end{bmatrix} , \tag{5.92}$$

$$G_4 = \begin{bmatrix} 0 & 1 \\ -1 & 0 \end{bmatrix} , \quad G_5 = \begin{bmatrix} 1 & 1 \\ -1 & 1 \end{bmatrix} . \tag{5.93}$$

Matrices (5.92) correspond to canonic systems, whereas matrices (5.93) correspond to noncanonic ones.

Let $\Phi(t)$ be a stationary normal process with the correlation function (5.72). As in (5.29), let us introduce the nondimensional time $t' = \omega_1 t$ and nondimensional parameters

$$\omega_2/\omega_1 = h \ , \quad \varepsilon_1/\omega_1 = \gamma_1 \ , \quad \varepsilon_2/\omega_2 = \gamma_2 \ , \quad \Theta/\omega_1 = \theta \ , \quad \alpha/\omega_1 = \rho \ . \tag{5.94}$$

We shall analyze the stability of the trivial solution of system (5.89) with respect to mathematical expectations and moment functions of the four-dimensional process $\underset{\sim}{x}(t)$ in the corresponding phase space. We shall apply here the modified method of moment functions [116].

System (5.89), with the correlation function of the process $\Phi(t)$ in the form (5.72) in the extended phase space $\underline{U} \oplus \underline{V}$, is equivalent to Ito's system of stochastic equations for the six-dimensional process $\underset{\sim}{z}(t)$ with the components $z_1 = u_1$, $z_2 = u_2$, $z_3 = \dot{u}_1$, $z_4 = \dot{u}_2$, $z_5 = \Phi$, $z_6 = \dot{\Phi}$:

$$
\begin{aligned}
dz_1 &= z_3 dt \ , \\
dz_2 &= z_4 dt \ , \\
dz_3 &= -(2\gamma_1 z_3 + z_1)dt - \mu z_5(g_{11}z_1 + g_{12}z_2)dt \ , \\
dz_4 &= -(2h\gamma_2 z_4 + h^2 z_2)dt - \mu h^2 z_5(g_{21}z_1 + g_{22}z_2)dt \ , \\
dz_5 &= z_6 dt \ , \\
dz_6 &= -(2\rho z_6 + \theta^2 z_5)dt + \sqrt{s}dw \ .
\end{aligned}
\tag{5.95}
$$

Here, the prime in t' is omitted, and for $\sigma_\phi^2 = 1$ it is to be assumed that $s = 4\rho\theta^2$.

The Kolmogorov equation for the probability density $p(z,t)$ of the six-dimensional process $\underset{\sim}{z}(t)$ has the form

$$
\begin{aligned}
\frac{\partial p}{\partial t} = {} & 2(\gamma_1 + h\gamma_2 + \rho)p - z_3\frac{\partial p}{\partial z_1} - z_4\frac{\partial p}{\partial z_2} + \\
& [z_1 + 2\gamma_1 z_3 + \mu z_5(g_{11}z_1 + g_{12}z_2)]\frac{\partial p}{\partial z_3} + \\
& [h^2 z_2 + 2h\gamma_2 z_4 + \mu z_5(g_{21}z_1 + g_{22}z_2)]\frac{\partial p}{\partial z_4} - \\
& z_6\frac{\partial p}{\partial z_5} + (2\rho z_6 + \theta^2 z_5)\frac{\partial p}{\partial z_6} + \frac{1}{2}s\frac{\partial^2 p}{\partial z_6^2} \ .
\end{aligned}
\tag{5.96}
$$

Using this equation, we shall construct a system of differential equations for moment functions of the process $z(t)$. This system will be infinite and coupled. The system of equations is closed by means of the hypothesis of quasi-normality. After the closure of the system, there remain 27 moment functions, five of which are directly expressed in terms of parameters of the process $\Phi(t)$. Thus, we arrive at a system of 22 nonlinear differential equations for moment functions of first and second order, including eight joint moment functions of the type $m_{15}(t)$, $m_{25}(t)$, etc. The stability of a trivial solution of this system is equivalent (accurate up to the hypothesis of quasi-normality) to the stability of the stochastic system (5.95) in relation to a set of mathematical expectations and moment functions of the second order of the four-dimensional process $\underset{\sim}{x}(t)$.

Let us denote the column-matrix made up of the remaining moment functions by $\underset{\sim}{m}_1^2(t)$. Consider the linearized system of equations (5.70), where H is a square matrix whose dimensionality is 22. In order to find the boundaries of the stability region in the parameter space, a numerical method utilizing V. Zubov's stability criterion is used. The matrix H_1 raised to the power 2^p is calculated from the formula (5.77). Actually the amount of computations required for obtaining a reliable result on stability varies from $p = 8$ to $p = 12$, depending on damping. For small damping, eigenvalues of the matrix H_1 are located close to the boundary of the unit circle. Therefore, the matrix H_1 had to be raised to a sufficiently large power 2^p before its elements became small in magnitude.

Results of Numerical Analysis

In Figures 41 - 46, the modulation parameter μ is laid off along the horizontal axis, and the ratio of the carrying frequency of the process Θ versus the principal natural frequency of the system, i.e., $\theta = \Theta/\omega_1$, along the vertical axis. Solid lines show the boundaries of instability regions constructed in approximation of the second moment functions. Stability regions are located on the left. All the calculations were made for the case $h = \frac{3}{2}$. The values of damping coefficients were assumed to be $\gamma_1 = h\gamma_2 = \gamma = 0.025;\ 0.1;\ 0.25;\ 0.5$. In all the graphs except Figure 44, the parameter of process correlation was assumed to be $\rho = \alpha/\omega_1 = 0.02$. In Figure 44, this parameter is varied ($\rho = 0.02;\ 0.1;\ 0.25;\ 0.5;\ 1.0$) for the constants $\gamma_1 = h\gamma_2 = \gamma = 0.025$.

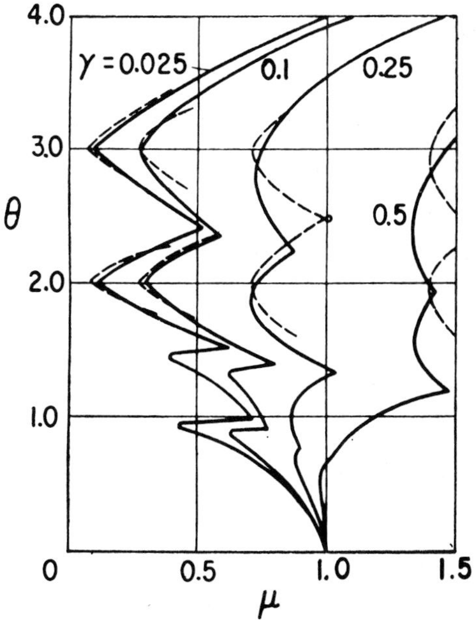

Figure 41

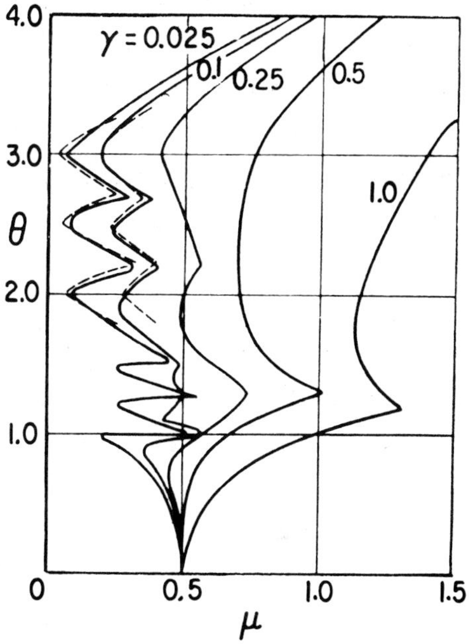

Figure 42

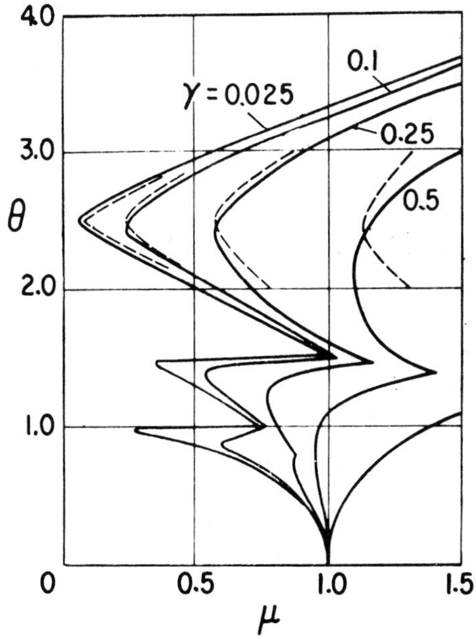

Figure 43

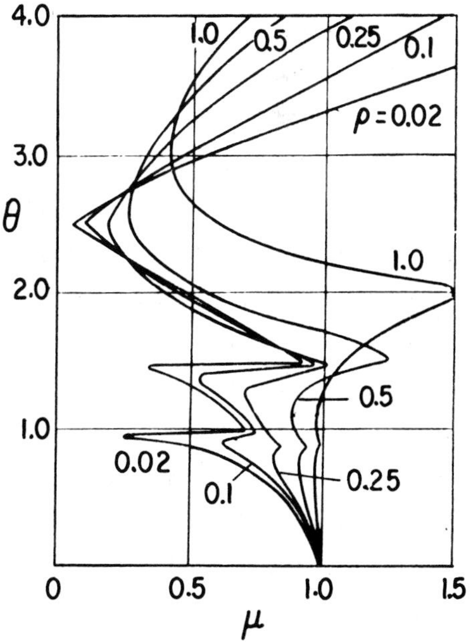

Figure 44

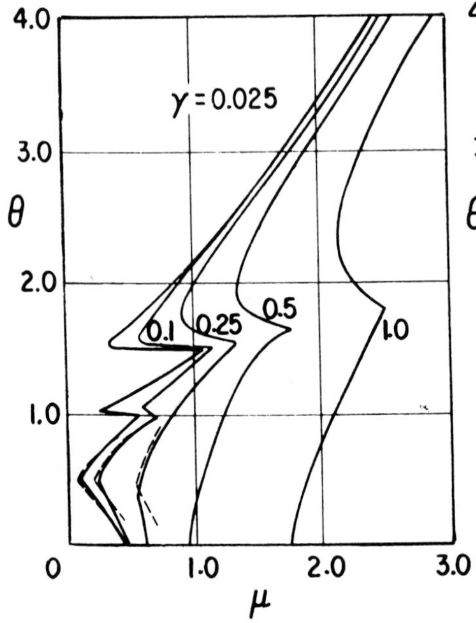

Figure 45

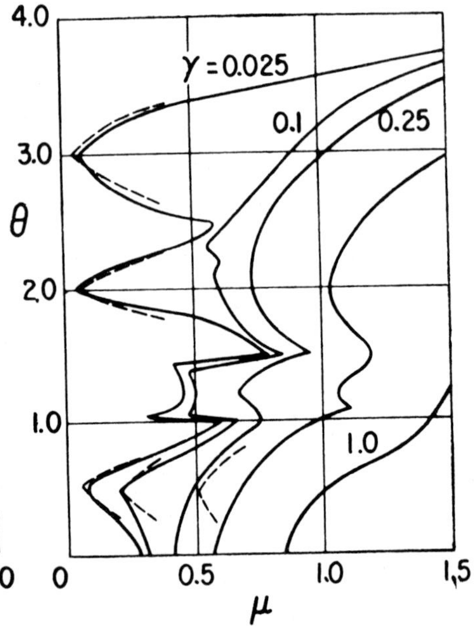

Figure 46

Figure 41 shows the case of $G = G_1$, where the matrix G_1 is determined from (5.92). For small values of γ, the instability regions have four wedges. The wedges located close to $\theta = 2$ and $\theta = 3$ correspond to principal parametric resonances $\Theta = 2\omega_1$ and $\Theta = 2\omega_2$. The wedges at $\theta = 1.0$ and $\theta = 1.5$ can be interpreted as analogs of secondary resonances for $\Theta = \omega_1$ and $\Theta = \omega_2$. The combination resonsnaces of the type (5.91) do not occur in this case. The matrix is diagonal and, therefore, the system (5.89) can be split into two independent equations analogous to the Mathieu-Hill equation. The instability regions in Figure 41 can be found by superimposing instability regions for two independent equations (5.51). As the damping parameter γ increases, the wedges of the instability regions flatten, i.e., the selective character of stochastic instability decreases with respect to the carrying frequency.

Figure 42 shows the results for the matrix $G = G_2$. The system (5.89) is in this case coupled. At small damping, two more wedges, in addition to those mentioned above, appear in the instability regions. One of them located close to $\theta = 2.5$ can be interpreted as an analog of the combination resonance $\Theta = \omega_1+\omega_2$. The other is located close to $\theta = 1.25$; it probably corresponds to the fractional combination resonance $\Theta = \frac{1}{2}(\omega_1+\omega_2)$. Starting from $\gamma = 0.25$, the selective character of instability with respect to the carrying frequency practically disappears.

For small values of θ, the transition from stability to instability takes place when the determinant of the matrix H becomes zero. Thus, for small θ, variation of moments with respect to time close to the boundary of an instability region is of a quasi-monotonical character. If the parameter θ is sufficiently large, the behaviour of moments close to the boundary will be oscillatory.

Figure 43 shows the case of the anti-diagonal matrix $G = G_3$. Here, there remain three wedges in the instability regions close to $\theta = 2.5$, $\theta = 1.5$, and $\theta = 1.0$. The first of them is an analog of the combination resonance $\Theta = \omega_1+\omega_2$. The second is possibly obtained from superimposition of the resonances $\Theta = \omega_2$ and $\Theta = \frac{1}{2}(\omega_1+\omega_2)$. The third wedge corresponds to the resonance $\Theta = \omega_1$. There are no analogs of principal parametric resonances in the vicinity of the frequency relations $\Theta = 2\omega_1$, and $\Theta = 2\omega_2$. It should be remembered that in the corresponding deterministic system, the width of the principal instability regions is of the order μ^2 as compared to the order for combination resonance regions at $\Theta = \omega_1+\omega_2$.

Figure 44 shows the influence of the correlation parameter of the process $\rho = \alpha/\omega_1$. The data are for the case $G = G_3$, $\gamma = 0.025$. Comparing Figure 44 to Figure 43, we see that the increase of the correlation parameter ρ transforms the boundaries of the instability regions approximately in the same way as the increase of the damping parameter γ does. At large values of ρ, the instability region has no typical resonance wedges.

Up to now, we have considered the analogs of canonic systems. Numerical results for noncanonic systems are given in Figures 45 and 46. Consider Figure 45 constructed for the case $G = G_4$, where G_4 is determined from (5.93). At small γ, the instability regions have three wedges, one of which (close to $\theta = 0.5$) corresponds to a combination resonance of a difference type, i.e., to the relation $\Theta = |\omega_1 - \omega_2|$. The other two wedges may correspond to the resonances $\Theta = \omega_1$ and $\Theta = \omega_2$. There are no principal resonances close to $\Theta = 2\omega_1$ and $\Theta = 2\omega_2$. As in the case of deterministic systems of an essentially noncanonic type, there is no combination resonance for the sum of frequencies.

The results of calculations for the case of the matrix $G = G_5$ are given in Figure 46. At small γ, the instability regions have up to five wedges. The analogs of principal parametric resonances $\Theta = 2\omega_1$ and $\Theta = 2\omega_2$, and the analog of the resonance of a difference type, can be clearly identified among them.

Comparison with Corresponding Deterministic System

Applying the limit transition for $\alpha \to 0$ to a stationary normal process with the correlation function (5.72), we do not obtain a sinusoidal process. We can only speak of equivalent sinusoidal processes in a certain sense, for example, $\Phi(t) = \sqrt{2}\,\sin(\Theta t + \psi)$, where Θ is a specified frequency, and the phase ψ is a stochastic variable. The factor $\sqrt{2}$ is introduced in order to satisfy the condition $\sigma_\Phi^2 = 1$. Henceforth, the comparison is made with instability regions for the system (5.89), in which the stochastic process $\Phi(t)$ is replaced by the function $\sqrt{2}\cos\Theta t$.

The approximate formulae relating the values of the frequency Θ and the parameter μ on the boundaries of dynamic instability regions can be found in the books mentioned in the footnote.* The formulae for the boundaries of the principal instability regions in terms of the notations (5.94) have the form

$$\theta = 2[1 \pm (1/8\ \mu^2 g_{11}^2 - \gamma_1^2)^{1/2}] , \tag{5.97}$$

$$\theta = 2h[1 \pm (1/8\ \mu^2 g_{22}^2 - \gamma_2^2)^{1/2}] .$$

For the boundaries of the instability region close to the resonances $\Theta = |\omega_1 \pm \omega_2|$ we have for a system of the canonic type the formula

$$\theta = 1+h \pm \frac{\gamma_1 + h\gamma_2}{(\gamma_1\gamma_2)^{1/2}} (1/8\ \mu^2 g_{12} g_{21} - \gamma_1\gamma_2)^{1/2} . \tag{5.98}$$

For essentially noncanonical systems, (for $g_{12}g_{21} < 0$), we have, respectively,

$$\theta = |1-h| \pm \frac{\gamma_1 + h\gamma_2}{(\gamma_1\gamma_2)^{1/2}} (1/8\ \mu^2 |g_{12} g_{21}| - \gamma_1\gamma_2)^{1/2} . \tag{5.99}$$

Formulae (5.97) - (5.99) are derived on the assumption that the parameters μ, γ_1, and γ_2 are sufficiently small as compared to unity. The results of the calculations using these formulae are shown in dashed lines in Figures 41 - 46. In the vicinity of the wedges, these results agree well with the data which relate to a stochastic system (especially for small μ and γ).

To conclude this section, we shall consider the points of intersection $\mu = \mu_*$ of the boundaries of instability regions with the line $\theta = 0$. The case $\theta \to 0$ corresponds to the quasi-static variation of function $\Phi(t)$. Therefore, we can naturally expect the critical values of μ_* to be found by considering some

---

*Bolotin, V.V., *Dynamic Stability of Elastic Systems*, Gostekhizdat, Moscow, 1956; and Schmidt, G., *Parametric Vibrations*, Mir, Moscow, 1978.

equivalent deterministic system with constant parameters. Such a system is:

$$\frac{d^2u_1}{dt^2} + 2\varepsilon_1 \frac{du_1}{dt} + \omega_1^2[u_1+\mu(g_{11}u_1+g_{12}u_2)] = 0 ,$$
$$\frac{d^2u_2}{dt^2} + 2\varepsilon_2 \frac{du_2}{dt} + \omega_2^2[u_2+\mu(g_{21}u_1+g_{22}u_2)] = 0 . \qquad (5.100)$$

And, indeed, for the matrices G_1, G_2 and G_3, the values of μ_* coincide with the roots of the characteristic equation $\det(I-\mu G) = 0$; that is $\mu_* = 1$ for $G = G_1$ and $G = G_3$ and if $G = G_2$, $\mu_* = \frac{1}{2}$. If the system (5.100) is not canonic for $\varepsilon_1 = \varepsilon_2 = 0$, then, generally, the instability of its trivial solution may be of an oscillating nature. The determinant of the matrix $I-\mu G$ is different from zero for any real μ, so, the critical values μ_* are to be determined from the stability condition of the zero solution of the system (5.100). For example, for $G = G_4$ we find that

$$\mu_*^2 = (1+h^2+4h\gamma_1\gamma_2)\frac{h\gamma_1+\gamma_2}{h(\gamma_1+h\gamma_2)} - \frac{(h\gamma_1+\gamma_2)^2}{(\gamma_1+h\gamma_1)^2} - 1 .$$

Calculation by means of this formula gives values for μ_* that practically coincide with those obtained from the analysis of the stability of a stochastic system.

The results given above have been obtained by means of an approximate numerical analysis of the problem using moment functions of the second order, and, therefore, are of an empiric nature. The question - how much the result will change if the system of moment functions is closed at a level of moment functions of a higher order - has not been answered. Unfortunately, in passing over to moment functions of the third order, difficulties in calculations increase drastically, i.e., the order of the matrix H becomes equal to 74.

5.6 COMPARISON WITH EXPERIMENTAL DATA

On Experimental Investigation of Instability Phenomena

Let us consider a physical system and its unperturbed motion. The task is to find a region in the space of the system's parameters (by varying them experimentally), in which the unperturbed motion is stable in the sense of one of definitions of stochastic stability. We shall consider a point in the space of parameters and try to answer the question of whether this point belongs to the stability region, or to the instability one, or to the boundary between these regions. In this connection some substantial technical difficulties do arise.

First of all, the experiment should involve the variation of initial data and the observation of the behaviour of solutions during an arbitrary long period of time.

Secondly, definitions of stability are of infinitesimal character, i.e., they involve numbers that should be made arbitrarily small during the experiment, which is practically unrealizable. For instance, the relation (5.14) involves three small positive numbers δ, ε and ρ characterizing the smallness of perturbations of initial data, the smallness of deviations of trajectories for $t > t_0$, and the smallness of deviations of probability from unity. Specification of the perturbations of the physical system and measurement of deviations from the unperturbed trajectory, starting from sufficiently small values, becomes impossible. Very small perturbations are not controlled, and are even undetectable. Nevertheless, they can still substantially influence the dynamic behaviour of the system.

Thirdly, the determination of regions of stochastic stability is in fact the verification of a system of statistical hypotheses. Even if we have any finite number of realizations of any duration, it is impossible to give an unambiguous answer to the question: to which of the three above mentioned sets does the considered point belong? It should be mentioned that the verification of the statistical hypotheses relating to small probabilities ρ requires a great amount of experimental data. Some experimental difficulties can be avoided if, instead of the definitions that involve relations of the type (5.14), (5.15), etc., we use definitions involving a finite time interval T and finite numbers ε, δ and ρ. However, such experimental data could not be compared with theoretical data (there are no corresponding theoretical results).

The difficulties listed above will take place in any experimental investigation where it is necessary to distinguish between stable solutions and unstable ones. As an example, we may take the experimental investigation of the stability of structures and their elements. Due to the indefiniteness of experimental data, it is generally impossible to uniquely determine the critical load of the moment in which the loss of stability of a structure occurs. For all of these reasons we cannot use strict definitions of stability. Instead, some "engineering" criteria are to be used that are based on easily classified properties of sampling realizations. Here, we give some results of investigations of stability of stochastic systems carried out on an analog computer. The method of electronic simulation gives a good approximation of a physical system to a mathematical model and allows us to vary the system's parameters easily.

To compare experimental data with theoretical results, the following method was used. Points in the space of parameters were classified according to the behaviour of the norm (or seminorm) $||\underset{\sim}{x}(t)||$ of realizations $\underset{\sim}{x}(t)$ within some sufficiently long time interval T. Uncontrolled fluctuations in an analog electronic device played the role of initial perturbations. A certain number x_* was selected that had the order of ten mean-square values from the level of natural fluctuations. If the inequalities were satisfied,

$$kx_* \leq \sup_{0 \leq t \leq T} \|\underset{\sim}{x}(t)\| < (k+1)x_* , \quad (k = 1,2,\ldots) , \tag{5.101}$$

the point in the parameter space fell into the class k.

If the time interval T and the level x_* are properly selected, the points of the class 0 can be interpreted as belonging to the stochastic stability region, and the points of a sufficiently high class (for instance, class 3) - as belonging to the instability region; the boundary of the stochastic stability region being diffuse. As a rule, there is a sufficiently large scatter of realizations, i.e., the results may vary noticeably when passing from one ensemble of realizations to another. Sometimes there is a self-crossing of the boundary, or, to be more exact, with a monotonical increase of the parameter, the class of points may vary non-monotonically. Nevertheless, it is worthwhile to compare the results obtained in this way with the theoretical boundary of the stochastic stability region.

Description of the Experiment

The input processes were created with the help of a noise generator and formative filters, and the investigated systems were set up on an analog electronic device. The input processes were registered by an electronic oscillograph and a recording potentiometer. The output processes were recorded by a digital attachment whose function is the calculation of the number of overshoots of the norm of the process beyond the prescribed level. The attachment allowed a simultaneous registration of overshoots beyond 15 levels. All the experiments were carried out for systems with frequencies of natural oscillations of the order of $1s^{-1}$. The standard duration of realization was assumed to be T = 400s, i.e., it has the order of 100 typical system periods. The perturbations acting on the system were determined only by uncontrolled fluctuations whose level had a value not exceeding 3 percent of the root mean values of the processes at the system's input.

Results of the Experiment

Consider a stochastic analog of the Mathieu-Hill equation (5.51). The comparison of experimental and theoretical results is shown in Figures 47 - 49. The scales of simulation were selected so that $\sigma_{\Phi} = 1$. The value $|u(t)|$ was taken as the norm (more exactly - a semi-norm) in the space $\underline{U}$. The level x_* from the inequality of the type (5.101) was taken to be equal to ten mean-square values of the uncontrollable fluctuations in the simulator. Points of class 0, i.e., points for which no realization overshots of the band $|u(t)| < x_*$ were ever observed, are plotted by light circles, points of class 1 by light circles with a dot, points of class 2 by white and black circles, and points of class 3 by black circles.

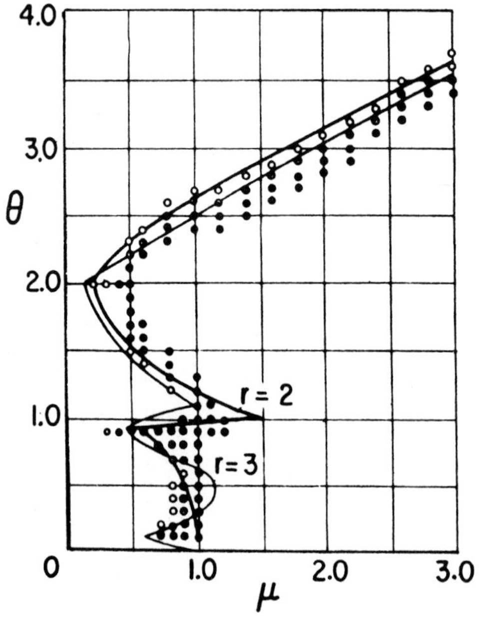

Figure 47

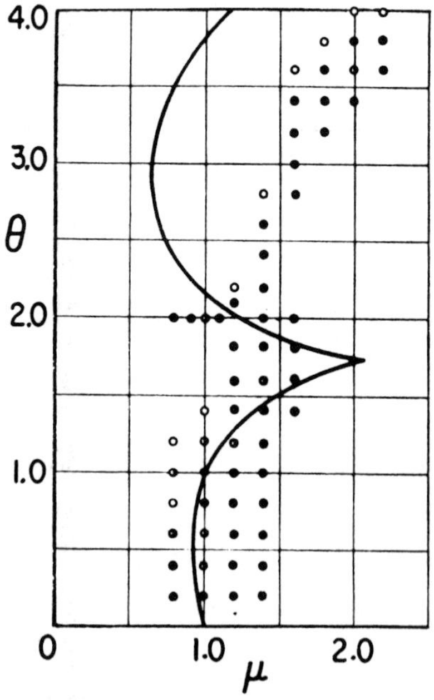

Figure 48

Figure 47 shows the results for the case $\rho = 0.02$, $\gamma = 0.025$, which corresponds to a narrow-band input process and to relatively small damping in the system. Comparing experimental results with calculations, we find that they agree well in the neighbourhood of the analog of the principal parametric resonance $\theta = 2$, i.e., $\Theta = 2\omega_0$. In addition, the experiment locates an analog of the second parametric resonance close to $\theta = 1$. There is a displacement of the point of high class in the direction of small μ for small ratios of the frequencies θ. This agrees with the results obtained in the approximation of the third and fourth moment functions.

The calculations given in Section 5.5 show that with the increase of damping in the system or with the input process growing more wide-band, the selective character of stochastic instability smoothes off. This is confirmed by the experimental data, shown in Figures 47, 48 and 49, with the latter two relating to the cases $\rho = 1.0$, $\gamma = 0.025$ and $\rho = \gamma = 1.0$, respectively.

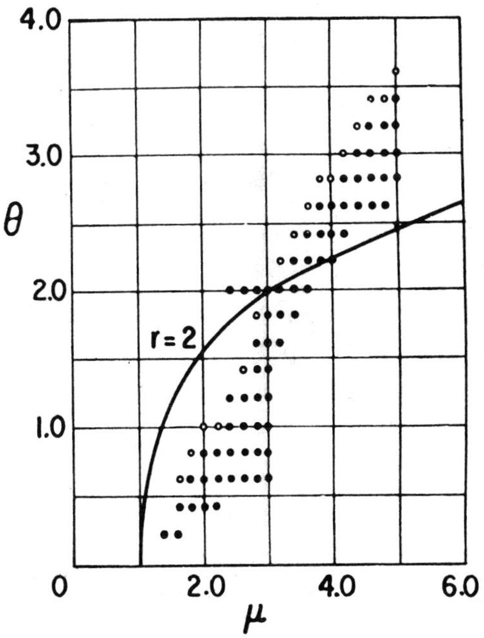

Figure 49

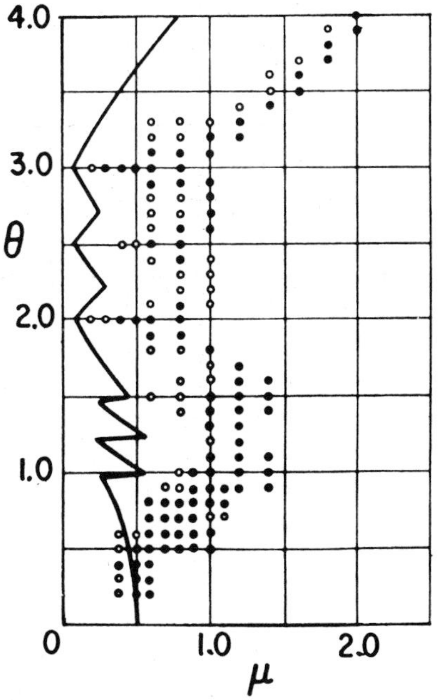

Figure 50

In Figures 50 - 52, the results of the simulation of system (5.89) with a correlation function of the process $\Phi(t)$ of the type (5.72) are given. We take the semi-norm $||\underset{\sim}{x}(t)||$ to be used in the inequality (5.101) as

$$||\underset{\sim}{x}(t)|| = \max\{|u_1(t)|, |u_2(t)|\} , \tag{5.102}$$

i.e., we classify the points in the parameter space depending on the behaviour of the realizations of a two-dimensional process $\{u_1(t), u_2(t)\}$ with respect to the square with sides $2x_*$. The classification and denotation of these points will be the same as before.

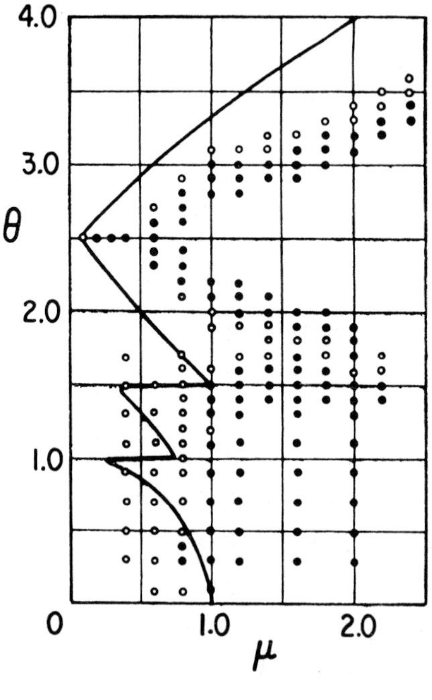

Figure 51

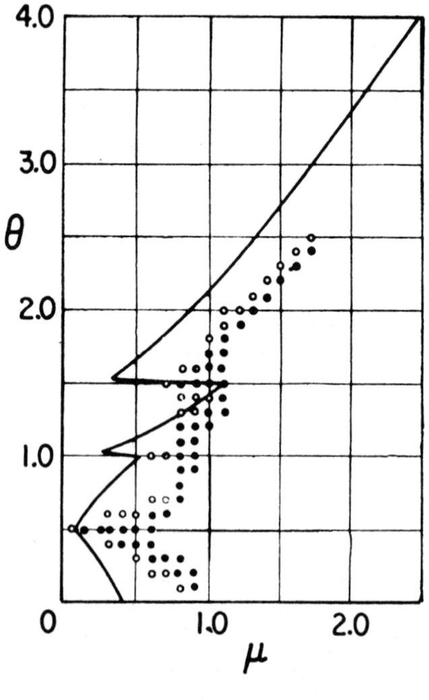

Figure 52

Figure 50 gives the results for the case $\rho = 0.02$, $\gamma = 0.025$, $G = G_2$, where the matrix G_2 is determined according to

(5.92). The agreement of the experimental data with the calculated ones is quite satisfactory, especially in the region of large values of $\theta = \Theta/\omega_1$. Specifically, the existence of analogs of parametric resonances close to the frequencies $\Theta = 2\omega_1$, $\Theta = 2\omega_2$ and $\Theta = \omega_1+\omega_2$ is confirmed. The picture becomes less clear with the decrease of θ. This is so because the level of vibration decreases and the typical period increases. Evidently, to obtain a better agreement with theory, it is necessary to lower the level x_*, or to increase the observation time T. The data shown in Figure 51 relate to the case $G = G_3$ with the same numerical values for the parameters. At the frequency $\Theta = \omega_1+\omega_2$, the combination resonance predominates. There are no principal resonances for $\Theta = 2\omega_1$ and $\Theta = 2\omega_2$, which agrees with the calculated results (Section 5.5). Figure 52 corresponds to the antisymmetric matrix $G = G_4$. The parametric resonance located close to the frequency $\Theta = |\omega_1-\omega_2|$ predominates. There is also a small wedge in the instability region close to $\Theta = \omega_2$.

5.7 SOME PROBLEMS OF PARAMETRIC VIBRATIONS OF CONTINUOUS SYSTEMS

The Elastic Rod under the Action of a Longitudinal Force which is a Stochastic Time Function

The problem of parametrically excited lateral vibrations of an elastic rod subjected to a longitudinal force was formulated in Section 5.1. If the cross-section of the rod is constant over the length, and if the rod's ends are simply supported, the generalized coordinates in the linearized equations of the perturbed motion can be separated. Moreover, the equations (5.4) for each generalized coordinate have the same form as equation (5.51). Therefore, all the results related to the latter equation can be used for investigating parametrically excited vibrations of an elastic rod.

Let the longitudinal force in equation (5.4) be given by the expression $N(t) = N_0\xi(t)$, where $\xi(t)$ is a stationary normal white noise of unit intensity, and N_0 is a numerical parameter. Comparing equations (5.4) and (5.42), we find that in order to make them identical, it is sufficient to replace μ in (5.42) by N_0/N_k, and ω_0 by ω_k. Then, from (5.46), we obtain the conditions for the parameter N_0 which ensure stability of each generalized coordinate in the mean square sense:

$$(N_0/N_k)^2 < 4\varepsilon/\omega_k\,, \qquad (k = 1,2,\ldots)\,.$$

Since N_k and ω_k are proportional to k^2, the stability condition of a straight rod is reduced to

$$N_0/N_1 < 2(\varepsilon/\omega_1)^{1/2}\,. \qquad (5.103)$$

The Elastic Spherical Shell Subjected to Stochastic Pressure

Consider a closed elastic spherical shell with the radius of the median surface R and thickness h. The shell is subjected to an external pressure $q(x,t)$, which is a stochastic function of the coordinates on the median surface and of time. Proceeding from the equations of a spherical shell in V. Vlasov's form, we shall introduce into them terms that take into account the normal components of the inertia forces and the forces of resistance proportional to velocity, and further additional terms that take account of initial stresses in the median surface caused by external pressure. As a result, we arrive at the equation for the normal deflection,*

* Bolotin, V.V., *Dynamic Stability of Elastic Systems*, Gostekhizdat, Moscow, 1956; and Schmidt, G., *Parametric Vibrations*, Mir, Moscow, 1978.

$$\left[\frac{h^2}{12(1-\nu^2)R^2}(\Delta+1)^2+1\right](\Delta+2)w+\frac{q(t)R}{2Eh}(\Delta+1-\nu)(\Delta+2)w+$$

$$\frac{\rho hR^2}{Eh}(\Delta+1-\nu)\left(\frac{\partial^2}{\partial t^2}+2\varepsilon\frac{\partial}{\partial t}\right)w=0\ . \qquad (5.104)$$

Here, Δ is the Laplace operator on the sphere, ν is Poisson's ratio, ρh is the surface mass density, and ε is the damping coefficient. We shall use the geographical coordinates ψ, β (ψ is the latitude, β is the longitude). We seek the solution of equation (5.104) in the form

$$w(\psi,\beta,t)=\sum_{k=1}^{\infty}u_k(t)\phi_k(\psi,\beta)\ , \qquad (5.105)$$

where $u_k(t)$ are generalized coordinates, and $\phi_k(\psi,\beta)$ are natural modes. For a closed uniform shell, these modes coincide with surface spherical functions. Ordering the natural modes by means of two indices characterizing the numbers of nodal parallels and nodal meridians, we have

$$\phi_{mn}(\psi,\beta)=P_m^{(n)}(\cos\psi)\begin{matrix}\sin m\beta\\ \cos m\beta\end{matrix},\qquad (m=0,1,2,\ldots;\ n=0,1,\ldots,m)\ , \qquad (5.106)$$

where $P_m^{(n)}(z)$ are the adjoint Legendre polynomials.

If the expressions (5.105) and (5.106) are substituted into equation (5.104), we arrive at a sequence of independent equations for the generalized coordinates $u_k(t)$ of the type (5.42). Thus, in this problem too, the results from Sections 5.3, 5.4 and 5.5 can be used.

Let us restrict ourselves to the case when the pressure q is uniform over the entire surface of the shell [72]. Let q(t) be a stationary normal process with the correlation function (5.72). In Figure 53, the stability region (on the left) is shown, calculated in the approximation of second moments for the values $\varepsilon=0.025\omega_R$, and $\alpha=0.02\Theta$. Here, ω_R is a natural frequency of the shell, calculated by means of formula (4.74), α and Θ are the correlation constant and a characteristic frequency in formula (5.72). A nondimensional parameter of the loading, i.e., $\mu = \sigma_q/q_*$ (where q_* is the critical pressure in the corresponding static problem) is plotted on the abscissa. Along the ordinate axis, the ratio $\theta = \Theta/\omega_R$ of the characteristic frequency of loading versus the natural frequency of the shell is plotted. The instability region is obtained by a multiple mapping of the instability region from Figure 38 onto the plane μ, Θ for various natural modes. Altogether, 22 values of the index m in expression (5.106) for natural modes (some of these values are shown in Figure 53) are taken into account in the diagram.

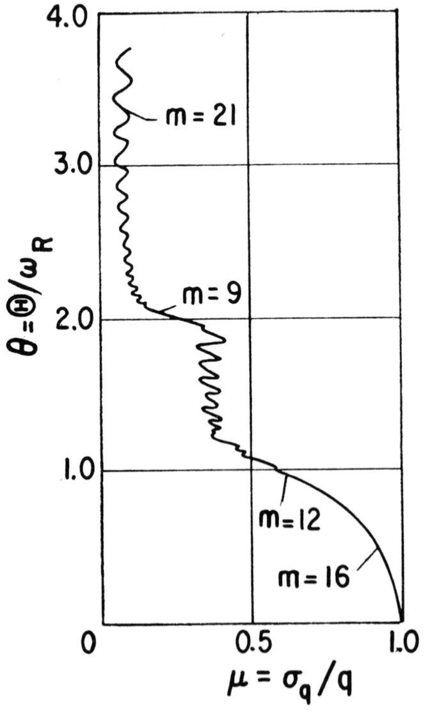

Figure 53

Chapter 6
Random Vibrations of Nonlinear Systems

6.1 GENERAL CHARACTERISTICS OF NONLINEAR PROBLEMS IN THE THEORY OF RANDOM VIBRATIONS

Preliminary Remarks

Problems of nonlinear systems in the theory of random vibrations are much more difficult than those of linear systems because the principle of superposition of solutions is generally not applicable to nonlinear systems as is the case in the classical theory of vibrations. In deterministic problems, this is manifested by the appearance of fractional, multiple, and combinative resonances when large-amplitude vibrations are sustained at a ratio of frequencies substantially different from the condition of simple resonance. The solutions to problems involving steady-state forced vibrations become multi-valued, which makes it necessary to investigate the problem of finding stable branches and to evaluate the influence of the system's history on its behaviour, even in the case of steady-state conditions for forced vibrations. From the mathematical point of view, the theory of nonlinear vibrations is rather complicated. In this case, approximate methods prevail, only a few of which can be strictly substantiated. Most of them are based on physical intuition and experiment. This holds true for stochastic problems as well. For linear systems, we have reliable exact methods (the method of differential equations for moment functions, the method of impulse unit-step functions, the spectral method, etc.). An attempt to extend these methods to nonlinear systems meets with great difficulties. If, for example, the method of moment functions leads for linear systems to independent systems of linear differential equations for the moment functions of the given order, we have for nonlinear systems an infinite system of nonlinear differential equations, in which moment functions of arbitrarily high order are encountered.

Let an equation of the system be specified in the form (2.2), where, for simplicity, we assume that $q(t)$ and $u(t)$ are scalar functions of time. If L is a nonlinear operator, it is not permutable with the operation of averaging. Isolating the linear part of L_0 from L, we shall write equation (2.2) as

$$L_0 u + \mu L_1 u = q \ . \qquad (6.1)$$

Here, L_1 is a nonlinear operator and μ is a deterministic numerical parameter (not necessarily small). Averaging (6.1) over a set of realizations, we obtain

$$L_0 \langle u \rangle + \mu \langle L_1 u \rangle = \langle q \rangle \ , \qquad (6.2)$$

instead of (2.6).

The equation (6.2), along with $\langle u \rangle$, contains the mathematical expectation of $L_1 u$. Let $L_1 u$ be a polynomial in u. Then, equation (6.2) will contain moment functions of $u(t)$ for coinciding time moments. For the moment functions of the second order, we obtain, analogously, an equation of the type (2.7):

$$\underset{t_1 t_2}{L_0 L_0}\langle u(t_1)u(t_2)\rangle + \mu \underset{t_1}{L_0}\langle u(t_1)\underset{t_2}{L_1} u(t_2)\rangle + \mu \underset{t_2}{L_0}\langle u(t_2)\underset{t_1}{L_1} u(t_1)\rangle +$$

$$\mu^2 \langle \underset{t_1}{L_1} u(t_1)\underset{t_2}{L_1} u(t_2)\rangle = \langle q(t_1)q(t_2)\rangle \ .$$

The moment functions of the third, fourth, and possibly of a higher order, enter into this equation. In general, an attempt to arrive at closed-form equations leads to an infinite system of equations that decouple only in the case when L is a linear operator.

Analogous difficulties arise when Green's function method is used. Let us denote Green's operator, corresponding to the linear part of the operator L, by H. Instead of (2.1), we have: $u = H(q-\mu L_1 u)$. Calculation of moment functions of u(t) leads again to relations in which, along with the moment functions of the given order, higher moment functions are involved. It is impossible to reduce this system to a system of a finite order without using additional assumptions. And, here, we obtain only an approximate solution, whose exactness can not always be verified.

In the statistical dynamics of nonlinear systems, approximate methods of nonlinear mechanics, such as the method of a small parameter, the linearization method, the stochastic analog of Krylov-Bogolubov-Mytropolski's method and others, are used. First, we shall consider the application of the small parameter method in its standard form.

Expansion in Terms of a Small Parameter

If μ is a number sufficiently small in magnitude, it is only natural to think of applying the small parameter method [122]. As a result, we obtain an infinite system of linear equations solvable successively. We shall represent the solution of equation (6.2) as a series in powers of the parameter μ:

$$u(t) = u_0(t) + \mu u_1(t) + \mu^2 u_2(t) + \cdots . \tag{6.3}$$

We shall also expand the nonlinear term in (6.1) as a series in powers of μ in order to obtain

$$L_1 u = L_1 u_0 + \mu \frac{\partial}{\partial u_0} (L_1 u_0) u_1 + \cdots . \tag{6.4}$$

Here, $u_0(t)$ is some generating solution. Substituting the series (6.3) and (6.4) into the equation (6.1), and equating terms which contain identical powers of μ, we arrive at the equations

$$\begin{aligned} L_0 u_0 &= q , \\ L_0 u_1 &= -L_1 u_0 , \\ L_0 u_2 &= -\frac{\partial}{\partial u_0} (L_1 u_0) u_1 , \\ &\cdots\cdots\cdots\cdots \end{aligned} \tag{6.5}$$

It can be seen from the first equation that the generating solution coincides with the solution of the corresponding linear system. Each of the solutions is linear, and their right sides depend only on the functions found in the preceding stage of calculations. As before, we shall designate $H \equiv L_0^{-1}$. The solution of system (6.5) takes the form

$$\begin{aligned} u_0 &= Hq , \\ u_1 &= -HL_1 u_0 , \\ u_2 &= -H\left[\frac{\partial}{\partial u_0} (L_1 u_0) u_1\right], \\ &\cdots\cdots\cdots\cdots \end{aligned}$$

Substituting these results into (6.3), multiplying the sums thus obtained, and averaging over a set of realizations, we find approximate expressions for the moment functions as power series in μ. For example, the first terms for the mathematical expectation will be

$$\langle u(t)\rangle = \langle u_0(t)\rangle - \mu \underset{\tau}{\overset{t}{H}} \langle L_1 u_0(\tau)\rangle + \cdots \, . \tag{6.6}$$

For the moment function of the second order, we obtain, analogously,

$$\langle u(t_1)u(t_2)\rangle = \langle u_0(t_1)u_0(t_2)\rangle - \mu \underset{\tau_1}{\overset{t_1}{H}} \langle u_0(t_2)L_1 u_0(\tau_1)\rangle - \mu \underset{\tau_2}{\overset{t_2}{H}} \langle u_0(t_1)L_1 u_0(\tau_2)\rangle + \cdots \, . \tag{6.7}$$

Example: Duffing's Stochastic Equation

In order to illustrate the considered methods, we shall use as an example the stochastic analog of Duffing's system, the vibrating system with one degree of freedom with a cubic nonlinearity, i.e.,

$$\frac{d^2u}{dt^2} + 2\varepsilon \frac{du}{dt} + \omega_0^2 u + \gamma u^3 = q(t) \, . \tag{6.8}$$

Here, ω_0 is the natural frequency of the linear system, ε is the damping coefficient, γ is the coefficient of the nonlinear function, and $\omega_0 > \varepsilon$, $\varepsilon > 0$, $\gamma \gtrless 0$. Unless otherwise specified, the external action $q(t)$ is assumed to be a stationary normal process with mathematical expection equal to zero and with the spectral density $S_q(\omega)$.

According to (6.3), we can represent the solution of equation (6.8) as a series

$$u = u_0(t) + \gamma u_1(t) + \gamma^2 u_2(t) + \cdots \, . \tag{6.9}$$

For the functions $u_k(t)$, we obtain the sequence of linear equations

$$\begin{aligned} &\frac{\partial^2 u_0}{dt^2} + 2\varepsilon \frac{du_0}{dt} + \omega_0^2 u_0 = q(t) \, , \\ &\frac{d^2 u_1}{dt^2} + 2\varepsilon \frac{du_1}{dt} + \omega_0^2 u_1 = -u_0^3 \, , \\ &\frac{d^2 u_2}{dt^2} + 2\varepsilon \frac{du_2}{dt} + \omega_0^2 u_2 = -3u_0^2 u_1 \, , \\ &\cdots\cdots\cdots\cdots\cdots \end{aligned} \tag{6.10}$$

We shall denote Green's function for the linear system by $h(t-\tau)$. Introducing the notations $t_1-\tau_1 = \Theta_1$, $t_2-\tau_2 = \Theta_2$, $t_2-\tau_1 = \tau$ we obtain for the correlation function $K_u(\tau)$ of the process $u(t)$:

$$K_u(\tau) = K_{u_0}(\tau) - \gamma \int_0^\infty h(\Theta_1)\langle u_0(\tau)u_0^3(-\Theta_1)\rangle d\Theta_1 -$$

$$\gamma \int_0^\infty h(\Theta_2)\langle u_0(0)u_0^3(\tau-\Theta_2)\rangle d\Theta_2 + \cdots . \tag{6.11}$$

Here, $K_{u_0}(\tau)$ is the correlation function of the zero approximation, i.e.,

$$K_{u_0}(\tau) = \int_0^\infty \int_0^\infty h(\Theta_1)h(\Theta_2)K_q(\tau+\Theta_1-\Theta_2)d\Theta_1 d\Theta_2 . \tag{6.12}$$

Since the process $u_0(t)$ is normal, the moments of fourth order, contained in the relation (6.11), can be expressed in terms of the correlation function (6.12) by means of the formulae (5.61):

$$\langle u_0(\tau)u_0^3(-\Theta_1)\rangle = 3K_{u_0}(0)K_{u_0}(\tau+\Theta_1) ,$$

$$\langle u_0(0)u_0^3(\tau-\Theta_2)\rangle = 3K_{u_0}(0)K_{u_0}(\tau-\Theta_2) .$$

Substituting these values into formula (6.11), we obtain

$$K_u(\tau) = K_{u_0}(\tau) - 3\gamma K_{u_0}(0) \int_0^\infty h(\Theta)[K_{u_0}(\tau-\Theta)+K_{u_0}(\tau+\Theta)]d\Theta + O(\gamma^2) . \tag{6.13}$$

We shall perform calculations for the case when $q(t)$ is stationary normal white noise of intensity s. Then, $S_q(\omega) = s/2\pi =$ const, and

$$K_{u_0}(\tau) = \frac{s}{4\varepsilon\omega_0^2} e^{-\varepsilon|\tau|}\left(\cos \omega_\varepsilon \tau + \frac{\varepsilon}{\omega_\varepsilon} \sin \omega_\varepsilon |\tau|\right) ,$$

where $\omega_\varepsilon^2 = \omega_0^2-\varepsilon^2$. The substitution of this expression into formula (6.13) gives the following formula for the second central moment $m_{11} \equiv K_u(0)$:

$$m_{11} = \frac{s}{4\varepsilon\omega_0^2}\left[1 - \frac{3}{4}\frac{\gamma s}{\varepsilon\omega_0^4} + O(\gamma^2)\right] . \tag{6.14}$$

If $\gamma s \ll \varepsilon\omega_0^4$, the correction for nonlinearity will be small. But if $\gamma s \approx \varepsilon\omega_0^4$, the correction will not be small; in this case, the written terms are not sufficient for finding an approximate solution. In general, for large nonlinearities, convergence of the method is doubtful. Such a situation is also typical for other problems in which the standard method of a small parameter is used, i.e., in those cases where this method is applicable, corrections for nonlinearity are small; yet in cases where corrections are not small, the method is inapplicable or requires the construction of approximations of too high an order. Therefore, further below, we shall concentrate on methods which are more suitable for the solution of nonlinear problems in the theory of random vibrations.

6.2 APPLICATION OF METHODS OF THE THEORY OF MARKOV PROCESSES

General Considerations

Consider a vibrating system whose evolution is described by the equation

$$\frac{d\underset{\sim}{x}}{dt} = \underset{\sim}{f}(\underset{\sim}{x},\underset{\sim}{q},t) . \tag{6.15}$$

Here, $\underset{\sim}{x}(t)$ is the phase vector in the space $\underline{U} = \underline{R}^n$, $\underset{\sim}{q}(t)$ is the vector of the input process, and $\underset{\sim}{f}(\underset{\sim}{x},\underset{\sim}{q},t)$ is a vector-function analytical in $\underset{\sim}{x}$, linear in $\underset{\sim}{q}$ and continuous in t.

Let the input process $\underset{\sim}{q}(t)$ be a stochastic function resulting from the passage of multi-dimensional white noise through some linear filters whose parameters can be continuous functions of time. Let us introduce the extended phase space $\underline{U} \oplus \underline{V}$, where the phase space $\underline{U}$ is extended so that the elements $\underset{\sim}{z}(t) \in \underline{U} \oplus \underline{V}$ are an N-dimensional Markov diffusion (Section 2.4). The process $\underset{\sim}{z}(t)$ satisfies Ito's stochastic differential equation (2.97)

$$d\underset{\sim}{z} = g(\underset{\sim}{z},t)dt + G(\underset{\sim}{z},t)d\underset{\sim}{w} , \tag{6.16}$$

where $g(\underset{\sim}{z},t)$ is a vector, $G(\underset{\sim}{z},t)$ is a matrix (analytical functions of $\underset{\sim}{z}$ and continuous functions of t), and $\underset{\sim}{w}(t)$ is the Wiener vector process with independent components.

Calculating the intensities of the process $\underset{\sim}{z}(t)$ from the formulae (2.102) - (2.103), we find the coefficients of the Kolmogorov equations (2.90) and (2.91). Further analysis is then reduced either to constructing the solutions of these equations, i.e., probability densities $p(\underset{\sim}{z};t)$, or to the calculation of moments of the process $\underset{\sim}{z}(t)$. General characteristics of the appropriate methods were given in Section 2.4, and the application of these methods to parametrically excited stochastic systems was given in the preceding chapter.

Stationary Vibrations in Duffing's System

Let us consider a special case of Duffing's stochastic system

$$\frac{d^2u}{dt^2} + 2\varepsilon \frac{du}{dt} + \omega_0^2 u + \gamma u^3 = \xi(t) , \tag{6.17}$$

where $\xi(t)$ is a stationary normal white noise of intensity s. Introducing the phase variables $x_1 = u$ and $x_2 = \dot{u}$, we obtain the corresponding Ito equations

$$dx_1 = x_2 dt ,$$

$$dx_2 = -(2\varepsilon x_2 + \omega_0^2 x_1 + \gamma x_1^3)dt + \sqrt{s}dw .$$

As before, w(t) denotes a Wiener process of unit intensity, i.e., $\langle w(t)w(t')\rangle = \min\{t,t'\}$. The formulae (2.102) and (2.103) lead to identical expressions for drift coefficients and the diffusion of the process $\underset{\sim}{x}(t)$:

$$\chi_1 = x_2 \ , \qquad \chi_2 = -(2\varepsilon x_2 + \omega_0^2 x_1 + \gamma x_1^3) \ , \qquad \chi_{22} = s \ ,$$

(the remaining coefficients of the diffusion $\chi_{\alpha\beta}$ are equal to zero). The Kolmogorov equation (2.90) takes the form

$$\frac{\partial p}{\partial t} = -x_2 \frac{\partial p}{\partial x_1} + \frac{\partial}{\partial x_2} [(2\varepsilon x_2 + \omega_0^2 x_1 + \gamma x_1^3) p] + \frac{1}{2} s \frac{\partial^2 p}{\partial x_2^2} \ . \qquad (6.18)$$

Its solution $p(\underset{\sim}{x};t)$ should satisfy the initial condition at $t = t_0$, the normalization condition and the conditions for the behaviour at infinity.

The solution to equation (6.18), independent of time, corresponds to stationary random vibrations. This solution has for $\gamma \geq 0$ the form

$$p(\underset{\sim}{x}) = \frac{1}{J_1} \exp\left[- \frac{2\varepsilon}{s}\left(\omega_0^2 x_1^2 + \frac{1}{2} \gamma x_1^4 \right)\right] \frac{1}{J_2} \exp\left(- \frac{4\varepsilon}{s} \frac{x_2^2}{2} \right) . \qquad (6.19)$$

Here, J_1 and J_2 are normalization factors. They are expressed in terms of elementary functions and in terms of the parabolic cylinder function $D_\mu(\zeta)$:

$$J_1 = \int_{-\infty}^{\infty} \exp\left[- \frac{2\varepsilon}{s} \left(\omega_0^2 x^2 + \frac{1}{2} \gamma x^4 \right)\right] dx = \left(\frac{s}{2\varepsilon\gamma}\right)^{1/4} \sqrt{\pi} \exp\left(\frac{\omega_0^4 \varepsilon}{2\gamma s}\right) D_{-1/2}\left(\omega_0^2 \sqrt{\frac{2\varepsilon}{\gamma s}}\right) , \qquad (6.20)$$

$$J_2 = \int_{-\infty}^{\infty} \exp\left(- \frac{4\varepsilon}{s} \frac{x^2}{2} \right) dx = \sqrt{\frac{\pi s}{2\varepsilon}}$$

If $\gamma < 0$, there is no stationary solution to equation (6.18). This can be accounted for by the fact that the phase trajectories of the system will tend to infinity as $t \to \infty$ in the case of a "soft" nonlinearity of the restoring force.

This example will be used further below for comparing the accuracy and effectiveness of various approximate methods. The expressions for the moments of the process $\underset{\sim}{x}(t)$ will be needed for comparison. Let us write the values of these moments that do not become identically zero:

$$m_{1(2k)2(2\ell)} \equiv m_{\underbrace{11\ldots1}_{2k}\underbrace{22\ldots2}_{2\ell}} \equiv \int_{-\infty}^{\infty}\int_{-\infty}^{\infty} x_1^{2k}x_2^{2\ell}p(x_1,x_2)dx_1dx_2 \; .$$

The substitution of expressions (6.19) and (6.20) here yields

$$m_{1(2k)2(2\ell)} = \left(\frac{s}{2\varepsilon\gamma}\right)^{k/2}\left(\frac{s}{4\varepsilon}\right)^{\ell}\frac{\Gamma(k+1/2)}{\Gamma(1/2)}(2\ell-1)!!\,\frac{D_{-(k+1/2)}(\zeta)}{D_{-1/2}(\zeta)} \; , \tag{6.21}$$

where the notation

$$\zeta = \omega_0^2\sqrt{2\;\varepsilon/\gamma s}$$

is used. Let us write the formulae for the moments of second order as

$$m_{11} = \frac{1}{2}\sqrt{\frac{s}{2\varepsilon\gamma}\frac{D_{-3/2}(\zeta)}{D_{-1/2}(\zeta)}} \; , \qquad m_{22} = \frac{s}{4\varepsilon} \; , \tag{6.22}$$

of the fourth order as

$$m_{1(4)} = \frac{3}{8}\frac{s}{\varepsilon\gamma}\frac{D_{-5/2}(\zeta)}{D_{-1/2}(\zeta)} \; , \qquad m_{2(4)} = \frac{3s^2}{16\varepsilon^2} \; , \tag{6.23}$$

and of the sixth order as

$$m_{1(6)} = \frac{15}{8}\left(\frac{s}{2\varepsilon\gamma}\right)^{3/2}\frac{D_{-7/2}(\zeta)}{D_{-1/2}(\zeta)} \; , \qquad m_{2(6)} = \frac{15s^3}{64\varepsilon^3} \; . \tag{6.24}$$

Some Generalizations

The stationary distribution (6.19) can be considered as a special case of wider class of distributions related to the Maxwell-Boltzmann distribution in statistical physics. For example, let us consider a vibrating system with one degree of freedom

$$\frac{d^2u}{dt^2} + 2\varepsilon \frac{du}{dt} + \frac{\partial \Pi(u)}{\partial u} = \xi(t) , \qquad (6.25)$$

where $\xi(t)$ is normal white noise, and where the function $\Pi(u)$ can be interpreted as the potential energy of the system. The corresponding Kolmogorov equation has the form

$$\frac{\partial p}{\partial t} = -x_2 \frac{\partial p}{\partial x_1} + \frac{\partial}{\partial x_2}\left\{\left[2\varepsilon x_2 + \frac{\partial \Pi(x_1)}{\partial x_1}\right] p\right\} + \frac{1}{2} s \frac{\partial^2 p}{\partial x_2^2} .$$

Its stationary solution will be

$$p(x_1,x_2) = \frac{1}{J} \exp\left\{- \frac{4\varepsilon}{s} [\Pi(x_1) + \frac{1}{2} x_2^2]\right\} , \qquad (6.26)$$

where the expression in square brackets, i.e.,

$$V(x_1,x_2) = \Pi(x_1) + \frac{1}{2} x_2^2 \qquad (6.27)$$

has the sense of the total mechanical energy of the system. It is possible to obtain a closed stationary solution to the Kolmogorov equation for a more general case of the system [120]

$$\frac{d^2u}{dt^2} + 2\varepsilon f\left[\Pi(u) + \frac{1}{2}\left(\frac{du}{dt}\right)^2\right] \frac{du}{dt} + \frac{\partial \Pi(u)}{\partial u} = \xi(t) .$$

Here,

$$p(x_1,x_2) = \frac{1}{J} \exp\left[- \frac{4\varepsilon}{s} \int_0^{V(x_1,x_2)} f(\eta)\, d\eta\right] ,$$

where $V(x_1,x_2)$ is, as before, the total mechanical energy (6.27).

If certain restrictions are imposed on the relations between the intensities of white noises and damping coefficients, these results can be extended to systems with multiple degrees of freedom which are under the action of potential positional forces. For example, let us consider a system with n degrees of freedom

$$\frac{d^2u_j}{dt^2} + 2\varepsilon_j \frac{du_j}{dt} + \frac{\partial \Pi(u_1, u_2, \ldots, u_n)}{\partial u_j} = \xi_j(t) , \qquad (j = 1,2,\ldots,n) , \tag{6.28}$$

where $\Pi(\underset{\sim}{u})$ is the potential energy of the system, and the $\xi_j(t)$ are independent white noises with intensities s_j. If the condition

$$\varepsilon_j / s_j = \text{const.} \tag{6.29}$$

is satisfied, the joint probability density of the stationary distribution of the generalized coordinates and the generalized velocities will have the form

$$p(\underset{\sim}{u}, \dot{\underset{\sim}{u}}) = \frac{1}{J} \exp \left\{ - \frac{4\varepsilon}{s} \left[\Pi(\underset{\sim}{u}) + \frac{1}{2} \sum_{j=1}^{n} \left(\frac{du_j}{dt} \right)^2 \right] \right\} . \tag{6.30}$$

Here, ε/s is the constant in relation (6.29). This relation leads to a uniform distribution of energy over all degrees of freedom.*

A review of other nonlinear problems for which it has been possible to find solutions of the Kolmogorov equation in a closed form is given in [96].

Application of Methods of Nonlinear Mechanics

If external actions are processes with a finite variance (for example, processes with latent periodicity), the dimensionality N

---

*Bolotin, V.V., "On Stationary Distributions in Statistical Dynamics of Elastic Systems", *Coll. Works: Voprosi Dynamiki i Dinamicheskoi Prochnosti*, (Problems of Dynamics and Dynamic Strength), No. 10, Academy of Science of Latvia, Riga, 1963.

of the extended phase space $\underline{U} \oplus \underline{V}$ may turn out to be too large. Let us consider some approximate methods [93], which make use of the well-known concepts of nonlinear mechanics, and which make it possible to avoid an excessive increase of the dimensionality of the phase space.

Let the system's behaviour be of the kind that can be described by the truncated equations of the Krylov-Bogolubov-Mitropolski method. Here, the amplitudes and phases of the generalized coordinates will vary slowly not only when compared to the carrying periodic process but also when compared to the input process. If, besides, the correlation time of the latter is not large as compared to the natural periods of the system, the joint evolution of amplitudes and phases can be treated as a multidimensional Markov process. For example, for the two-dimensional process $\underset{\sim}{x}(t) = \{u(t), \dot{u}(t)\}$ at the output of the system described by

$$\frac{d^2u}{dt^2} + \omega_0^2 u + \mu f\left(u, \frac{du}{dt}\right) = q(t) \tag{6.31}$$

to be approximately considered as a Markov process, it is necessary for the correlation time τ_c of the input process $q(t)$ to be small as compared to the natural period of the system $\tau_0 = 2\pi/\omega_0$. Thus, it is necessary that $\tau_c \ll \tau_0$. Now let us consider the case of a narrow-band process $q(t)$ whose amplitude and phase vary sufficiently slowly so that their typical variation time τ_1 is large as compared to τ_0. Let then $\tau_0 \sim \tau_c$. Then, the input process, in relation to the slowly varying amplitude and phase, can be treated as white noise. Thus, the application of the method of slowly varying amplitudes and phases extends the range of application of the theory of Markov processes to the cases in which the conditions

$$\tau_c \sim \tau_0 \ll \tau_1 \tag{6.32}$$

are satisfied.

If the nonlinearity and the nonstationarity of the system are sufficiently small, the solution of the equation (6.31) will differ only slightly from the harmonic motion with a frequency ω_0 . So we assume that

$$u = a_1(\tau) \sin \omega_0 t + a_2(\tau) \cos \omega_0 t , \tag{6.33}$$

where $a_1(\tau)$ and $a_2(\tau)$ are functions of the "slow" time τ. When differentiating, we shall consider the derivatives with respect to τ to be small as compared to the derivatives with respect to t. When integrating with respect to t within the interval τ_0, the "slow" time will be considered as a parameter. Let us expand the nonlinear function $f(u,\dot{u})$ and the external action q(t) into Fourier series and truncate all the terms containing harmonics:

$$\begin{gathered} f(u,\dot{u}) \approx F_1(a_1,a_2,\tau) \cos \omega_0 t + F_2(a_1,a_2,\tau) \sin \omega_0 t , \\ q(t) \approx q_1(\tau) \cos \omega_0 t + q_2(\tau) \sin \omega_0 t . \end{gathered} \tag{6.34}$$

Substituting these expressions into equation (6.31), we shall obtain (after averaging with respect to time on the interval $[0,2\pi/\omega_0]$) the truncated equations for the functions of "slow" time $a_1(\tau)$ and $a_2(\tau)$

$$\begin{gathered} 2\omega_0 \frac{da_1}{d\tau} = -\mu F_1(a_1,a_2,\tau) + q_1(\tau) , \\ -2\omega_0 \frac{da_2}{d\tau} = -\mu F_2(a_1,a_2,\tau) + q_2(\tau) . \end{gathered} \tag{6.35}$$

Equations (6.35) formally coincide with the equations of first approximation in the Krylov-Bogolubov's method. Here, however, $q_1(\tau)$ and $q_2(\tau)$ are random functions. If the conditions (6.32) are satisfied, the external actions $q_1(\tau)$ and $q_2(\tau)$ can be approximately treated as white noises, and, thus, one can write the

Kolmogorov equation for the joint probability density $p(a_1,a_2;\tau)$ of the amplitudes $a_1(\tau)$ and $a_2(\tau)$. The next objective is to express the intensities of these white noises in terms of probability characteristics of the process $q(t)$.

Let us, for example, consider the correlation function

$$\langle q_1(\tau_1)q_2(\tau_2)\rangle = \frac{1}{\tau_0^2}\int_{\tau_1}^{\tau_1+\tau_0}\int_{\tau_2}^{\tau_2+\tau_0}\langle q(\Theta_1)q(\Theta_2)\rangle \cos\omega_0\Theta_1 \cos\omega_0\Theta_2 \, d\Theta_2 d\Theta_1 . \tag{6.36}$$

Since the output process is assumed to be narrow-band, its properties are essentially determined by the value of the spectral density $S_q(\omega)$ of the input process $q(t)$ corresponding to the frequency ω_0. Therefore, when calculating the integral, the real process can be replaced by white noise of intensity $s = 2\pi S_q(\omega_0)$. Substituting this expression into the right side of the formula (6.36) and integrating, we obtain

$$\langle q_1(\tau_1)q_1(\tau_2)\rangle \approx \frac{1}{2}\,\omega_0\, S_q(\omega_0) .$$

The formula is valid if τ_1 and τ_2 fall into one period $\tau_0 = 2\pi/\omega_0$. Otherwise, the right side should be assumed as equal to zero. An analogous expression is obtained for the correlation function of the process $q_2(\tau)$. Thus,

$$\langle q_1(\tau_1)q_1(\tau_2)\rangle = \langle q_2(\tau_1)q_2(\tau_2)\rangle \approx \begin{cases} \frac{1}{2}\,\omega_0 S_q(\omega_0) & (|\tau_1-\tau_2| \le \tau_0) \\ 0 & (|\tau_1-\tau_2| > \tau_0) . \end{cases} \tag{6.37}$$

It is not difficult to see that the joint correlation function of the processes $q_1(\tau)$ and $q_2(\tau)$ can be assumed to be equal to zero. But, from the viewpoint of the slowly varying processes $a_1(\tau)$ and $a_2(\tau)$, the nature of the correlation of the processes $q_1(\tau)$ and $q_2(\tau)$ within one period is not of great importance. So, without a significant error, it can be assumed that

$$\langle q_1(\tau_1)q_1(\tau_2)\rangle = \langle q_2(\tau_1)q_2(\tau_2)\rangle = \pi S_q(\omega_0)\delta(\tau_1-\tau_2) \ , \tag{6.38}$$

i.e., that the processes $q_1(\tau)$ and $q_2(\tau)$ are uncorrelated white noises with intensities

$$s = \pi S_q(\omega_0) \ . \tag{6.39}$$

The Kolmogorov equation for the joint density $p(a_1,a_2;\tau)$ takes the form

$$\frac{\partial p}{\partial \tau} = \frac{\mu}{2\omega_0}\left[\frac{\partial}{\partial a_1}(F_1 p) - \frac{\partial}{\partial a_2}(F_2 p)\right] + \frac{\pi S_q(\omega_0)}{8\omega_0^2}\left(\frac{\partial^2 p}{\partial a_1^2} + \frac{\partial^2 p}{\partial a_2^2}\right) . \tag{6.40}$$

The functions $F_1(a_1,a_2,\tau)$ and $F_2(a_1,a_2,\tau)$ are determined according to the first formula (6.34) as the Fourier coefficients of the function $f(u,\dot{u})$. For example, in the case of equation (6.8),

$$F_1(a_1,a_2) = \frac{3}{4}(a_1^2+a_2^2)a_2 \ , \qquad F_2(a_1,a_2) = \frac{3}{4}(a_1^2+a_2^2)a_1 \ .$$

6.3 METHOD OF STATISTICAL LINEARIZATION

The Concept of the Method

The method of statistical linearization is analogous to the method of harmonic linearization in the theory of nonlinear vibrations. According to the method of harmonic linearization, nonlinear functions entered into the differential equation of a system are replaced by certain equivalent linear functions over a set of sinusoidal solutions with unknown amplitude and phase or unknown amplitude and frequency. Solutions of the linear equation obtained are dependent on these unknown values as parameters. To determine these unknown values, algebraic, transcendental or functional equations are obtained. In the method of statistical linearization, transition to the equivalent linear equation is performed over a certain set of random functions. As a rule, the hypothesis of stationarity and normality of the output process is used, the coefficients of the equivalent equation depending on mathematical expectation and on moments of the second order of the output process. By applying the averaging operation to the equivalent equation, we arrive at the equations for the probabilistic characteristics of the output process.

Let us illustrate the concept of the method by using the example of a scalar equation such as

$$L_0 u + \mu f(u) = q(t) , \tag{6.41}$$

where L_0 is a linear operator and $f(u)$ is a deterministic nonlinear function. Let the input process $q(t)$ be stationary, centralized and have symmetrical distribution, and let the function $f(u)$ be odd. Then, the output process $u(t)$ will also be stationary, centralized and have symmetrical distribution. Let us replace the function $f(u)$ by the linear function

$$f(u) \sim cu , \tag{6.42}$$

where c is a deterministic constant. This constant will be chosen according to the condition that variances of both parts of the relation (6.42) should be equal. Hence,

$$c = \left[\frac{\langle f^2(u) \rangle}{\langle u^2 \rangle} \right]^{1/2} . \tag{6.43}$$

Another method for determining the constant is given by the criterion of the minimum of the standard deviation of function $f(u)$ from cu:

$$c = \frac{\langle f(u)u \rangle}{\langle u^2 \rangle} . \tag{6.44}$$

Generally, applying either the formulae (6.43) or (6.44), we obtain different values for the constant c. In order to apply these formulae, it is necessary to know the one-dimensional probability density p(u). In this case,

$$\langle f^2(u)\rangle = \int_{-\infty}^{\infty} f^2(u)p(u)du ,$$

$$\langle f(u)u\rangle = \int_{-\infty}^{\infty} f(u)up(u)du .$$

As a result, the constant c is dependent on the parameters, accurate to the degree to which the distribution of the process is specified. For example, if we assume that u(t) is a normal process, then,

$$p(u) = \frac{1}{\sqrt{2\pi}\sigma_u} \exp\left(- \frac{u^2}{2\sigma_u^2}\right) , \tag{6.45}$$

where σ_u^2 is the unknown variance of the process u(t), and the coefficient c in the linearized equation

$$L_0 u + \mu c u = q(t) \tag{6.46}$$

depends on σ_u. Calculating the variance of the output process for the equation (6.46), and equating it to the value σ_u^2, we obtain the equation for determining the single unknown parameter of the output process. Let q(t) be a stationary random process with the spectral density $S_q(\omega)$. The spectral density $S_u(\omega)$ of the stationary response of the linearized system to this process is determined from formula (2.41) as

$$S_u(\omega) = \frac{S_q(\omega)}{|L_0(i\omega)+\mu c(\sigma_u)|^2} .$$

In order to obtain the equation for σ_u, we shall use the relation between the variance of the stationary random process and its spectral density. Hence,

$$\sigma_u^2 = \int_{-\infty}^{\infty} \frac{S_q(\omega)d\omega}{|L_0(i\omega)+\mu c(\sigma_u)|^2} . \qquad (6.47)$$

If the integral on the right side is calculated in closed form, we obtain for determining σ_u an algebraic or a transcendental equation. For the general case, equation (6.47) is solved numerically or by the method of successive approximations.

Example

We shall consider the application of formulae (6.43), (6.44) and (6.47) to the stochastic analog of Duffing's equation (6.8). Assuming that the distribution p(u) does not significantly differ from the normal one, we find from formulae (6.43) and (6.44) that $c = \alpha\sigma_u^2$. The coefficient α turns out to be equal to $\sqrt{15}$ if linearization is performed according to the condition of equality of variances. The criterion of the minimum of the standard deviation yields $\alpha = 3$. Equation (6.47) takes the form

$$\sigma_u^2 = \int_{-\infty}^{\infty} \frac{S_q(\omega)d\omega}{|\omega_0^2+2i\varepsilon\omega-\omega^2+\alpha\gamma\sigma_u^2|^2} .$$

The integral on the right side is, for instance, easily calculated in the case when S_q = const. Then, we obtain for determining σ_u^2 the quadratic equation

$$\sigma_u^2 = \frac{\pi S_q}{2\varepsilon(\omega_0^2+\alpha\gamma\sigma_u^2)} . \qquad (6.48)$$

Equation (6.48) can be applied in the case when the spectral density $S_q(\omega)$ is a slowly varying function of ω and the damping in the system is so small that for an approximate calculation of the integral, a method that leads to the formula (2.49) can be used. Then, it is sufficient to substitute the value of the spectral density $S_q(\omega)$, instead of the constant S_q, into equation (6.48) for $\omega = \sqrt{\omega_0^2+\alpha\gamma\sigma_u^2}$.

Let us compare the results obtained from the equation (6.48) with the formula (6.14), which was derived by applying the method of a small parameter. Assume that statistical linearization is performed using the condition of the minimum of the standard deviation of the nonlinear function and its linear equivalent. Then, if the notations of formula (6.14) are used, equation (6.48) takes the form

$$m_{11} = \frac{s}{4\varepsilon(\omega_0^2 + 3\gamma m_{11})} . \tag{6.49}$$

The solution of this equation coincides with the right-hand side of the formula (6.14) up to terms of the order γ. We shall deal with the exact solution in the next section.

A More General Scheme of the Method of Statistical Linearization

Instead of (6.41), we shall consider a more general equation

$$L_0 u + \mu f(u,\dot{u}) = q(t) , \tag{6.50}$$

where $u(t)$ and $q(t)$ are nonstationary random functions with non-zero mathematical expectations. We shall separate the fluctuating components $u_1(t)$ and $q_1(t)$ from the functions $u(t)$ and $q(t)$:

$$u_1(t) = u(t) - \langle u(t)\rangle , \qquad q_1(t) = q(t) - \langle q(t)\rangle .$$

Averaging the equation (6.50) over a set of realizations, and subtracting the result from the equation (6.50), we obtain

$$L_0 u_1 + \mu[f(u,\dot{u}) - \langle f(u,\dot{u})\rangle] = q_1(t) . \tag{6.51}$$

Let us apply linearization to equation (6.51). The equivalent linear equation has the form

$$L_0 u_1 + \mu(c_1 u_1 + c_2 \dot{u}_1) = q_1(t) \,, \tag{6.52}$$

where c_1 and c_2 are nonrandom constants. In order to find these constants, we shall use the criterion of the minimum of the standard deviation, i.e.,

$$\langle [f(u,\dot{u}) - \langle f(u,\dot{u})\rangle - (c_1 u_1 + c_2 \dot{u}_1)]^2 \rangle \to \min_{c_1, c_2} \,.$$

Hence, we obtain a system of two equations for c_1 and c_2

$$\begin{aligned} \langle u_1^2 \rangle c_1 + \langle u_1 \dot{u}_1 \rangle c_2 &= \langle f(u,\dot{u}) u_1 \rangle \,, \\ \langle u_1 \dot{u}_1 \rangle c_1 + \langle \dot{u}_1^2 \rangle c_2 &= \langle f(u,\dot{u}) \dot{u}_1 \rangle \,, \end{aligned} \tag{6.53}$$

in which time t is considered as a parameter. If the process u(t) is stationary, $\langle u_1 \dot{u}_1 \rangle = 0$, and we arrive at a formulae of the type (6.44). More details about the method of statistical linearization and its applications can be found in [79].

6.4 APPLICATION OF THE METHOD OF MOMENT FUNCTIONS

Features of the Application of the Method of Moment Functions to Nonlinear Problems

If the nonlinearity is analytical (for instance, polynomial), the method of moment functions, as it was presented in Section 5.4, is convenient for calculating probabilistic characteristics of random vibrations. Let us introduce (as we did in Section 5.4) the extended phase space with the elements $\underset{\sim}{z}(t)$. Consider a set of one-point moment functions of the process $\underset{\sim}{z}(t)$

$$m_{jk\ell\ldots}(t) = \langle z_j(t) z_k(t) z_\ell(t) \ldots \rangle \,. \tag{6.54}$$

Without considering the moments of the input process, and moments identically equal to one another, we shall form vector functions consisting of moments of the same order ρ:

$$\underset{\sim}{m}_{\rho}(t) = \{m_{\underbrace{11\ldots}_{\rho}}(t), m_{\underbrace{11\ldots2}_{\rho}}(t), \ldots\} , \quad (\rho = 1,2,\ldots) . \tag{6.55}$$

Differential equations for $\underset{\sim}{m}_{\rho}(t)$ can be obtained either by directly averaging the system's equations, or by using Ito's principle of differentiation and by the operation of averaging, or by applying the appropriate Kolmogorov equation. As a result, we arrive at equations of the type (5.40),

$$\frac{d\underset{\sim}{m}_{\rho}}{dt} = \Phi_{\rho}(\underset{\sim}{m}_1, \ldots, \underset{\sim}{m}_{\rho}, \ldots) , \quad (\rho = 1,2,\ldots) , \tag{6.56}$$

whose right sides are analytical vector-functions of the moments $\underset{\sim}{m}_{\rho}$ and continuous functions of time t.

By this method of moment functions, a stochastic problem is reduced to the system of equations (6.56).* The equation (6.56) can easily be solved numerically, and in some cases analytically, after being reduced to a finite system. The advantage of the method is that, firstly, it admits an investigation of the stability of the solution (which is especially important for nonlinear vibrating systems), and, secondly, it makes the reconstruction possible of the one-point distribution of probabilities of the vector $\underset{\sim}{x}(t)$ by moment functions. The method also allows one to obtain partial information on multipoint distributions (more exactly, at contiguous moments of time, which is sufficient for the solution of problems in the theory of level crossings).

Each equation of the infinite system (6.56) contains in its right side moment functions of higher order. As a consequence

---

*Bolotin, V.V., "On the Application of the Method of Moment Functions in Statistical Dynamics of Non-Linear Systems", Coll. Works: *Rascheti na Prochnost, (Strength Analysis)*, No. 19, Mashinostroenie, Moscow, 1978.

of this, there arises the problem of the closure of the system (6.56), i.e., of its reduction to a closed system of a finite number of equations. Here, as in Section 5.4, we shall use the quasi-normality hypothesis for closure. Restricting ourselves to moments up to the order r (inclusive), we shall assume all the cumulant functions of higher order to be equal to zero. In terms of central moments, the closure condition of the order r is given by the formulae (5.61), and in terms of initial moments by formulae (5.62) - (5.65).

Closure at the Level of Moments of Second Order

Assuming that all the cumulant functions of the order $r \geq 3$ are equal to zero, we obtain the simplest approximation. The only distribution with cumulant functions identically equal to zero for $r \geq 3$ is the normal distribution. Therefore, it is natural to expect that such a distribution will be equivalent to a variant of the method of statistical linearization based on the normal distribution. This can be verified by direct calculations. We shall consider a few examples, restricting ourselves to stationary solutions.

As in the first example, we shall consider a stochastic analog of Duffing's oscillating system, where $\xi(t)$ is a stationary normal white noise of intensity s.

The Kolmogorov equation (6.18) for the probability density $p(\underset{\sim}{x};t)$ of the phase vector $\underset{\sim}{x} = \{u,\dot{u}\}$ corresponds to equation (6.17). In order to obtain equations for moment functions of the vector $\underset{\sim}{x}(t)$, we shall multiply the equation (6.18) by x_1, x_2, x_1^2, x_1x_2, x_2^2,... termwise, and integrate over the whole space $\underline{U}$. Using wherever necessary, integration by parts, and taking into consideration the behaviour of the function $p(\underset{\sim}{x};t)$ at infinity, we obtain the system of equations

$$\frac{dm_1}{dt} = m_2 \ ,$$

$$\frac{dm_2}{dt} = -(2\varepsilon m_2 + \omega_0^2 m_1 + \gamma m_{111}) \ ,$$

$$\frac{dm_{11}}{dt} = 2m_{12} \ , \qquad (6.57)$$

$$\frac{dm_{12}}{dt} = m_{22} - (2\varepsilon m_{12} + \omega_0^2 m_{11} + \gamma m_{1111}) \ ,$$

$$\frac{dm_{22}}{dt} = -2(2\varepsilon m_{22} + \omega_0^2 m_{12} + \gamma m_{1112}) + s \ ,$$

. . . .

The equations for the mathematical expectations m_1 and m_2 contain the moment of the third order m_{111}, the equations for the moments of the second order contain the moments of the fourth order m_{1111} and m_{1112}, etc. In order to obtain an approximation at the level of the moments of the second order, the moments of higher order should be excluded from the equations written in (6.57). Application of the formulae (5.62) and (5.63) gives

$$m_{111} = 3m_{11}m_1 - 2m_1^3 \ ,$$

$$m_{1111} = 3m_{11}^2 - 2m_1^4 \ ,$$

$$m_{1112} = -2m_1^3 m_2 + 3m_{11}m_{12} \ .$$

From this, we obtain the truncated system corresponding to the infinite series (6.57):

$$\frac{dm_1}{dt} = m_2 \ , \qquad \frac{dm_2}{dt} = -(2\varepsilon m_2 + \omega_0^2 m_1) - \gamma m_1(3m_{11} - 2m_1^2) \ ,$$

$$\frac{dm_{11}}{dt} = 2m_{12}, \qquad \frac{dm_{12}}{dt} = m_{22} - (2\varepsilon m_{12} + \omega_0^2 m_{11}) - \gamma(3m_{11}^2 - 2m_1^4) \ , \qquad (6.58)$$

$$\frac{dm_{22}}{dt} = -2(2\varepsilon m_{22} + \omega_0^2 m_{12}) - 2\gamma(-2m_1^3 m_2 + 3m_{11}m_{12}) + s \ .$$

For stationary random vibrations, all the moments of the process $\underset{\sim}{x}(t)$ are constant. It is sufficient to assume the left sides of equations (6.58) to be equal to zero to find the corresponding values of the moments. As a result, we obtain $m_1 = m_2 = m_{12} = 0$,

$$m_{11} = \frac{\omega_0^2}{6\gamma}\left[\left(1 + \frac{3\gamma s}{\varepsilon\omega_0^4}\right)^{1/2} - 1\right], \quad m_{22} = \frac{s}{4\varepsilon} . \tag{6.59}$$

The right side of the first formula coincides exactly with the positive root of equation (6.49), i.e., with the result that is obtained by the method of statistical linearization, provided that the linearization is performed by means of the condition of the minimum of the mean square deviation of the nonlinear function from its linear equivalent.

As a second example, we shall consider a self-oscillating system whose equation is a stochastic analog of Rayleigh's equation

$$\frac{d^2u}{dt^2} - 2\varepsilon\frac{du}{dt} + 2\beta\left(\frac{du}{dt}\right)^3 + \omega_0^2 u = \xi(t) . \tag{6.60}$$

The notations used here are the same as in the equation (6.17). Stable self-oscillating regimes in this system are possible if $\varepsilon > 0$ and $\beta > 0$. Let us write the first five equations of the system (6.56):

$$\frac{dm_1}{dt} = m_2 ,$$

$$\frac{dm_2}{dt} = 2\varepsilon m_2 - \omega_0^2 m_1 - 2\beta m_{222} ,$$

$$\frac{dm_{11}}{dt} = 2m_{12} , \tag{6.61}$$

$$\frac{dm_{12}}{dt} = m_{22} + 2\varepsilon m_{12} - \omega_0^2 m_{11} - 2\beta m_{1222} ,$$

$$\frac{dm_{22}}{dt} = 2(2\varepsilon m_{22} - \omega_0^2 m_{12} - 2\beta m_{2222}) + s .$$

In order to close the system (6.61), we shall assume, taking into consideration formulae (5.62) and (5.63), that

$$\begin{aligned} m_{222} &= 3m_{22}m_2 - 2m_2^3 , \\ m_{1222} &= -2m_1m_2^3 + 3m_{12}m_{22} , \\ m_{2222} &= 3m_{22}^2 - 2m_2^4 . \end{aligned} \qquad (6.62)$$

Substituting (6.62) into (6.61), and restricting ourselves to the stationary solution m_{11} = const, m_{22} = const, $m_1 = m_2 = m_{12} = 0$, we obtain

$$m_{11} = \frac{\varepsilon}{6\beta\omega_0^2}\left[\left(1 + \frac{3\beta s}{\varepsilon^2}\right)^{1/2} + 1\right], \qquad m_{22} = \omega_0^2 m_{11} . \qquad (6.63)$$

Finally, let us consider the nonlinear parametric system

$$\frac{d^2u}{dt^2} + 2\varepsilon\frac{du}{dt} + \omega_0^2[1+\mu\xi(t)]u + \gamma u^3 = 0 , \qquad (6.64)$$

where we assume that $\varepsilon > 0$, $w_0 > 0$, $\gamma > 0$. The system is excited by stationary white noise $\xi(t)$ of intensity s. The Kolmogorov equation has the form

$$\frac{\partial p}{\partial t} = -x_2\frac{\partial p}{\partial x_1} + \frac{\partial}{\partial x_2}[(2\varepsilon x_2+\omega_0^2x_1+\gamma x_1^3)p] + \frac{1}{2}\omega_0^4\mu^2 s x_1^2\frac{\partial^2 p}{\partial x_2^2} . \qquad (6.65)$$

We shall obtain equations for moments by multiplying equation (6.65) termwise by x_1, x_2, and by their products and calculating the corresponding integrals. The first five equations have the form

$$\frac{dm_1}{dt} = m_2 , \qquad \frac{dm_2}{dt} = -(2\varepsilon m_2+\omega_0^2 m_1+\gamma m_{111}) ,$$

$$\frac{dm_{11}}{dt} = 2m_{12}, \qquad \frac{dm_{12}}{dt} = m_{22}-(2\varepsilon m_{12}+\omega_0^2 m_{11}+\gamma m_{1111}) , \qquad (6.66)$$

$$\frac{dm_{22}}{dt} = -2(2\varepsilon m_{22}+\omega_0^2 m_{12}+\gamma m_{1112})+\omega_0^4\mu^2 s m_{11} .$$

We seek a stationary solution to the system (6.66), closing it by means of the relations (5.62) and (5.63). Since in this case $m_2 = m_{12} = m_{1112} = 0$, the equations (6.66) are reduced to the form

$$\begin{aligned} &\omega_0^2 m_1 + \gamma m_1(3m_{11} - 2m_1^2) = 0 , \\ &m_{22} - \omega_0^2 m_{11} - \gamma(3m_{11}^2 - 2m_1^2) = 0 , \\ &-4\varepsilon m_{22} + \omega_0^4 \mu^2 m_{11} s = 0 . \end{aligned} \tag{6.67}$$

The system (6.67) has the trivial solution $m_1 = m_{11} = m_{22} = 0$. Consider a nontrivial solution. The first equation is satisfied identically for $m_1 = 0$, $m_{11} \neq 0$. From the two others, we obtain

$$m_{11} = \frac{\omega_0^2}{3\gamma}\left(\frac{\mu^2\omega_0^2 s}{4\varepsilon} - 1\right), \qquad m_{22} = \frac{\mu^2\omega_0^4 s}{4\varepsilon} m_{11} . \tag{6.68}$$

The solution (6.68) loses its meaning if the expression in parenthesis becomes negative, i.e., for

$$\mu^2 s < 4\varepsilon/\omega_0^2 . \tag{6.69}$$

Condition (6.69) coincides with the stability condition (5.46) of the corresponding linear system. Thus, if $\gamma > 0$, the variance of the stationary solution, which is equal to zero when condition (6.69) is satisfied, increases monotonically after passing the stability threshold.

Construction of Higher Approximations

For the approximate determination of moments of higher order, as well as for making results pertaining to lower order moments more precise, equations should be closed on a higher level. But here some substantial difficulties arise which were discussed earlier

(Section 5.4). The assumption that all the cumulant functions of an order higher than r are equal to zero is equivalent to breaking off the Taylor's series for the logarithm of the characteristic functional. It may turn out that there exist no distributions for which, with values of lower moments specified, all the cumulant functions of an order higher than r are equal to zero. Approximations based on the hypothesis of quasi-normality may be of an asymptotic character (in the sense that for each problem there exists a relatively low level of closure giving the best approximation). This is confirmed by applying the method of moment functions to parametrically excited systems. As it was shown in Section 5.5, increasing the level r does not result in greater accuracy in the region dependent on the properties of lower moments. It does, however, allow one to find new phenomena that were not described in the scope of lower approximations.

An analogous phenomenon is found in nonlinear stochastic systems. As an example, consider Duffing's stochastic system for which it is easy to obtain an exact expression for the stationary distribution of probabilities (6.19) as well as exact expressions for moments (6.21).

Figure 54 shows the results of calculations of the moment m_{11} for the case when $\varepsilon = 0.05\omega_0$ and $\gamma = 0.1\omega_0^2$. The solid line corresponds to the exact solution (6.22), and the dashed line corresponds to the solution which is obtained by statistical linearization from the condition of equality of variances. The continuous lines were obtained by the method of moment functions for various values of the closure level r. The method of statistical linearization using the condition of the minimum of the mean square deviation, as it has already been mentioned, gives the same results as the method of moment functions for $r = 2$. Note that the diagram in Figure 54 corresponds to fairly large nonlinearities: for example, for $s = \omega_0^3$ the exact value of the moment m_{11} is nearly five times smaller than the corresponding value for a linear system.

For comparison, dotted lines are plotted in Figure 54 which correspond to the closure of equations by means of the hypothesis of higher central moments being equal to zero (this hypothesis can be formalized by introducing the corresponding small parameter). Application of this hypothesis leads to unsatisfactory results, even when noise levels are not high.

Analogous graphs for the moments $m_{1(4)}$ and $m_{1(6)}$ are given in Figure 55. Here

$$m_{1(k)} = m_{\underbrace{11\cdots}_{k}} \, .$$

In order to approximately calculate the moment $m_{1(4)}$, closure should be performed at least at the level $r = 4$, and, in order to obtain the moment $m_{1(6)}$, at the level $r = 6$. As can be seen in the graphs, the method of moment functions gives satisfactory approximation for higher moments even at a significant deviation from a normal distribution.

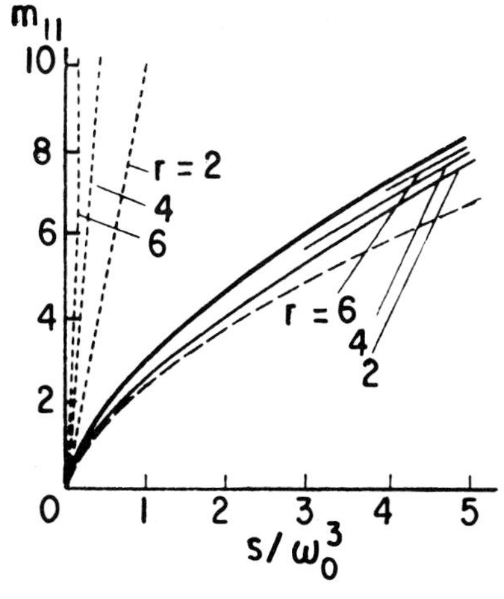

Figure 54

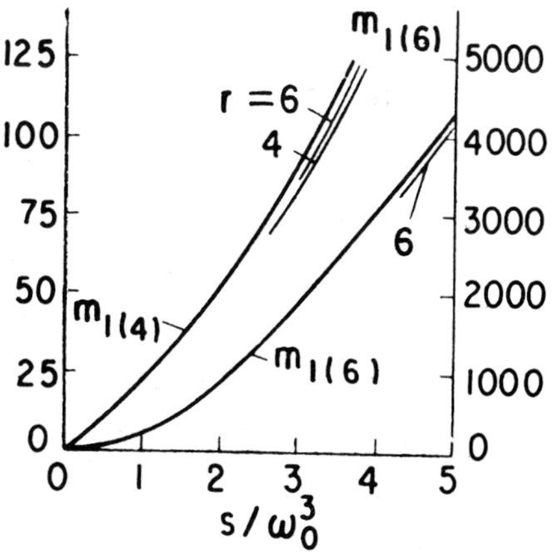

Figure 55

Application of the method for high approximations requires a lot of calculations, which makes it necessary to use a computer. For example, the dimensionality of the vector $\underset{\sim}{m}$ in the basic equation (6.56) equals 5 for the considered problems, if closure is performed at $r = 2$; equals 14, if $r = 4$; and equals 27, if $r = 6$. This implies that for the analysis of transient vibrations or of stability it is necessary to consider the system of 5, 14 and 27 simultaneous nonlinear differential equations, respectively. The analysis of probabilistic characteristics of steady vibration leads to a system of algebraic equations whose degree of the resultant is lower than the order of the system of differential equations, but is still relatively high. In this case, it is also necessary to separate all the branches of real solutions, choosing the stable branches from among them, which again requires using the system of differential equations. But, if the method of stabilization is applied for determining stationary solutions, the equations (6.56) being used as stabilization equations, the stable branches are chosen in the solution process.

A more serious verification for the method of moment functions is the problem where the corresponding autonomous systems have more than one state of stable equilibrium. Equation (6.17) with $\omega_0^2 < 0$ and $\gamma > 0$ can be taken as a model of such a system. Another example is a stochastic analog of the equation

$$\frac{d^2u}{dt^2} + 2\varepsilon \frac{du}{dt} + \omega_0^2 u + \beta u^3 + \gamma u^3 = \xi(t) \ , \qquad (6.70)$$

where a system with two stable focuses and one inflection point can be obtained by an appropriate choice of the coefficients β and γ. This is shown in Figure 56, where the graph for the potential energy $\Pi(u)$ of the system for $\gamma = 0.1\omega_0^2$ and various values of β/ω_0^2 is given.

The system (6.70) is often used as the simplest model of a "buckling" panel - a thin elastic shallow shell that can vibrate with "buckling" subjected to random fluctuations of pressure (see Section 6.5). If $\xi(t)$ is stationary normal white noise, it is not difficult to obtain a formula, analogous to (6.19), for a stationary distribution of probability. Calculated from the formula, stationary moments of the process u(t) and its first derivative will be used for an estimate of the effectiveness of the method of moment functions.

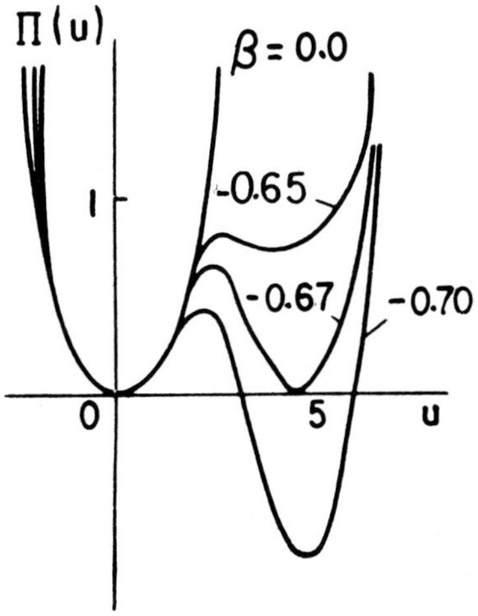

Figure 56

In Figure 57, the results of calculations for $\varepsilon = 0.05\omega_0$, $\beta = -0.65\omega_0^2$, and $\gamma = 0.1\omega_0^2$ are shown. The continuous heavy line corresponds to exact values of the moment m_{11}, and thin lines correspond to an approximation by means of the method of moments for $r = 2$, $r = 4$ and $r = 6$. In the approximation of second moments, there is a lack of uniqueness of the moment m_{11}. This also occurs when the method of statistical linearization is applied, which indicates that the method is imperfect. At sufficiently high levels of noise (for $s > 0.1\omega_0^3$ in the considered example) the dependence becomes unique again. With the increase of the closure level r, the results of calculations approach the exact solution in the wide range of ratio s/ω_0^3. This is also confirmed by the data in Figure 58, where analogous dependences for the fourth moment $m_{1(4)} = m_{1111}$ are shown.

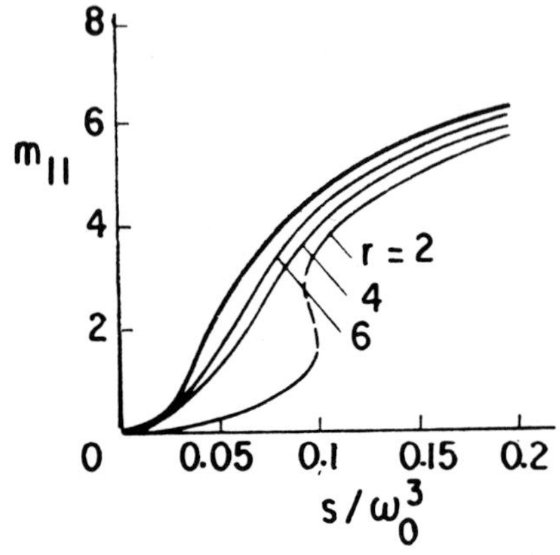

Figure 57

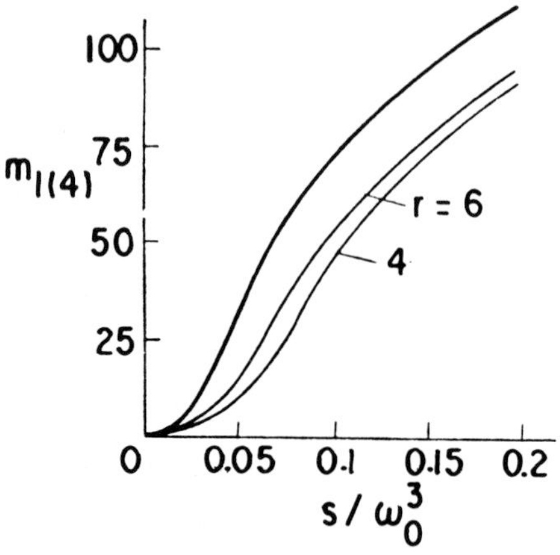

Figure 58

6.5 RANDOM VIBRATIONS OF NONLINEAR CONTINUOUS SYSTEMS

Application of the Method of Generalized Coordinates to Nonlinear Problems

Most of the approximate methods of statistic dynamics for systems with a finite number of degrees of freedom which were described above can be extended to continuous systems. An example is the method of closing the equations for moment functions by means of hypotheses about the relations between higher moment functions. Further examples are the method of a small parameter and the method of statistical linearization which also admit of an extension to problems which are described by partial differential equations.

One of the most effective approximate methods is that of approximate reduction to systems with a finite number of degrees of freedom by expanding in terms of appropriate base functions;

for example, in terms of natural modes of the corresponding linear system. Representing the solution of a nonlinear problem as a finite series, and using a variational method, we arrive at a system of ordinary differential equations for the generalized coordinates $u_k(t)$. Further investigation is carried out by means of methods in the theory of random vibrations developed for systems with a finite number of degrees of freedom.

In order to illustrate the method of generalized coordinates, we shall consider a relatively simple problem. Suppose a straight rod of constant cross-section and of linearly elastic material is vibrating under a transverse loading with amplitudes of the order of its cross-sectional height, the latter being small as compared to the rod's length. The function describing the rod's deflection will be denoted by $w(x,t)$. If the rod's ends at $x = 0$ and $x = \ell$ are fixed so as to prevent axial displacements, the axial force N appearing under vibrations is the basic nonlinear factor. This force is determined from the approximate formula

$$N \approx \frac{EF}{2\ell} \int_0^\ell \left(\frac{\partial w}{\partial x}\right)^2 dx \,, \qquad (6.71)$$

where F is the cross-section area of the rod. Substitute the expression (6.71) into the equation of flexural vibrations of the rod, complemented by the terms which take into account the force N. As a result, we obtain the equation

$$EI \frac{\partial^4 w}{\partial x^4} + m \frac{\partial^2 w}{\partial t^2} + 2m\varepsilon \frac{\partial w}{\partial t} - \frac{EF}{2\ell} \frac{\partial^2 w}{\partial x^2} \int_0^\ell \left(\frac{\partial w}{\partial x}\right)^2 dx = q \,, \qquad (6.72)$$

where I is the moment of inertia of the cross-section, m is the rod's mass per unit length, ε is the damping coefficient, and $q(x,t)$ is the intensity of the transverse loading. Boundary conditions for the function $w(x,t)$ are

$$w = \partial^2 w/\partial x^2 = 0 \,, \qquad (x = 0,\ x = \ell) \,. \qquad (6.73)$$

The solution of equation (6.72) satisfying the conditions (6.73) is sought in the form

$$w(x,t) = \sum_{k=1}^{n} u_k(t) \sin \frac{k\pi x}{\ell} . \tag{6.74}$$

Substituting this series into equation (6.72) leads, after standard calculations, to the system of ordinary nonlinear differential equations

$$\frac{d^2u_j}{dt^2} + 2\varepsilon \frac{du_j}{dt} + \omega_j^2 u_j + f_j(u_1,u_2,\ldots,u_n) = Q_j(t) , \quad (j = 1,2,\ldots,n). \tag{6.75}$$

Here, the following notations are used:

$$\omega_j = \frac{j^2\pi^2}{\ell^2}\left(\frac{EI}{m}\right)^{1/2} , \qquad Q_j(t) = \frac{2}{m\ell}\int_0^\ell q(x,t) \sin \frac{j\pi x}{\ell}\, dx , \tag{6.76}$$

$$f_j(u_1,u_2,\ldots,u_n) = \frac{j^2\pi^4 EF}{4m\ell^4} u_j \sum_{k=1}^{n} k^2 u_k^2 . \tag{6.77}$$

On the Application of Methods in the Theory of Markov Processes

Consider finite-dimensional systems that are obtained as a result of an approximation of vibration fields in continuous systems by means of expansions of the type

$$\underset{\sim}{u}(\underset{\sim}{x},t) = \sum_{k=1}^{n} u_k(t) \underset{\sim}{\phi}_k(\underset{\sim}{x}) . \tag{6.78}$$

The corresponding generalized forces are defined as

$$Q_k(t) = \frac{(\underset{\sim}{q},\underset{\sim}{\phi}_k)}{(A\underset{\sim}{\phi}_k,\underset{\sim}{\phi}_k)} . \tag{6.79}$$

For the generalized forces (6.79) to be white noises, it is necessary and sufficient for the external action $\underset{\sim}{q}(\underset{\sim}{x},t)$ to be delta-correlated in time. In this case, the evolution of the 2n-dimensional phase vector with the components $u_1(t), u_2(t), \ldots, \dot{u}_n(t)$ will be a Markov process. If the external action $q(\underset{\sim}{x},t)$ is a time-stationary process with a rational time spectral density, then, reduction to a Markov process is achieved by an appropriate extension of the phase space.

The problem of establishing conditions under which finite-dimensional systems of the type (6.75) will have the Maxwell-Boltzmann stationary distribution (6.30) requires special consideration. If for any finite n, the stationary distribution has the form (6.30), it can be expected to be fulfilled as $n \to \infty$, which would make it possible to obtain the exact solution for a continuous system as a result of the passage to this limit. To satisfy the relations (6.28) and (6.29), the following conditions are to be met:

$$\langle Q_j(t)Q_k(t')\rangle = s_j\delta(t-t')\delta_{jk} , \qquad (6.80)$$

$$s_j/\varepsilon_j = \text{const} . \qquad (6.81)$$

As follows from (6.79), (6.80) and (6.81), these conditions include certain requirements on the load $q(\underset{\sim}{x},t)$, on the natural modes $\phi_k(\underset{\sim}{x})$, and on the distribution of mass and damping in the system.

For the sake of definiteness, we assume that $q(\underset{\sim}{x},t)$ and $u(\underset{\sim}{x},t)$ are scalar functions of the coordinates $\underset{\sim}{x} \in V$, and the action of the operator A is reduced to that of a multiplication by the mass density $m(\underset{\sim}{x})$. Then, from formula (6.79), we obtain

$$\langle Q_j(t)Q_k(t')\rangle = \frac{\int_V\int_V \langle q(\underset{\sim}{x},t)q(\underset{\sim}{x}',t')\rangle\phi_j(\underset{\sim}{x})\phi_k(\underset{\sim}{x}')d\underset{\sim}{x}d\underset{\sim}{x}'}{\int_V m(\underset{\sim}{x})\phi_j^2(\underset{\sim}{x})d\underset{\sim}{x}\int_V m(\underset{\sim}{x})\phi_k^2(\underset{\sim}{x})d\underset{\sim}{x}} . \qquad (6.82)$$

In order to satisfy condition (6.80), it is sufficient to assume that

$$\langle q(\underset{\sim}{x},t)q(\underset{\sim}{x}',t')\rangle = s\delta(\underset{\sim}{x}-\underset{\sim}{x}')\delta(t-t') , \qquad (6.83)$$

i.e., that the load $\underset{\sim}{q}(x,t)$ is time and space delta-correlated.* Then,

$$s_j = s \int_V \phi_j^2(\underset{\sim}{x})d\underset{\sim}{x} \left[\int_V m(\underset{\sim}{x}) \phi_j^2(\underset{\sim}{x}) d\underset{\sim}{x} \right]^{-2} . \tag{6.84}$$

Using relations (6.81) and (6.84), it is not difficult to formulate the conditions imposed on the damping coefficients ε_j. In particular, if $s_1 = s_2 = \cdots = s_n$, condition (6.81) will be satisfied for $\varepsilon_1 = \varepsilon_2 = \cdots = \varepsilon_n$.

As an example, we again consider a rod undergoing vibrations with finite amplitudes. Equations (6.75) are reduced to the form (6.28) for

$$\Pi(\underset{\sim}{u}) = \frac{1}{2} \sum_{k=1}^{n} \omega_k^2 u_k^2 + \frac{\pi^4 EF}{8m\ell^4} \left(\sum_{k=1}^{n} k^2 u_k^2 \right)^2 . \tag{6.85}$$

Let the load q(x,t) be of the type (6.83). For $\phi_j(x) = \sin(j\pi x/\ell)$, formula (6.84) yields $s_j = 2s/(m^2\ell)$. Thus, the condition (6.81) is satisfied, which leads to the stationary distribution

$$p(\underset{\sim}{u},\underset{\sim}{\dot{u}}) = \frac{1}{J} \exp\left\{ - \frac{4\varepsilon}{s} \left[\Pi(\underset{\sim}{u}) + \frac{1}{2} \sum_{k=1}^{n} \dot{u}_k^2 \right] \right\} , \tag{6.86}$$

valid for any finite n. The expression in the square bracket of formula (6.86) is equal, accurate up to a constant, to the total mechanical energy of the system

$$V = \frac{1}{2} \int_0^\ell EI \left(\frac{\partial^2 w}{\partial x^2} \right)^2 dx + \frac{EF}{8\ell} \left[\int_0^\ell \left(\frac{\partial w}{\partial x} \right)^2 dx \right]^2 + \frac{1}{2} \int_0^\ell m \left(\frac{\partial x}{\partial t} \right)^2 dx ,$$

if, instead of the deflection w(x,t), we substitute its expansion in terms of natural modes.

---

*Bolotin, V.V., "On Stationary Distributions in Statistical Dynamics of Elastic Systems", *Coll. Works: Voprosi Dynamiki i Dinamicheskoi Prochnosti*, (Problems of Dynamics and Dynamic Strength), No. 10, Academy of Science of Latvia, Riga, 1963.

The Problem of Panel "Buckling"

A great number of publications [28,29,34,40] are devoted to the following problem. A thin elastic panel is fixed along its contour and is subjected to a normal load $q(\underset{\sim}{x},t)$ which is a random function of coordinates and time. Deflections of the panel are considered as small in comparison to the panel's dimensions, but are comparable with its thickness and rise. Such panels can have more than one stable form of equilibrium, one of which corresponds to a nonbuckled panel and at least one more corresponding to a buckled panel. If the panel is subjected to a time fluctuating load, stemming, for example, from pulsations in the turbulent boundary layer or from a source of random acoustic waves, the panel will undergo vibrations of various types: vibrations in relation to a nonbuckled form of equilibrium or about the buckled form, and containing both these states of equilibrium. In connection with this situation there arise problems of the distribution of probabilities of the amplitudes of these vibrations, of the average number of "snap-through bucklings" per unit time, of sporadic "snap-through bucklings", etc.

In the simplest form, nonlinear equations of the buckling panel read

$$D\Delta\Delta w = \kappa_2 \frac{\partial^2\psi}{\partial x_1^2} + \kappa_1 \frac{\partial^2\psi}{\partial x_2^2} + L(w,\psi) - m\frac{\partial^2 w}{\partial t^2} - 2m\varepsilon\frac{\partial w}{\partial t} + q(x_1,x_2,t) \ , \tag{6.87}$$

$$\frac{1}{Eh}\Delta\Delta\psi = -\frac{1}{2}L(w,w) - \left(\kappa_2\frac{\partial^2 w}{\partial x_1^2} + \kappa_1\frac{\partial^2 w}{\partial x_2^2}\right) .$$

Here, x_1 and x_2 are the principal curvilinear coordinates on the median surface, κ_1 and κ_2 are the principal curvatures, $w(x_1,x_2,t)$ is the deflection function, $\psi(x_1,x_2,t)$ is the force function in the median surface, and $L(w,\psi)$ is the bilinear operator

$$L(w,\psi) = \frac{\partial^2 w}{\partial x_1^2}\frac{\partial^2\psi}{\partial x_2^2} + \frac{\partial^2 w}{\partial x_2^2}\frac{\partial^2\psi}{\partial x_1^2} - 2\frac{\partial^2 w}{\partial x_1\partial x_2}\frac{\partial^2\psi}{\partial x_1\partial x_2} .$$

Other notations have the same meaning as before. For an approximate calculation, the method [27] developed by P. Papkovich is used, according to which the deflection function is represented as

$$w(x_1,x_2,t) = u(t)\phi(x_1,x_2) \ , \tag{6.88}$$

where $u(t)$ is a generalized coordinate, and the function $\phi(x_1,x_2)$ satisfies the boundary conditions given for $w(x_1,x_2,t)$. The expression (6.88) is substituted into the right side of the second equation (6.87) which is solved with tangential boundary conditions being approximately satisfied. The expression obtained from here for $\psi(x_1,x_2,t)$, together with (6.88), is substituted into the first equation, and after this, the Bubnov-Galerkin method is used. As a result, the continuous system is replaced in the first approximation by a system with one degree of freedom, the equation for the generalized coordinate $u(t)$ coinciding in form with equation (6.70). The polynomial nonlinearity of this equation allows one to describe the phenomenon of random "snap-through buckling" of a panel under the action of a fluctuating normal load.

Chapter 7
Reliability and Longevity under Random Vibrations

7.1 BASIC CONCEPTS OF RELIABILITY THEORY

Introduction

Methods for stress analysis and for the analysis of deformations and displacements in mechanical systems are given by the theory of vibrations and by mechanics of deformable bodies. The final judgement regarding the suitability of a system under design or a system in use should be based on the theory of reliability. From the point of view of mechanics that part of reliability theory is of greatest interest in which failure is treated as the result of time changes in the system's parameters (general, physical or parametric theory of reliability). To a great extent, this theory is based on the analysis of fluctuations of random processes and fields describing the system's behaviour.

Failure is defined as complete or partial loss of the ability of the system to fulfil its functions. Failure can be caused either by the development of defects which have already been in the system by the time it came into use, or by accumulation of damages and irreversible changes occurring in the process of the system's operation. The initial distribution of defects, operating conditions, and interaction of the system with the environment are generally of random nature. Therefore, failures should be considered as random phenomena.

In the theory of reliability it is customary to distinguish between sudden failures and gradual ones. This distinction depends primarily on a fundamental understanding and description of the phenomena leading to one or another class of failures. If the description of processes resulting in failure presents difficultie[s] or is not expedient, the notion of sudden failure is used. In this case, an empirical-statistical approach is used for the analysis of reliability, the reliability characteristics being found from mass reliability tests. Mass tests of mechanical systems (structures, mechanisms and their members) are, as a rule, unacceptable from an economical point of view and are often unrealizable. On the other hand, the behaviour of these systems during their interaction with the environment can be analyzed with sufficient accuracy and in detail. Due to this, it is expedient to include interactions between a mechanical system and the environment into the general scheme of reliability theory [13,15].

Consider the behaviour of some system under external actions. The equation of the system will be taken in the form

$$L\underset{\sim}{u} = \underset{\sim}{q} \,, \tag{7.1}$$

where $\underset{\sim}{q}$ is an element of the input parameter space $\underline{Q}$, $\underset{\sim}{u}$ is an element of the output parameter space $\underline{U}$ and L is the operator of the system. The space $\underline{U}$ is chosen with a view to characterizing, by its elements $\underset{\sim}{u}$, any state of the system, and is called the state space. The time evolution of states for $0 \leq t < \infty$ is described by the functions $\underset{\sim}{u}(t)$, the trajectories in the state space $\underline{U}$ being their geometrical image.

Introduce the space $\underline{V}$ describing the quality of the system. Let the element $\underset{\sim}{v}(t)$ correspond to a property of the system, the time t playing the role of a parameter. To each trajectory $\underset{\sim}{u}(t)$ in the space $\underline{U}$ there corresponds a certain trajectory $\underset{\sim}{v}(t)$ in the quality space $\underline{V}$. The relation between the elements of these spaces and their trajectories is given by the operator equation

$$\underset{\sim}{v} = M\underset{\sim}{u} \ . \qquad (7.2)$$

The operator M can be, in particular, an identity operator. In some cases, the space $\underline{V}$ turns out to be a subspace of $\underline{U}$.

The set of the system's states which are admissible from the point of view of quality forms a region of admissible states Ω in the quality space $\underline{V}$. Henceforth, the region Ω is assumed to be an open set. The boundary of region Ω corresponds to limit states. This boundary will be referred to as the limit surface and will be designated by Γ. If $\underset{\sim}{v} \in \Omega$, it implies that the parameters of the system's quality remain within the specified tolerances. The first crossing of the limit surface Γ by the vector $\underset{\sim}{v}(t)$ in the direction of the external normal (positive crossing of the limit surface Γ, excursion from the region Ω of admissible states) corresponds to the failure of the system.

The notions introduced become especially clear if the system is finite-dimensional. In this case, the spaces $\underline{Q}$, $\underline{U}$ and $\underline{V}$ are Euclidean spaces. Figure 59 shows the trajectories q(t), $\underset{\sim}{u}(t)$ and $\underset{\sim}{v}(t)$ for the case when the spaces $\underline{Q}$, $\underline{U}$ and $\underline{V}$ are three-dimensional Euclidean spaces.

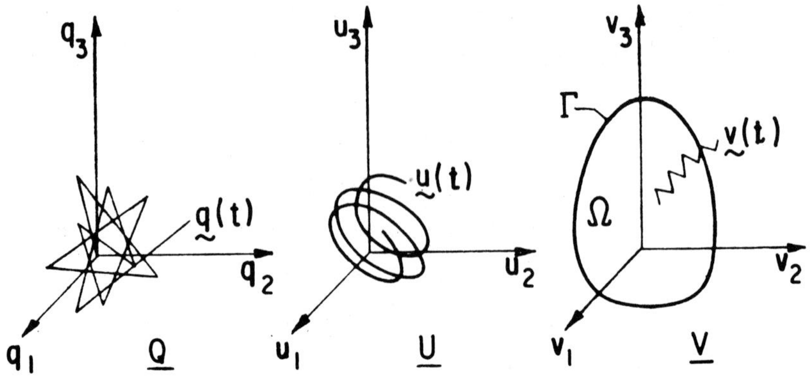

Figure 59

Reliability Function

Let the external action $\tilde{q}(t)$ and/or the operator L of the system be stochastic. In this case, the trajectories $\tilde{v}(t)$ in the quality space $\underline{V}$ will also be stochastic, and failure will be a random event. The reliability function is defined as the probability of failure-free operation within the interval [0,t], i.e., as the probability of the vector $\tilde{v}(\tau)$ being in the admissible region Ω during this time interval:

$$R(t) = \tilde{P}\{\tilde{v}(\tau) \in \Omega;\ \tau \in [0,t]\} . \qquad (7.3)$$

The definition of the reliability function in the form (7.3) is easily extended to the case when repeated failures are admissible, i.e., when provision is made for repair, reconstruction, etc. The parameter t may represent not only the physical time, but also the operating time, the number of production cycles, the amount of output, or other parameters suitable for systems of the given type.

The region of admissible states can be stochastic. For instance, it can vary stochastically when passing from one element of a set of systems to another. If the stochastic properties of a system and the external action to which it is subjected are characterized by a finite number of stochastic parameters, it then becomes expedient to determine the reliability function in two stages. In the first stage, the reliability function is constructed for the system with fixed parameters. This function is actually the probability of the system staying within the admissible region, provided the parameters of the system $\underset{\sim}{r}$ and the parameters of the external action $\underset{\sim}{s}$ are fixed:

$$R(t|\underset{\sim}{r},\underset{\sim}{s}) = P\{\underset{\sim}{v}(\tau|\underset{\sim}{r},\underset{\sim}{s}) \in \Omega(r);\ \tau \in [0,t]\} . \qquad (7.4)$$

By analogy with conditional probability, we shall refer to the function $R(t|\underset{\sim}{r},\underset{\sim}{s})$ as the conditional reliability function. In the second stage, the formula of total probability is used, and the probability of failure-free operation for a randomly chosen system of a given set and for a randomly chosen action is calculated:

$$R(t) = \iint R(t|\underset{\sim}{r},\underset{\sim}{s})p(\underset{\sim}{r},\underset{\sim}{s})d\underset{\sim}{r}d\underset{\sim}{s} . \qquad (7.5)$$

Here, $p(\underset{\sim}{r},\underset{\sim}{s})$ is a joint probability density for the parameters $\underset{\sim}{r}$ and $\underset{\sim}{s}$.

The condition of reliability of the system has the form

$$R(t) \geq R_*(t) , \qquad t \in [0,T_*] , \qquad (7.6)$$

where $R_*(t)$ is a normative reliability function, and T_* is a normative lifetime. As a rule, condition (7.6) is made for $t = T_*$. Designating $R_*(T_*) \equiv R_*$, we shall rewrite (7.6) in the form

$$R(T_*) \geq R_* . \qquad (7.7)$$

For "mass-produced" systems, reliability indices admit of a statistical interpretation, and normative values can be determined by a statistical analysis of the operating experience with respect to analogous structures and mechanisms, or by some other statistical considerations. Another method of obtaining normative values is a probability-optimization approach, the normative reliability indices being determined from the condition of minimum of the mathematical expectation of a certain function of losses related to failure. If the system is unique, as in the case of very large structures and constructions, statistical interpretation is meaningless. However, reliability indices are still important characteristics of the quality of the system and can be used for comparing different engineering solutions or for the optimization of parameters. So, the best engineering solution for a system of vibration protection is the one (other conditions being equal or comparable) which gives maximum reliability. Thus, we arrive at the criterion of maximum reliability (Section 7.5).

Other Characteristics of Reliability

The reliability function is the basic reliability characteristic determining the system's ability to operate without failures within a specified time interval. Other reliability indices are derived from the reliability function as follows:

(1) probability of at least one failure occurring in the interval $[0,t]$

$$Q(t) = 1 - R(t) \; ; \tag{7.8}$$

(2) probability density of time intervals before the first failure

$$f(t) = -R'(t) \; ; \tag{7.9}$$

(3) rate of failures (probability density of time intervals before the first failure on a set of systems which have not failed before the instant t)

$$\nu(t) = -R'(t)/R(t) . \qquad (7.10)$$

The relationship between the reliability function and the rate of failures is given by the formula

$$R(t) = R(0) \exp\left[- \int_0^t \nu(\tau)d\tau \right] . \qquad (7.11)$$

T, the time before the first failure (lifetime, resource), is a stochastic value with the probability density (7.9) and the probability distribution function

$$F(T) = 1 - R(T) . \qquad (7.12)$$

The average lifetime (average resource) is determined as

$$\langle T \rangle = \int_0^\infty TF'(T)dT = \int_0^\infty R(t)dt . \qquad (7.13)$$

Thus, the characteristics of longevity are contained in the reliability function R(t). Taking into account repairs, reconstructions, regular inspections, etc. requires the introduction of special characteristics of longevity, consideration of which is beyond the scope of this book [33].

7.2 STOCHASTIC MODELS OF FAILURES

Elementary Models of Failures

The properties of the stochastic process $\underset{\sim}{v}(t)$ - stationarity or nonstationarity, relation to pre-history, ergodicity or nonergodicity, special distribution characteristics (for example - normality of the process or the absence of it), etc. can serve as a basis for the classification of stochastic models. Further below, we shall consider some special stochastic models that are the most suitable for describing failures of mechanical systems and for reliability prediction at the design stage. Quality spaces for the analysis of structures, mechanisms and construction are multidimensional, and admissible regions can have a complicated configuration. From this point of view, it may be of interest to develop stochastic models of failures that allow one to obtain an exact or at least approximate solution to the problem that would show the whole complexity of interaction between the systems and their environment.

In the first papers on reliability of structures, models based on the concepts of elementary probability theory were widely used. As an example, we can take a simple model where the structural strength is determined by the relation between two stochastic values - the resistance r and the load s. The admissible region is specified as

$$\Omega = \{r, s: \quad s < r\} . \tag{7.14}$$

If the joint probability density p(r,s) is known, the system's reliability is calculated as

$$R = \int\int_{s<r} p(r,s)\,dr\,ds . \tag{7.15}$$

This model admits of various generalizations, even if considerations of the theory of stochastic processes are not involved. Suppose, for example, that loading is performed at the times $t_1, t_2, \ldots, t_n$, with the distributions of the load parameter q(t) for various loadings being independent and, generally speaking, different. The resistance r is generally a slowly varying stochastic function of time. The probability of failure for the k-th loading is determined by a formula of the type (7.15), say

$$Q_k = \underset{\sim}{P}\{q(t_k) > r(t_k)\} = \int\int_{q>r} p(r,q;t_k)\,dr\,dq \ . \qquad (7.16)$$

Introduction of multidimensional quality spaces opens up another way for generalization. Their components can be both loads and resistance, and the differences of loads and respective resistance. And, finally, the calculation of reliability in the case of models with single loadings reduces to using the formula

$$R = \iint_{\Omega} p(\underset{\sim}{r},\underset{\sim}{s})\,d\underset{\sim}{r}\,d\underset{\sim}{s} \qquad (7.17)$$

or its extension of the type (7.16) in the case of multiple loadings.

Markov Models

Suppose the system's evolution in the quality space $\underline{V}$ is an n-dimensional Markov diffusion process. At the initial time t_0, the system is at the point $\underset{\sim}{v}_0$. The behaviour of the system for $t > t_0$ is described by the transition probability density $p(\underset{\sim}{v},t|\underset{\sim}{v}_0,t_0)$. This density as a function of the variables $\underset{\sim}{v}$, t satisfies the forward Kolmogorov equation (2.94). If we consider $p(\underset{\sim}{v},t|\underset{\sim}{v}_0,t_0)$ as a function of the variables $\underset{\sim}{v}_0$, t_0, then, instead of (2.94), we obtain Kolmogorov's conjugate (backward) equation (2.95).

Henceforth, we shall only consider systems whose intensities χ_α and $\chi_{\alpha\beta}$ are explicitly independent of time, the transient probability density $p(\underset{\sim}{v},t|\underset{\sim}{v}_0,t_0)$ depending on the difference $t-t_0$. As a result, equation (2.95) takes the form

$$\frac{\partial p}{\partial t} = \Lambda_0^* p \ , \qquad (7.18)$$

where the operator Λ_0^* is given by the formula

$$\Lambda_0^* = \sum_{\alpha=1}^{n} \chi_\alpha(\underset{\sim}{v}_0)\frac{\partial}{\partial v_\alpha^0} + \frac{1}{2}\sum_{\alpha=1}^{n}\sum_{\beta=1}^{n}\chi_{\alpha\beta}(\underset{\sim}{v}_0)\frac{\partial^2}{\partial v_\alpha^0 \partial v_\beta^0} \ . \qquad (7.19)$$

The solution to this equation is sought for the initial condition

$$p = \delta(\underset{\sim}{v}-\underset{\sim}{v}_0) \text{ for } t = t_0 \ .$$

Consider the conditional reliability function $R(t|\underset{\sim}{v}_0,t_0)$ as the probability that the system is at the point $\underset{\sim}{v}_0 \in \Omega$ when $t = t_0$ and will not go beyond the boundary Γ of the region Ω within the time interval $(t_0,t]$:

$$R(t|\underset{\sim}{v}_0,t_0) = P\{\underset{\sim}{v}(\tau) \in \Omega;\ \tau \in (t_0,t]\,|\,\underset{\sim}{v}(t_0) = \underset{\sim}{v}_0 \in \Omega\} \ . \tag{7.20}$$

In order to derive the differential equation which the function $R(t|\underset{\sim}{v}_0,t_0)$ should satisfy, note that the transition probability density $p_\Gamma(\underset{\sim}{v},t|\underset{\sim}{v}_0,t_0)$ defined over a set of realizations of the process $\underset{\sim}{v}(t)$ which are contained by the boundary Γ also satisfies the equation (7.18). The boundary conditions for $p_\Gamma(\underset{\sim}{v},t|\underset{\sim}{v}_0,t_0)$ are given as $p = 0$ for $\underset{\sim}{v} \in \Gamma$, $t \geq t_0$. Then

$$R(t|\underset{\sim}{v}_0,t_0) = \int_\Omega p_\Gamma(\underset{\sim}{v},t|\underset{\sim}{v}_0,t_0)d\underset{\sim}{v} \ . \tag{7.21}$$

Integrating equation (7.18) for the function $p_\Gamma(\underset{\sim}{v},t|\underset{\sim}{v}_0,t_0)$ with respect to $\underset{\sim}{v} \in \Omega$, we find that the conditional reliability function also satisfies the equation (7.18), i.e.,

$$\frac{\partial R}{\partial t} = \Lambda_0^* R \tag{7.22}$$

with the initial condition

$$R = 1 \ , \qquad (t = t_0,\ \underset{\sim}{v}_0 \in \Omega) \tag{7.23}$$

and with the boundary condition

$$R = 0 \ , \qquad (t \geq t_0,\ \underset{\sim}{v}_0 \in \Gamma) \ . \tag{7.24}$$

For many applications it is necessary to distinguish failures of various types. These correspond to excursions of the process $\underset{\sim}{v}(t)$ from the region Ω through different parts of the surface Γ. The probability of failure of type α in the time interval $(t_0,t]$ is introduced as

$$Q_\alpha(t|\underset{\sim}{v}_0,t_0) = P\{\underset{\sim}{v}(\tau)\uparrow\Gamma_\alpha;\ \tau \in (t_0,t]|\underset{\sim}{v}(t_0) = \underset{\sim}{v}_0 \in \Omega\} , \tag{7.25}$$

where the symbol $\underset{\sim}{v}\uparrow\Gamma_\alpha$ stands for the process $\underset{\sim}{v}(t)$ leaving the region Ω through a part of the surface Γ_α. The probability of failure $Q_\alpha(t|\underset{\sim}{v}_0,t_0)$ as a function of the variables $\underset{\sim}{v}_0$, t_0 also satisfies equation (7.18), i.e.,

$$\frac{\partial Q_\alpha}{\partial t} = \Lambda_0^* Q_\alpha . \tag{7.26}$$

The initial condition has the form

$$Q_\alpha = 0 , \qquad (t = t_0,\ \underset{\sim}{v}_0 \in \Omega) , \tag{7.27}$$

and the boundary conditions on Γ_α and on the remaining part of the surface Γ are different:

$$\begin{aligned} Q_\alpha &= 1 , \qquad (t \geq t_0,\ \underset{\sim}{v}_0 \in \Gamma_\alpha) , \\ Q_\alpha &= 0 , \qquad (t \geq t_0,\ \underset{\sim}{v}_0 \in \Gamma\backslash\Gamma_\alpha) . \end{aligned} \tag{7.28}$$

For a large dimensionality of the space $\underline{V}$, the solution of the boundary value problems (7.22) and (7.26) presents considerable difficulties. It will be less difficult if, instead of a calculation of the conditional reliability function, we consider the problem of determining those times at which the boundary Γ is reached. The time required to reach the boundary Γ is a stochastic value with the conditional probability density

$$p(T|\underset{\sim}{v}_0,t_0) = - \left.\frac{\partial R(t|\underset{\sim}{v}_0,t_0)}{\partial t}\right|_{t=T} . \qquad (7.29)$$

The moments of this stochastic value are introduced as

$$T_k(\underset{\sim}{v}_0,t_0) = \int_{t_0}^{\infty} p(T|\underset{\sim}{v}_0,t_0)T^k dT , \qquad (7.30)$$

and satisfy the sequence of equations

$$\Lambda_0^* T_k = -kT_{k-1} , \qquad (k = 1,2,\dots) , \qquad (7.31)$$

for which it is necessary to assume that $T_0 = 1$. Equation (7.31) for $k = 1$ determines the mathematical expectation of the time at which the boundary is reached. It was first determined by L. Pontriagin [2]. Equations (7.31) for the nonsingular matrix of the diffusion coefficients $\chi_{\alpha\beta}$ are of the elliptical type. The boundary conditions for the time at which the boundary is reached have the form

$$T_k(\underset{\sim}{v}_0,t_0) = 0 , \qquad (\underset{\sim}{v}_0 \in \Gamma;\ k = 1,2,\dots) . \qquad (7.32)$$

In many cases, when the process $\underset{\sim}{v}(t)$ is not a Markov process, it is possible to increase the dimensionality of the quality space $\underline{V} = \underline{R}^n$ in such a way that the process in the extended phase space $\underline{R}^{n+m}$ becomes a Markov process. The technique described is also fully valid for an extended phase space, the only difference being that the region of admissible states is taken as the cylinder $\Omega \times \underline{R}^m$.

Using a simple form of boundary conditions (7.24), (7.28) and (7.32), we can consider the following method for an approximate calculation of the reliability function, the failure probabilit and the time to the first failure. The solution is sought as a series in terms of coordinate functions $\phi_j(\underset{\sim}{v}_0)$ of the components of the vector $\underset{\sim}{v}_0$ which satisfy the above mentioned boundary conditions. The coefficients of these series are functions of time or, if moments of time to the first failure are sought, certain numerical coefficients. Applying Bubnov-Galerkin's variational method to the equations (7.22), (7.26) and (7.31), we obtain a system of ordinary linear differential equations for the desired functions or a system of linear algebraic equations for the desired coefficients. This method is especially effective if the region has a simple geometric form, for instance, that of a rectangular parallelepiped or a sphere. Examples of applications of this method can be found in [14].

Poisson Models

Consider the sequence of times $t_1, t_2, \ldots$ at which random events occur. This sequence is called a flow of events. Let the flow be ordinary in the sense that the probability of the occurrence of an event in an infinitely small time interval is proportional to the interval's length, and the probability of the repeated occurrence of the event in this interval has a higher order of smallness. If the flow is homogeneous, and the occurrence of events in two nonoverlapping time intervals is stochastically independent, the probability of occurrence of k events in the interval [0,t] follows Poisson's distribution

$$P_k(t) = \frac{(\nu t)^k}{k!} e^{-\nu t} , \qquad (7.33)$$

where the constant $\nu > 0$ is called the rate of the flow of events. Flows following the distribution (7.33) are called Poisson flows.

The model of a Poisson flow is widely used in statistical reliability theory, the probability of failure-free operation, i.e., the reliability function R(t), following from (7.33) with $k = 0$ as:

$$R(t) = e^{-\nu t} . \qquad (7.34)$$

Formula (7.34) describes the exponential reliability law.

The natural generalization of a Poisson flow is the following: the flow is assumed to be ordinary, as before, and, instead of assuming independence of the occurrence of failures, a weaker hypothesis of failures as rare events is used. Then, instead of (7.34), we obtain the approximate formula

$$R(t) \approx \exp[-\langle N(t)\rangle] \; . \tag{7.35}$$

Here, $N(t)$ is a random variable equal to the number of the excursions of process $\underset{\sim}{v}(t)$ from the region Ω in the time interval $[0,t]$. This formula can be written in another form:

$$R(t) \approx \exp\left[-\int_0^t \nu(\tau)d\tau\right] , \tag{7.36}$$

where $\nu(t)$ is the mathematical expectation of the number of excursions from the region Ω per unit time. Comparing (7.34) with (7.36), we see that $\nu(t)$ can be interpreted as the rate of failures. Henceforth, all models of failures based on representing the reliability function in terms of the number of excursions $N(t)$ will be referred to as models of the Poisson type.

Cumulative Models

Most failures of structures and mechanisms are caused by a gradual accumulation of damage in the system: plastic deformation, fatigue damage, wear, etc. Mathematically this can be represented by cumulative models of failures that describe a quasi-monotonic deterioriation of the parameters of the system's quality which takes place during the system's operation and interaction with the environment [13].

The concept of a cumulative model will be introduced in the following way. Let the admissible region Ω be a bounded, convex set in the space $\underline{V}$, so that the condition $c\underset{\sim}{v}_1+(1-c)\underset{\sim}{v}_2 \in \Omega$ is satisfied for any two elements $\underset{\sim}{v}_1$, $\underset{\sim}{v}_2 \in \Omega$ and with $0 < c < 1$. Let the point $\underset{\sim}{v}_0 = 0$ belong to the region Ω. In this case, the condition $\underset{\sim}{v} \in \Omega$ can be represented as

$$||\underset{\sim}{v}|| < 1 , \tag{7.37}$$

where $||\underset{\sim}{v}||$ is a properly chosen norm in the region $\underline{V}$. For example, the following method for constructing the norm $||\underset{\sim}{v}||$ can be used. Let the straight line drawn from the point $\underset{\sim}{v}_0 = 0$ through the point $\underset{\sim}{v} \neq 0$ intersect the surface Γ at the point $\underset{\sim}{v}_\Gamma$. Then,

$$||\underset{\sim}{v}|| = |\underset{\sim}{v}|/|\underset{\sim}{v}_\Gamma| .$$

For an admissible region in the form of a symmetric rectangular parallelepiped $|v_j| < v_{j*}$, $(j = 1,\ldots,n)$, we have

$$||\underset{\sim}{v}|| = \max_{1\leq j\leq n} \{|v_1|/v_{1*},\ldots,|v_n|/v_{n*}\} . \tag{7.38}$$

For a symmetric and symmetrically arranged ellipsoid with the semi-axes v_{j*}, we obtain analogously

$$||\underset{\sim}{v}|| = [(v_1/v_{1*})^2 + \cdots + (v_n/v_{n*})^2]^{1/2} ,$$

etc.

Now let us consider the random process $\underset{\sim}{v}(t)$ specified in the time interval T. This process will be referred to as cumulative in the interval T, provided the condition

$$||\underset{\sim}{v}(t_2)|| \geq ||\underset{\sim}{v}(t_1)|| , \qquad (t_2 > t_1) , \tag{7.39}$$

is satisfied for any t_1, $t_2 \in T$.

The basic characteristic of cumulative models follows from condition (7.39). According to this condition, the vector of the system's quality quasimonotonically approaches the boundary of the admissible region, so that the probability of its staying in the admissible region in any interval $[t_1, t_2]$ coincides with the probability of its staying in this region at the moment $t_2 > t_1$. Hence, the reliability function reads

$$R(t) = P\{\underset{\sim}{v}(\tau) \in \Omega;\ \tau \in [0,t]\} \equiv P\{\underset{\sim}{v}(t) \in \Omega\} . \qquad (7.40)$$

Therefore, determination of the characteristics of the reliability and of longevity turns out to be quite simple for many types of cumulative models. However, generally speaking, the use of cumulative models does not completely solve the problem of finding the probability of the first excursion. Suppose, for example, that the process of loading is nonmonotonic, and the value of the residual plastic deformation is taken as the measure of damage $\underset{\sim}{v}(t)$. This value depends substantially on the extremal values of the loading process $\underset{\sim}{q}(t)$ which are realized in the interval $[0,t]$. Thus, the processes $\underset{\sim}{q}(t)$ and $\underset{\sim}{v}(t)$ have to be analyzed in the whole interval in order to find the distribution of the damage measure at the end of this interval.

Further simplifications can be obtained by using the additive nature of the process $\underset{\sim}{v}(t)$. If the conditions of the generalized central limit theorem[*] for the process $\underset{\sim}{v}(t)$ are satisfied, the values of the damage measure are distributed normally asymptotically. Then, the value of the reliability function (7.40) is fully determined by the values of the mathematical expectation $<\underset{\sim}{v}(t)>$ and the correlation matrix of the process $K_v(t,t')$. The application of the central limit theorem will be given in Section 7.8.

───────────

[*] Ibragimov, I.A. and Linnik, Yu.V., *Independent and Stationarily Coupled Quantities*, Moscow, Nauka, 1965.

7.3 APPROXIMATE ESTIMATES FOR RELIABILITY FUNCTIONS

Two-Sided Estimates

Formulae (7.35) and (7.36) for Poisson's models of failure relate the reliability function R(t) to the mathematical expectation of the number of excursions from the region Ω in the specified time interval. Mainly following references [13,19], we shall give some other estimates of the reliability function in terms of moments of the number of excursions.

Let the system at t = 0 be with a probability equal to unity in the admissible region Ω. Let the flow of crossings of the limit surface Γ by the process $\underset{\sim}{v}(t)$ be ordinary. By $Q_1(t)$, $Q_2(t), \ldots$, we will designate the probabilities of single, double, etc. crossings of the surface Γ in the direction of the external normal to this surface in the time interval [0,t]. The reliability function (7.3) is expressed in terms of the probabilities $Q_k(t)$ in the following way:

$$R(t) = 1 - \sum_{k=1}^{\infty} Q_k(t) . \tag{7.41}$$

Along with the reliability function, we shall also consider moments of the number of positive crossings of the surface Γ, i.e.,

$$\langle N^{\alpha}(t)\rangle = \sum_{k=1}^{\infty} k^{\alpha} Q_k(t) . \tag{7.42}$$

Combining the relations (7.41) and (7.42), we obtain the following result: if the flow of crossings of the limit surface is ordinary, and R(0) = 1, the reliability function for t > 0 satisfies the inequalities

$$1 - \langle N(t)\rangle \leq R(t) \leq 1 - \frac{3}{2}\langle N(t)\rangle + \frac{1}{2}\langle N^2(t)\rangle . \tag{7.43}$$

To apply the upper estimate (7.43), it is necessary to know not only the mathematical expectation $\langle N(t)\rangle$, but also the mean square of the number of excursions. For example [95], if $v(t)$ is a one-dimensional random process, and the region Ω is given by the inequality $v < v_*$, where v_* is a nonrandom number, then

$$\langle N(t)\rangle = \int_0^t \int_0^\infty \dot{v} p(v_*, \dot{v}; \tau) d\dot{v} d\tau \, , \tag{7.44}$$

$$\langle N^2(t)\rangle = \int_0^t \int_0^t \int_0^\infty \int_0^\infty \dot{v}_1 \dot{v}_2 p(v_*, v_*, \dot{v}_1, \dot{v}_2; \tau_1, \tau_2) d\dot{v}_1 d\dot{v}_2 d\tau_1 d\tau_2 \, . \tag{7.45}$$

Here, $p(v,\dot{v};\tau)$ is the joint probability density of the process $v(t)$ and its first derivative $\dot{v}(t)$ at the coinciding time moments τ; and $p(v_1,v_2,\dot{v}_1,\dot{v}_2;\tau_1,\tau_2)$ is the joint probability density at two noncoinciding time moments τ_1 and τ_2. The estimates (7.43) can be improved by considering moments of higher order of the number of crossings. For example, by considering moments up to the fourth order, inclusive*

$$1 - \frac{11}{6}\langle N(t)\rangle + \langle N^2(t)\rangle - \frac{1}{6}\langle N^3(t)\rangle \le R(t) \le 1 - \frac{25}{12}\langle N(t)\rangle$$

$$+ \frac{35}{24}\langle N^2(t)\rangle - \frac{5}{12}\langle N^3(t)\rangle + \frac{1}{24}\langle N^4(t)\rangle \, .$$

The use of the formulae (7.44), (7.45), etc. requires a great deal of calculations. Figure 60 shows the results of computer calculations for the case when $v(t)$ is a one-dimensional stationary centred Gaussian process with the correlation function (1.18). The regions given by the inequalities (7.43) are hatched. The dashed lines are plotted according to formula (7.35). It should be noted that, for a stationary process $R(0) = 1 - F(v_*) \neq 1$,

---

* Bolotin, V.V., "Problems of Reliability Theory in Mechanics of Solids", *Coll. Works: Mekhanika Sploshnoi Sredi*, (Mechanics of Continua and Related Problems of Analysis), Nauka, Moscow, 1972.

where F(v) is the one-dimensional probability distribution function of the process v(t). Therefore, in this case, the inequalities (7.43) lose their exact meaning. They can be used as approximate (rough) estimates at such t for which the condition $1-R(t) >> F(v_*)$ is satisfied.

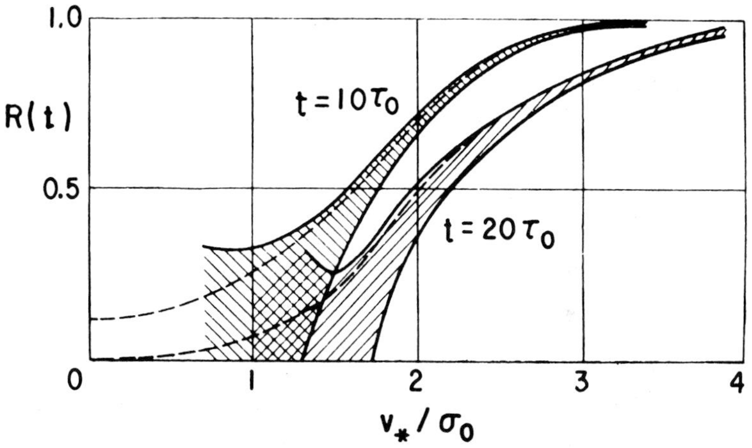

Figure 60

Approximate Estimates for Highly Reliable Systems

Suppose excursions from admissible regions are rare events, i.e., for all the considered $t > 0$,

$$\langle N(t)\rangle << 1 \, , \tag{7.46}$$

the moments of higher powers of the number of fluctuations being negligibly small as compared to lower moments. Systems satisfying these conditions will be referred to as highly reliable. From formulae (7.41) and (7.42) we will derive approximate relations expressing the function R(t) for highly reliable systems in terms of moments of the number of excursions. Thus, breaking off the series at k = 1, we obtain the approximate formula [7]:

$$R(t) \approx 1 - \langle N(t) \rangle \ . \tag{7.47}$$

This formula gives values coinciding with the lower bound (7.43). The formula

$$R(t) \approx 1 - \frac{3}{2} \langle N(t) \rangle + \frac{1}{2} \langle N^2(t) \rangle \ , \tag{7.48}$$

which is derived from relations (7.41) and (7.42) if the series are broken off at $k = 2$, can be expected to give a smaller error.

Another method of obtaining an approximate estimate for $R(t)$ is based on the hypothesis that the flow of crossings of the limit surface Γ is of the Poisson type. In this case, the reliability function is determined from formula (7.35), and the moments of the first and second order of the number of fluctuations are related by the expression

$$\langle N^2(t) \rangle = \langle N(t) \rangle + \langle N(t) \rangle^2 \ . \tag{7.49}$$

Taking formulae (7.35) and (7.49) into account, we find that, in the case of a Poisson flow of crossings, the estimate (7.48) gives the values of the first three terms of the series obtained by expanding the reliability function (7.35) into a power series. The linear estimate (7.47) corresponds to the retention of two terms of the series, and it is somewhat more convenient for calculations than the exponential estimate (7.35). In the general case of a non-Poisson flow of crossings, the estimate (7.47) is conservative (in the reliability margin).

An experimental verification of formula (7.47) is given in [57]. This verification was carried out by the method of electronic simulation on an installation consisting of a noise generator, filters and counters of the number of crossings. Figure 61 shows the sampling values of the reliability function for the stationary centred quasi-Gaussian random process with spectral density

$$S(\omega) = \begin{cases} \dfrac{\sigma_0^2}{\pi} \dfrac{\alpha}{\omega^2+\alpha^2} & (|\omega| \leq \omega_c) , \\ 0 & (|\omega| > \omega_c) . \end{cases} \tag{7.50}$$

Here, $\alpha > 0$, and ω_c is the cut-off frequency. In the simulation, the values $\alpha = 1s^{-1}$, $\omega_c = 30s^{-1}$ were assumed. For these numerical values, it is possible to approximately assume that the coefficient σ_0^2 in formula (7.50) coincides with the variance of the process $v(t)$. The region Ω is taken in the form $v < v_*$. Numbers at the curves designate the value of the ratio v_*/σ_0. Straight lines in Figure 61 correspond to the approximate estimate

$$R(t) \approx 1 - \nu t , \tag{7.51}$$

following from (7.74), and dashed lines correspond to the expo-ential estimate (7.34). Here, ν is the mathematical expectation of the number of fluctuations per unit time of the process $v(t)$ over the level v_*. The values ν were calculated from a formula of the type (7.44), i.e.,

$$\nu = \int_0^\infty \dot{v} p(v_*, \dot{v}) d\dot{v} . \tag{7.52}$$

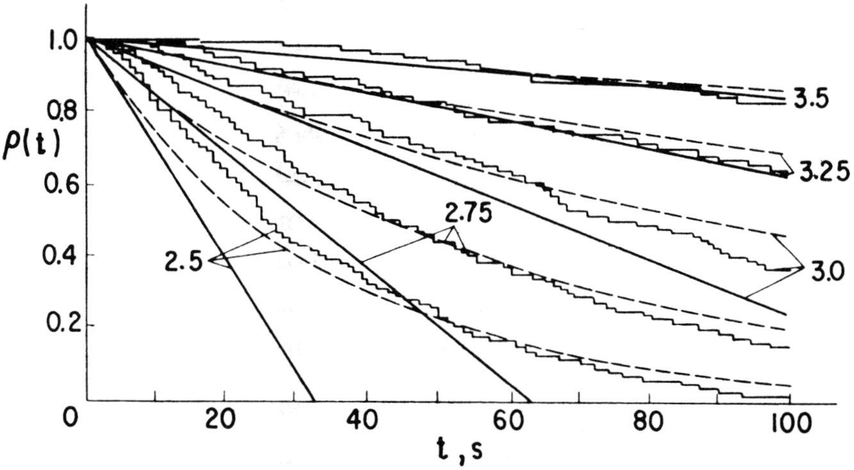

Figure 61

As can be seen from Figure 61, the estimate (7.51) gives good results in the region of rare fluctuations. In one case, the left inequality in (7.43) is violated. However, this inequality is formulated for probabilities, not for sampling characteristics. Note that estimates of the type (7.35), (7.47) and (7.48) are naturally extended to the case $R(0) \neq 1$. For example, instead of (7.35), we have

$$R(t) \approx R(0) \exp[-\langle N(t|\Omega)\rangle] ,$$

where the number of excursions $N(t|\Omega)$ is determined on a set of trajectories satisfying the condition $v(0) \in \Omega$.

Calculation of the Mathematical Expectation of the Number of Excursions

Below, some formulae for the mathematical expectation ν of the number of excursions from the admissible region Ω per unit time are given without derivation. Details can be found in [19].

Let $v(t)$ be a stationary normal process with the mathematical expectation a, variance σ^2, and spectral density $S(\omega)$. The region of admissible states is given by the inequality $v < v_*$. Application of the formula (7.52) gives

$$\nu = \frac{\omega_e}{2\pi} \exp\left[- \frac{(v_*-a)^2}{2\sigma^2}\right], \tag{7.53}$$

where ω_e is the effective frequency of the process given by the formula

$$\omega_e = \frac{\sqrt{2}}{\sigma} \left[\int_0^\infty S(\omega)\omega^2 d\omega\right]^{1/2} . \tag{7.54}$$

If $v(t)$ is a nonstationary normal process, the closed-form formula for $\nu(t)$ has a rather cumbersome form, namely,

$$\nu(t) = \frac{s}{2\pi\sigma}\left\{\sqrt{1-\rho^2}\exp\left(-\frac{1}{1-\rho^2}\frac{v_{**}^2}{2\sigma^2}\right) + \sqrt{2\pi}\,\frac{\rho v_{**}}{\sigma}\exp\left(-\frac{v_{**}^2}{2\sigma^2}\right)\Phi\left(\frac{\rho}{\sqrt{1-\rho^2}}\frac{v_*}{\sigma}\right)\right\}, \tag{7.55}$$

where $v_{**}(t) = v_* - \langle v(t)\rangle$, $\rho(t)$ is a correlation coefficient between $v(t)$ and $\dot{v}(t)$ in coinciding time moments, $s^2(t)$ is the variance of the derivative $\dot{v}(t)$, and $\Phi(u)$ is the normalized Gaussian distribution function, i.e.,

$$\Phi(u) = \frac{1}{\sqrt{2\pi}}\int_{-\infty}^{u} e^{-x^2/2}dx\ . \tag{7.56}$$

Consider the case when the quality space $\underline{V}$ is an n-dimensional Euclidean space. The general formula for the mathematical expectation of the number of excursions of the process $\underset{\sim}{v}(t)$ from the region Ω with piecewise smooth boundary Γ has the form

$$\nu(t) = \int_{\Gamma} d\Gamma \int_{\dot{v}_n>0} p(\underset{\sim}{v}_{\Gamma},\dot{\underset{\sim}{v}};t)\dot{v}_n d\dot{\underset{\sim}{v}}\ . \tag{7.57}$$

Here, $p(\underset{\sim}{v},\dot{\underset{\sim}{v}};t)$ is the joint probability density of the vector process $\underset{\sim}{v}(t)$ and its first derivative with respect to time, $\dot{\underset{\sim}{v}}(t)$ and $\dot{v}_n$ is the normal component of $\dot{\underset{\sim}{v}}(t)$ on the surface Γ. In calculating the internal integral in (7.75), it is assumed that $\underset{\sim}{v} = \underset{\sim}{v}_{\Gamma}$, where $\underset{\sim}{v}_{\Gamma}$ is taken on the surface Γ. Formula (7.57) can be treated as the extension of formula (7.52) to the multidimensional quality space.

Let the space $\underline{V}$ be two-dimensional, $\underset{\sim}{v}(t)$ be the stationary normal process $\{v_1(t), v_2(t)\}$, and the region Ω be specified by the inequalities

$$|v_1| \le v_{1*}\ , \qquad |v_2| \le v_{2*}\ . \tag{7.58}$$

Application of formula (7.57) to this case gives

$$\nu = \frac{\omega_{e1}}{2\pi}\left\{\exp\left[-\frac{(v_{1*}-a_1)^2}{2\sigma_1^2}\right][\Phi(u_{+2,+1})-\Phi(u_{-2,+1})] + \exp\left[-\frac{(v_{1*}+a_1)^2}{2\sigma_1^2}\right][\Phi(u_{+2,-1})-\Phi(u_{-2,-1})]\right\} + \frac{\omega_{e2}}{2\pi}\left\{\exp\left[-\frac{(v_{2*}-a_2)^2}{2\sigma_2^2}\right][\Phi(u_{+1,+2})-\Phi(u_{-1,+2})] + \exp\left[-\frac{(v_{2*}+a_2)^2}{2\sigma_2^2}\right][\Phi(u_{+1,-2})-\Phi(u_{-1,-2})]\right\}, \tag{7.59}$$

where the following notations are used:

$$u_{\pm j,\pm k} = \frac{1}{\sqrt{1-\rho^2}}\left(\frac{\pm v_{j*}-a_j}{\sigma_j} - \rho\,\frac{\pm v_{k*}-a_k}{\sigma_k}\right), \quad \omega_{e1} = \frac{s_1}{\sigma_1}, \quad \omega_{e2} = \frac{s_2}{\sigma_2}. \tag{7.60}$$

In formulae (7.59) and (7.60), a_j are the mathematical expectations of the components $v_j(t)$ of the process $\underset{\sim}{v}(t)$, σ_j^2 are their variances, s_j^2 are the variances of their first derivatives with respect to time, and ρ is the correlation coefficient of the processes $v_1(t)$ and $v_2(t)$ for coinciding moments of time.

Application of formula (7.59) requires quite a few calculations. If excursions are rare events, it is possible, instead of (7.59), to take the simpler relation

$$\nu \approx \frac{\omega_{e1}}{2\pi}\left\{\exp\left[-\frac{(v_{1*}-a_1)^2}{2\sigma_1^2}\right] + \exp\left[-\frac{(v_{1*}+a_1)^2}{2\sigma_1^2}\right]\right\} + \frac{\omega_{e1}}{2\pi}\left\{\exp\left[-\frac{(v_{2*}-a_2)^2}{2\sigma_2^2}\right] + \exp\left[-\frac{(v_{2*}-a_2)^2}{2\sigma_2^2}\right]\right\}. \tag{7.61}$$

On the right side, we have the sum of the mathematical expectations of the number of excursions from the strips $|v_1| \leq v_{1*}$ and $|v_2| \leq v_2$ Formula (7.61) allows an extension to any finite number of dimensions [19].

Application to Narrowband Processes

The estimates (7.35), (7.47) and (7.48) may turn out to be ineffective and even unfit in the case of insufficiently mixed processes. Narrowband processes describing random vibrations in systems with small dissipation can serve as a most important example for an application of such processes. The moments of the number of crossings of such processes generally do not decrease with the increase of α as fast as necessary for a truncation of the series (7.41) and (7.42). For example, if in the case of a one-dimensional process $v(t)$ the level v_* is not sufficiently high, the probability of two-fold, three-fold, etc. crossings on this level may not turn out to be small as compared to the probability of a one-fold crossing.

For systems whose vibrations are a narrowband random process, measures of fatigue damage [10,13] which are quasimonotonic random time functions can naturally be taken for quality parameters. Another approach is to take the envelopes of the corresponding narrowband processes for quality parameters, the reliability function $R(t)$ for highly-reliable systems being estimated either from a formula of the type (7.47)

$$R(t) \approx 1 - \langle \overline{N}(t) \rangle \,, \tag{7.62}$$

or from the formula of Poisson's approximation

$$R(t) \approx \exp[-\langle \overline{N}(t) \rangle] \,. \tag{7.63}$$

Here, $\overline{N}(t)$ is the number of excursions of the process $\underset{\sim}{\overline{v}}(t)$ whose components are the envelopes of the components of the specified process $v(t)$.

As an example, we shall consider the one-dimensional stationary narrow-band Gaussian process $v(t)$ with the mathematical expectation a, the variance σ^2, and the carrying frequency Θ. The mathematical expectation of the number of excursions of the envelope per unit time over the fixed level $v_* > a$ is determined from the formula [95]

$$\bar{\nu} = \frac{\bar{\omega}_e}{\sqrt{2\pi}} \frac{v_*}{\sigma} \exp\left[-\frac{(v_*-a)^2}{2\sigma^2}\right]. \tag{7.64}$$

Here, $\bar{\omega}_e$ is the effective frequency of the envelope and is expressed in terms of the spectral density $S(\omega)$ of the process $v(t)$ as follows:

$$\bar{\omega}_e = \frac{\sqrt{2}}{\sigma}\left[\int_0^\infty (\omega-\Theta)^2 S(\omega)\,d\omega\right]^{1/2}. \tag{7.65}$$

Extension of Reliability Theory to Continuous Systems

Basic concepts of reliability theory are applicable to both systems with a finite number of degrees of freedom and continuous systems.* In the latter case, certain scalar, vector, and tensor fields serve as quality elements $\underset{\sim}{v}(t)$. The space $\underline{V}$ is a functional phase space with elements parametrically dependent of time, and with the probability measure specified in it. As a rule, limitations on the parameters of the system are such that the admissible region Ω can be specified with the help of a certain norm in the space $\underline{V}$. Independent of the geometric nature of the elements $\underset{\sim}{v}(t)$, the norm will be a scalar time function. This follows from the fact that the quality of the system and its reliability measure are by their nature invariant with respect to transformations of the system of coordinates.

As an example, we shall consider a body subjected to the action of external forces. Suppose there appears a random field

---
*Bolotin, V.V., "Reliability Theory of Continuous Mechanical Systems" Izvestia Ac. of Sc., U.S.S.R., *MTT*, 1969, No. 6; and Bolotin, V.V., "Reliability Theory of Problems in Mechanics of Solids", *Coll. Works Mekhanika Sploshnoi Sredi*, (Mechanics of Continua and Related Problems of Analysis), Nauka, Moscow, 1972.

of stresses $\sigma_{jk}(\tilde{x},t)$ in the body on which the limitation is imposed that at no point of the region G occupied by the body is the intensity of the stresses $\sigma(\tilde{x},t)$ to exceed a certain specified value σ_*. The space $\underline{V}$ is defined in the following way:

$$\underline{V} = \{\tilde{v}(t);\sigma_{jk}(\tilde{x},t);\tilde{x} \in G, t \geq 0\} .$$

The admissible region Ω is specified by the condition

$$\Omega = \{\tilde{v}(t):\sup_{\tilde{x}\in G} \sigma(\tilde{x},t) < \sigma_*\} ,$$

where σ is expressed in terms of σ_{jk} as

$$\sigma^2 = \check{\sigma}_{jk}\check{\sigma}_{jk} , \quad \check{\sigma}_{jk} = \sigma_{jk} - \frac{1}{3}\sigma_{\ell\ell}\delta_{jk} , \tag{7.66}$$

and where the positive value of the root of σ^2 is taken. It can be seen from formula (7.60) that the upper bound of the intensity of stresses $\sigma(\tilde{x},t)$ in the region G can be interpreted as a norm (more exactly as a semi-norm) in the space $\underline{V}$. Thus,

$$\Omega = \{\tilde{v}(t):||\tilde{v}(t)|| < \sigma_*\} ,$$

where the notation

$$||\tilde{v}(t)|| = \sup_{\tilde{x}\in G} \sigma(\tilde{x},t)$$

is used.

The above mentioned approximate estimates for the reliability function are naturally extended to the case of the functional quality space. For simplicity, we restrict ourselves to the case when the space $\underline{V}$ is formed from realizations of the scalar field $v(\tilde{x},t)$ specified in the region G, and the region of admissible values is defined by the inequality $v(\tilde{x},t) < v_*$. Here,

v_* is a prescribed non-random number. The reliability function $R(t)$ is given as

$$R(t) = \underset{\sim}{P}\{\sup_{\underset{\sim}{x}\in G} v(\underset{\sim}{x},t) < v_*; \tau \in [0,t]\} . \qquad (7.67)$$

We introduce the following interpretation of an excursion from Ω. Consider the (n+1)-dimensional space $\underline{R}^{n+1}$ of the variables $\underset{\sim}{y} = \{\underset{\sim}{x},\tau\}$ (henceforth, we shall use the notation $n+1 = m$). The range of the function y is the cylinder $G_t = G\times[0,t]$ (see Figure 62, where $n = 2$). A connected set of points in G_t for which $v(\underset{\sim}{y}) > v_*$ corresponds to the excursion of the field $v(\underset{\sim}{x},t)$ from Ω. The number of excursions $N(t)$ in the region G_t is related to $N_{max}(t)$, the number of maxima of the field $v(\underset{\sim}{y})$ exceeding the level v_*, by the relation $N(t) \leq N_{max}(t)$. With levels of v_* sufficiently high and with the field $v(\underset{\sim}{y})$ sufficiently mixed, it is possible to assume that $N(t) \approx N_{max}(t)$. Then, instead of (7.35), we obtain the estimate

$$R(t) \approx \exp[-\langle N_{max}(t)\rangle] , \qquad (7.68)$$

and, instead of (7.47), the estimate

$$R(t) \approx 1 - \langle N_{max}(t)\rangle . \qquad (7.69)$$

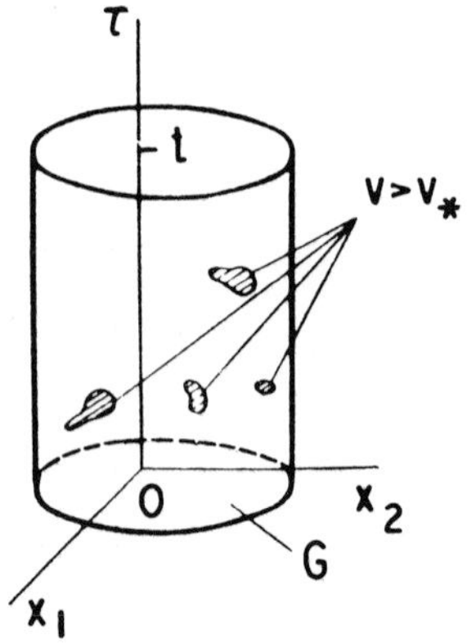

Figure 62

Let the field $v(\underset{\sim}{y})$ be twice differentiable with respect to all arguments, and let the probability of the excursion region crossing the boundary of region G_t be negligibly small. Then $\langle N_{max}(t)\rangle$ can be identified with the mathematical expectation of the number of analytical maxima in G_t. The method of calculating the latter characteristic is given in [6]. Let the joint probability density $p(v,\underset{\sim}{\phi},\underset{\sim}{\kappa};\underset{\sim}{y})$ of the field $v(\underset{\sim}{y})$, of its gradient $\underset{\sim}{\phi} = \{\partial v/\partial x_\alpha\}$ and of the column matrix $\underset{\sim}{\kappa}$ be given, with the elements of $\underset{\sim}{\kappa}$ being identically equal to noncoinciding elements of the Hessian matrix $H = \{\partial^2 v/\partial x_\alpha \partial x_\beta\}$. Then,

$$\langle N_{max}(t)\rangle = \int_{G_t} d\underset{\sim}{y} \int_{v_*}^{\infty} dv \int_{\underset{\sim}{\kappa}\in \underset{-}{K}_m} p(v,\underset{\sim}{0},\underset{\sim}{\kappa};\underset{\sim}{y})\,|\det \underset{\sim}{H}|\,d\underset{\sim}{\kappa}\ , \tag{7.70}$$

where the set of values of $\underset{\sim}{\kappa}$ for which the matrix H is negative-definite is denoted by $\underset{-}{K}_m$. A more accurate calculation of the mathematical expectation of the number of fluctuations of a random field, based on topological considerations, is given in the papers mentioned in the previous footnote. In the same papers, approximate methods of solving reliability problems for continuous systems based on a reduction to finite-dimensional subspaces can be found.

7.4 APPLICATION OF RELIABILITY THEORY TO PROBLEMS CONCERNED WITH PROTECTION FROM RANDOM VIBRATION

Basic Concepts of Problems Involving Vibration Protection

If a system is subjected to the action of a vibration that causes undesirable effects, the first thing to do is to eliminate such action or reduce it to a safe level. If this does not work, it is necessary to seek other ways. One of them is to change the system's parameters so as to make it less sensitive to vibration actions. Another is to introduce intermediate elements that would isolate the system from external action. For intermediate elements, one can use regular dampers, springs, rubber cords, or soft porous pads. In constructions and devices of some special importance, it is possible to use more complicated active (electromechanical, for instance) systems, in which vibration protection is effected on the principle of automatic control.

The purpose of the theory of vibration protection [49,55] is to establish the principles for the design of vibration protection systems and methods of computing their optimal characteristics. We shall illustrate the formulation of problems of vibration protection by a very simple example (Figure 63). Let an object of mass m be placed on a rigid horizontal base which is vibrating in the direction of the normal with displacement $u_0(t)$. If accelerations caused by this motion are undesirable from the point of view of the operation of the object, it is natural to try to decrease these accelerations by fixing the object with a linear visco-elastic element with a stiffness coefficient c and a viscosity coefficient b. The task is to choose the parameters c and b so as to reduce the acceleration of mass m to a desired level. As a characteristic of vibration protection, we generally use the coefficient of vibration isolation $k(\omega)$, which is equal to the modulus of the ratio of the base amplitude to the object's amplitude for a harmonic vibration with frequency ω, i.e.,

$$k(\omega) = \left|\frac{U_0(\omega)}{U(\omega)}\right| = \sqrt{\frac{(1-\omega^2/\omega_0^2)^2+4\gamma^2\omega^2/\omega_0^2}{1+4\gamma^2\omega^2/\omega_0^2}} \,. \tag{7.71}$$

Here, ω_0 is the object's natural frequency and γ is a nondimensional damping characteristic, i.e.,

$$\omega_0 = (c/m)^{1/2} \,, \qquad \gamma = b/2m\omega_0 \,. \tag{7.72}$$

In this case, the design of a vibration protection is reduced to the choice of stiffness parameters and that of damping subject to the condition

$$k(\omega) \geq k_* \,, \tag{7.73}$$

i.e., that the vibration isolation coefficient $k(\omega)$ be not smaller than a certain specified value k_*.

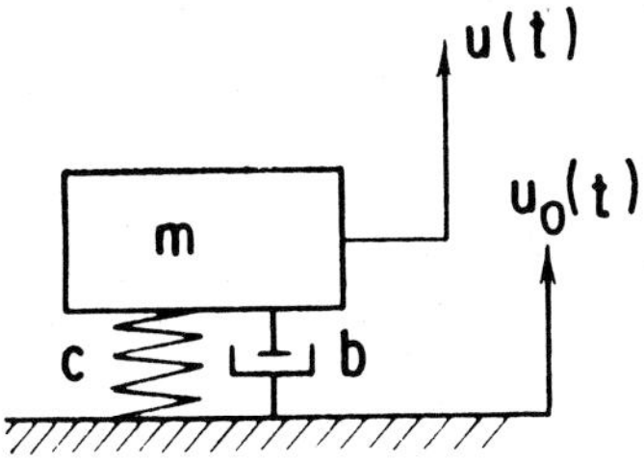

Figure 63

If the dampers' stiffness is chosen so that $\omega > \omega_0\sqrt{2}$, the condition $k(\omega) > 1$ is satisfied for all possible values of damping. To increase the vibration isolation coefficient, it is necessary to decrease the natural frequency further. However, conditions of physical implementation, strength conditions with respect to quasi-static overloads, conditions imposed on the free deflection of the damper, etc. put a lower bound on the dampers' stiffness. In practice, the dampers' stiffness is, as a rule, chosen so that $\omega \in [2.5\omega_0, 5.0\omega_0]$ at moderate damping. Although damping here reduces slightly the effectiveness of vibration protection, it plays a beneficial role when the system passes through resonance, and, generally, when low frequency components appear in the spectrum.

The approach described above admits of an extension to linear systems with many degrees of freedom and to linear continuous systems. The general statement of the problem is the following: the equation of the system's motion is written in the operator form

$$L\underset{\sim}{u} = L_0\underset{\sim}{u}_0 , \tag{7.74}$$

where $\underset{\sim}{u}_0(x,t)$ is the external kinematic action, $u(x,t)$ is the displacement field of the system, and L_0 and L are linear operators, some parameters of which are chosen from an admissible region so as to satisfy the specified conditions of vibration protection. Consider solutions to equation (7.74) to be in the class

$$\underset{\sim}{u}_0(\underset{\sim}{x},t) = \underset{\sim}{U}_0(\underset{\sim}{x},\omega)e^{i\omega t}, \qquad \underset{\sim}{u}(\underset{\sim}{x},t) = \underset{\sim}{U}(\underset{\sim}{x},\omega)e^{i\omega t} .$$

Observing some additional considerations (caused by vibration protection requirements), we introduce a norm in the joint region of definition of the functions $\underset{\sim}{U}_0(\underset{\sim}{x},\omega)$ and $\underset{\sim}{U}(\underset{\sim}{x},\omega)$. Analogously to (7.71), it is natural to define the coefficient of vibration isolation as the ratio of complex amplitudes' norms at the input to those at the output of the system [137], i.e.,

$$k(\omega) = \frac{||\underset{\sim}{U}_0(\underset{\sim}{x},\omega)||}{||\underset{\sim}{U}(\underset{\sim}{x},\omega)||} . \tag{7.75}$$

The analysis of the vibration protection system is reduced to the choice of parameters appearing in the operators L_0 and L such that the vibration isolation coefficient (7.75) satisfies the inequality (7.73)

Formulations of Problems in the Theory of Protection from Random Vibrations

The classical theory of vibration protection, the fundamentals of which were given above, extends naturally to stationary random processes. As an example, let us consider the system in Figure 63, in which mass displacement $u(t)$ and base displacement $u_0(t)$ are represented as

$$u(t) = \int_{-\infty}^{\infty} U(\omega)e^{i\omega t}d\omega , \qquad u_0(t) = \int_{-\infty}^{\infty} U_0(\omega)e^{i\omega t}d\omega .$$

Here, $U(\omega)$ and $U_0(\omega)$ are generalized stationary random functions of the frequency ω related to the spectral densities of the processes $u(t)$ and $u_\omega(t)$ by the expressions

$$\langle U^*(\omega)U(\omega')\rangle = S_u(\omega)\delta(\omega-\omega') ,$$
$$\langle U_0^*(\omega)U_0(\omega')\rangle = S_{u0}(\omega)\delta(\omega-\omega') . \tag{7.76}$$

Introduce the vibration isolation coefficient $k(\omega)$, equal to the positive value of the square root of the ratios of these spectral densities, i.e.,

$$k(\omega) = \sqrt{S_{u0}(\omega)/S_u(\omega)} . \tag{7.77}$$

It is easy to see that for the system plotted in Figure 63 the value of $k(\omega)$ is given by formula (7.71). In the general case of the operator equation (7.74), the vibration isolation coefficient $k(\omega)$ is expressed in terms of a properly chosen norm of the operator $L_0^{-1}(i\omega)L(i\omega)$. The parameters of vibration protection are then determined based on the criterion (7.73).

The approach described above will be referred to as the spectral approach. This approach may prove to be effective if external vibrations are a narrow-band random process. In this case, the parameters of vibration protection are chosen from the condition (7.73) at the frequency ω equal to the carrying frequency of the external action Θ. Here, the spectral approach leads to the same results as the classical theory of vibration protection for periodic action at the frequency Θ. In the case of broadband action whose spectrum is bounded by the frequencies ω_* and ω_{**}, it is natural to replace the criterion (7.73) by

$$k(\omega) \geq k_* , \qquad \omega \in [\omega_*,\omega_{**}] . \tag{7.78}$$

The area of practical application of the criterion (7.78) is determined by processes that are bounded below by sufficiently high frequencies.

In the theory of random vibration protection, criteria of the type of minimum standard deviation are most extensively used [61,64,98]. One of them is for example the criterion of minimum variance of the object's acceleration with a restriction imposed on the object's displacements with respect to the base. Applied to a system with one degree of freedom, this criterion has the form

$$\langle\tilde{a}^2\rangle \to \min , \qquad \langle\tilde{u}_1^2\rangle < u_*^2 , \tag{7.79}$$

where $a(t)$ is the mass acceleration, $u_1(t) = u(t)-u_0(t)$ is the relati displacement of the mass, and u_*^2 is the limiting value of the displacement variance. Instead of (7.79), the criterion

$$\langle\tilde{u}_1^2\rangle \to \min , \qquad \langle\tilde{a}^2\rangle < a_*^2 \tag{7.80}$$

is often used. Here, a_*^2 is the limiting value of the variance of vibration acceleration. If the external action has a quasi-static component, an additional restriction is imposed on the mathematical expectation of the vibration acceleration.

Variances of stationary random processes are quadratic functionals over sets of phase trajectories of the system. For systems with many degrees of freedom whose behaviour is characterized by the stationary random functions $v_j(t)$, $(j = 1,2,\ldots,n)$, the criteria (7.79) and (7.80) are generalized in the following way:

$$I = \sum_{j=1}^{n} \sum_{k=1}^{n} \lambda_{jk}\langle\tilde{v}_j(t)\tilde{v}_k(t)\rangle \to \min . \tag{7.81}$$

Here, some coefficients λ_{jk} are prescribed numbers, and the remaining ones are Lagrangian multipliers and are found from the restrictions imposed on some moments of second order of the output parameters. A natural extension of the criterion (7.79) to non-stationary random processes has the form

$$\sup_{0\le t\le T} \langle\tilde{a}^2(t)\rangle \to \min , \qquad \sup_{0\le t\le T} \langle\tilde{u}_1^2(t)\rangle < u_*^2 . \tag{7.82}$$

Here, T is the specified lifetime. The criteria (7.80) and (7.81) are extended analogously.

Criteria of the type (7.79) - (7.82) are extensively used in the theory of automatic control and in the theory of communications. Moreover, along with parametric optimization, problems in which the operators' structure (control loops, ripple filters, etc.) is not preassigned are formulated and solved. An analogous statement in the theory of vibration protection generally leads to systems containing active elements [56]. Such systems cannot be realized by means of purely mechanical links, i.e., elastic and dissipative constraints. Physical realization of these systems requires turning to measuring devices, feedback loops, servomechanisms, etc. Henceforth, we shall restrict ourselves to mechanical means of vibration protection.

Criterion of Maximum Reliability in the Theory of Vibration Protection

Spectral criteria of the type (7.78) and criteria of the mean-square type (7.79) - (7.82), are convenient in that they make it possible to use the well-developed technique for solving corresponding optimization problems. However, these criteria have a number of grave disadvantages. One of them is that these criteria cannot be clearly interpreted from the physical point of view, and it is difficult to determine the values of parameters entered in them based on engineering (operational) considerations. For example, the criterion (7.79) expresses the requirement of minimum vibration overloads for restrictions imposed on the free displacement of a damper. However, the minimum of variance of vibration acceleration still does not imply a good protection of a system from vibrations. From the point of view of fatigue damages, not only the value of stresses' variance is important, but their spectrum as well. On the other hand, restriction of the variance of relative displacement still allows significant fluctuations of the system, which are accompanied by severe shocks against the stopping device. Although the magnitude of the ultimately admissible variance is connected with the admissible free motion, it cannot be prescribed without resorting to a more detailed theoretical-probabilistic or experimental-statistical analysis.

The difficulties listed above increase if the system has many degrees of freedom, and if restrictions are imposed on a large number of dynamic parameters. For example, criteria of the type (7.81) do not generally allow a physical interpretation. It is difficult to make a justified choice of the coefficients λ_{jk}. Additional difficulties arise in considering nonstationarity of random processes, conflicting requirements resulting from conditions of vibration protection, shock protection and protection from quasi-static overloads, and in considering statistical scatter of mechanical characteristics of dampers, etc.

These difficulties are not inherent in the theory of vibration protection. They actually result from a wrong choice of optimization criteria. The engineer wants the vibration protection system to provide a trouble-free operation for the object. The conditions of trouble-free operation are characterized to some extent by the ratios of the spectral densities appearing in (7.77), as well as by variances of vibration accelerations and vibration displacements appearing in the criteria (7.79), (7.80), etc. But, from the engineering point of view, it would be more appropriate to minimize the probability of the random event which consists of the object's parameters exceeding the admissible values at least once during its operation. This is equivalent to requiring the reliability measure of the system to take on the maximum value. Reliability optimization, being more natural and justified, obviates the difficulties arising from applying spectral and mean-square criteria. This optimization method can be applied to both linear and non-linear systems, the number of degrees of freedom of the system and the number of parameters entered in optimization criterion can be arbitrary, and no restrictions are imposed on the stochastic nature of external actions. Below, following mainly reference [18], we shall describe the application of the maximum reliability criterion to the analysis of vibration protection.

Let the system perform random vibrations under the action of kinematic or force excitation. In the course of this process there appears in the system a field of vibrational displacements $\underset{\sim}{u}(\underset{\sim}{x},t)$, a field of accelerations $\underset{\sim}{a}(\underset{\sim}{x},t)$, etc. Suppose that, from the condition of normal operation of the system, some restrictions are imposed on the parameters of the vibration fields which can be represented in terms of certain quality elements $\underset{\sim}{v}(t)$ in this way:

$$\underset{\sim}{v}(t) \in \Omega \; . \tag{7.83}$$

Here, Ω is an admissible region in the quality space $\underline{V}$. The elements $\underset{\sim}{v}(t)$ are formed from the vibration field characteristics. Following formula (7.3) introduce the reliability function $R(t)$ for the vibrating system. Let T_* be the specified time of operation of the system. We shall consider the vibration protection system optimal if, with restrictions imposed on vibration protection devices, the index of the reliability of the system for $t = T_*$ takes on the maximum value. Thus, the optimization criterion in the analysis of a vibration protection system takes the form

$$R(T_*) \to \max \; . \tag{7.84}$$

In some applications, instead of the criterion (7.84), related criteria requiring a maximum (minimum) of a certain functional of the reliability function $R(t)$ are used. Such modified criteria arise, in particular, when an engineering-economical approach to reliability analysis is used. The simplest form of a criterion of this type takes the form

$$\int_0^{T_*} \rho(t)\,|R'(t)|\,dt \to \min , \qquad (7.85)$$

where $\rho(t)$ is some non-negative function of time. This function takes into account nonequivalence of the failures occurring in the interval $[0,T_*]$. The function $\rho(t)$ can be interpreted as a cost function.

Examples Showing the Formulation of the Problem of Optimal Vibration Protection

One of the problems in the theory of vibration protection is the choice of the parameters for a suspension of a perfectly rigid body inside a container executing arbitrary random motions. Let $\tilde{u}(\tilde{x},t)$ be the field of displacements of the body with respect to the container, and $\tilde{a}(\tilde{x},t)$ the field of absolute accelerations. In view of an operational condition, the maximum vibration overloads should be limited, the modulus by the value a_*, and the minimal relative displacements by the value u_*. The region of admissible values is determined by the conditions

$$\max_{\tilde{x}\in G} ||\tilde{u}(\tilde{x},t)|| < u_* , \qquad \max_{\tilde{x}\in G} ||\tilde{a}(\tilde{x},t)|| < a_* . \qquad (7.86)$$

Here, G is the volume occupied by the body; the double lines denote the Euclidean norm (vector modulus). From kinematical considerations it follows that maximal values of accelerations and displacements are reached on the body's surface; if the surface has the form of a polyhedron, they are found on its vertices. In the latter case, instead of the conditions (7.86), we obtain

$$||\tilde{u}(\tilde{x}_j,t)|| < u_* , \quad ||\tilde{a}(\tilde{x}_j,t)|| < a_* , \quad (j = 1,2,\ldots,s) \qquad (7.87)$$

where the $\tilde{x}_j$ are the coordinates of the polyhedron's vertices.

Let the displacements $\underset{\sim}{u}(\underset{\sim}{x},t)$ be sufficiently small as compared to the typical dimensions of the body. In this case, the fields $\underset{\sim}{u}(\underset{\sim}{x},t)$ and $\underset{\sim}{a}(\underset{\sim}{x},t)$ are approximately represented as linear combinations of six generalized coordinates $u_k(t)$ and of six generalized accelerations $a_k(t)$. In this case, it is convenient to take the 12-dimensional space of displacements and accelerations as a quality space. Suppose, for example, that the body is attached to the container by means of six linear constraints, each of which is characterized by a stiffness coefficient c_j and a damping coefficient b_j. In the general case, the location of each constraint is fixed by four coordinates (two curvilinear coordinates on the body's surface and two curvilinear coordinates on the inner surface of the container). Thus, in this case, variation is performed over 36 parameters. In a specific situation, the number of varied parameters can be significantly reduced due to restrictions imposed on the choice of dampers and their fixation points, and from symmetry considerations.

As a second example we shall consider a vibration protection design for devices installed on an elastic plate. Let the chassis (to which the plate is attached) undergo random vibrations in the direction orthogonal to the plate's plane. The vibration field in the system is characterized by the deflection function of the plate $w(\underset{\sim}{x},t)$ and by the absolute displacements of instruments $u_j(t)$, $(j = 1,2,\ldots,s)$. Analogously to (7.87), restrictions will be imposed on vibration overloads to each instrument separately, and on displacements of instruments with respect to the plate. The admissible region is given by the inequalities

$$|\ddot{u}_j(t)| < a_{*j} \ , \qquad |u_j(t)-w(\underset{\sim}{x}_j,t)| < u_{*j} \ , \qquad (j = 1,2,\ldots,s) \ , \tag{7.88}$$

where a_{*j} and u_{*j} are limiting values of accelerations and relative displacements. For reliability optimization, we have available the parameters of the dampers attaching the instruments to the plate and, also, if possible, the parameters of the plate and its fixation on the chassis.

Along with the displacements of instruments with respect to the base, restrictions can be imposed on the displacements of instruments relative to one another. For example, if two instruments are connected by a constraint that might be damaged or broken due to relative displacements, the following conditions are added to those of the type (7.88):

$$|u_j - u_k| < u_{*jk} \; . \tag{7.89}$$

A similar condition is used in the case when shock-absorbing members are positioned with a clearance, and it is required that under vibrations this clearance should be maintained. As an example, take a system with two degrees of freedom (Figure 68). Consider random vibrations in a horizontal plane. The clearance between the members m_1 and m_2 at the initial state equals u_*. Reducing the stiffnesses c_1 and c_2, we reduce the vibration overloads on the members m_1 and m_2. But, in this case, displacements of the members with respect to the chassis increase, and so does the danger of their collision. The admissible region Ω is given by the inequalities

$$|a_1| < a_{*1} \,, \qquad |a_2| < a_{*2} \,, \qquad u_1 - u_2 < u_* \; . \tag{7.90}$$

Along with the reliability of the vibration isolation system, the reliability of the dampers should be ensured. Let $u_j(t)$ be the relative displacement of the supporting planes of the damper along the principal axis, $F_j(u_j, \dot{u}_j)$ the force acting on the damper, and F_{j*} its limiting value. Conditions for the normal operation of the dampers may have the form

$$|F_j(u_j, \dot{u}_j)| < F_{j*} \,, \qquad (j = 1, 2, \ldots) \; . \tag{7.91}$$

For example, in the case of linear dampers with viscous friction, we have

$$F_j = -(c_j u_j + b_j \dot{u}_j) , \qquad (7.92)$$

where c_j are stiffness coefficients, and b_j damping coefficients.

Up to this point, conditions for the normal operation of a shock-protected object and of a system of dampers were expressed either in terms of kinematic parameters or in terms of damper forces. However, failures are often caused by accumulation of fatigue damage in the members of a system which are under the greatest stress. The quality of a system, and, hence, the quality of the shock absorbing system, can be expressed in terms of the parameters characterizing the degree of fatigue damage (Sections 7.6 - 7.8).

Optimization problems are solved with prescribed restrictions imposed on the parameters of the dampers. Among them are, primarily, conditions of physical realizability. For example, for linear dampers whose forces are determined by the relations (7.92), the conditions of physical realizability are

$$c_j > 0 , \qquad b_j > 0 . \qquad (7.93)$$

In practice, stricter conditions than conditions (7.93) or conditions similar to them should be satisfied. For instance, when standard dampers are used, stiffness coefficients are limited by typical dimensions, and the value of the relative energy dissipation or loss tangent is determined by the structure of the given series [49].

7.5 SOME PROBLEMS OF VIBRATION ISOLATION

Modified Criterion of Maximum Reliability

Approximate and two-sided estimates (7.35), (7.43) and (7.47) make it possible to express the criterion of maximum reliability in terms of the mathematical expectation of the number of excursions of the system away from the region of admissible states. So, using the approximate relation (7.35), we shall represent the criterion (7.84) in the form

$$\langle N(T_*)\rangle \to \min . \qquad (7.94)$$

Thus, the vibration protection system should be chosen so that, with prescribed restrictions, the mathematical expectation of the number of excursions from the admissible region during the operation would be minimal. The "weighted" criterion (7.85) is transformed analogously. Noting that $R'(t) \approx -\nu(t)$, we obtain

$$\int_0^{T_*} \rho(t)\nu(t)dt \to \min . \qquad (7.95)$$

If the process $\underset{\sim}{v}(t)$ is stationary, the criterion of maximum reliability can be replaced by the criterion of the minimum of the number of excursions from the admissible region per unit time. Thus, we arrive at the criterion

$$\nu \to \min . \qquad (7.96)$$

Since ν has the meaning of the rate of failures, the criterion of maximum reliability in the form (7.96) can be called the criterion of minimum failure rate.

The mathematical expectations of the numbers of excursions are nonlinear and, generally, nonconvex and nonconcave functions of varied parameters. Therefore, an application of well-developed methods for the solution of optimal problems encounters certain difficulties. It is expedient to start an analysis of these problems with a global review without imposing essential restrictions on the parameters.

Example: Linear System with One Degree of Freedom

Let us take a simple example of an application of the criterion of maximum reliability. We shall again consider a linear system with one degree of freedom (Figure 63). Let the base perform vibrations with acceleration $a_0(t)$, which is a stationary normal process. The equation of motion for the mass m is reduced to the form

$$\frac{d^2u_1}{dt^2} + 2\gamma\omega_0 \frac{du_1}{dt} + \omega_0^2 u_1 = -a_0(t) \,, \tag{7.97}$$

where the notations in (7.72) are used and where $u_1(t)$ is the relative displacement of the mass.

Let the conditions for the normal operation of the system be specified as inequalities of the type (7.87), i.e.,

$$|u_1(t)| < u_* \,, \qquad |a(t)| < a_* \,. \tag{7.98}$$

Here, a(t) is the absolute acceleration of mass m. The quality space $\underline{V}$ is thus two-dimensional, and the admissible region Ω is rectangular in shape (Figure 64). The criterion of maximum reliability leads to (7.96). Minimization of the mathematical expectation of the number of fluctuations ν is performed over the parameters ω_0 and γ.

Now we shall give some numerical results from [85]: assume that the acceleration $a_0(t)$ is a normal exponentially-correlated process with a mathematical expectation equal to zero.

Its spectral density has the form (1.17). Spectral densities of the processes $u_1(t)$ and $a(t)$ are found from formulae of the type (2.41) with regard to equation (7.97). The mathematical expectation of the number of fluctuations per unit time is determined from formula (7.59). The continuous line in Figure 65 shows the dependence of the ratio ν/α on the nondimensional natural frequence ω_0/ω_*. Here ω_* is

$$\omega_* = (a_*/u_*)^{1/2} . \tag{7.99}$$

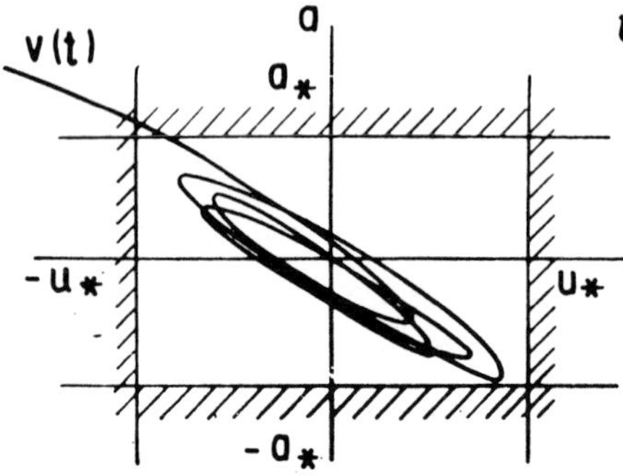

Figure 64

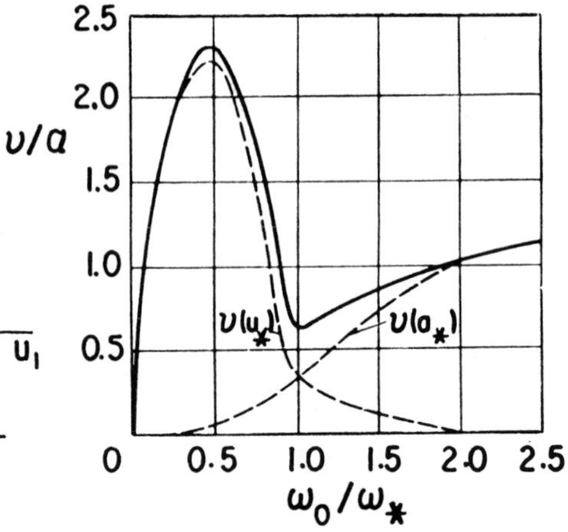

Figure 65

The remaining parameters have the following values: $\gamma = 0.5$, $a_* = 2\sigma_0$, $u_*\alpha^2 = 10^{-4}\sigma_0$.

As can be seen in Figure 65, the absolute minimum of the mathematical expectation of the number of excursions corresponds to the natural frequency ω_0 approaching zero. However, due to restrictions resulting from physical realizability, this minimum is of no practical significance. A relative minimum is observed close to the frequency ω_* determined from formula (7.99). The dashed lines in Figure 65 correspond to excursions from the bands $|u_1| < u_*$ and $|a| < a_*$.

The abscissa of the intersection point of these lines is close to the abscissa of the relative minimum. This fact becomes comprehensible if we consider narrowband vibrations of the mass m. For these vibrations, we can assume that $a = a_0 + \ddot{u}_1 \approx a_0 - \omega_0^2 u_1$. If $\omega_0 = \omega_*$, the phase point on the plane $\{u_1, a\}$ will move along trajectories which are close to the diagonal of the rectangle in Figure 64. Thus, the minimum of the mathematical expectation of the number of fluctuations from this rectangle is reached under conditions providing approximately equal danger of excursions from the bands $|u_1| < u_*$ and $|a| < a_*$.

Dependence of the mathematical expectation of the number of excursions on both varied parameters (the natural frequency ω_0 and the damping parameter γ) is shown in Figure 66. The curves in this graph correspond to equal values of ν/α. In the considered region of the variation of parameters, there are no isolated minima. Variation of damping over a wide range does not significantly influence the number of excursions. Dependence of this number on the natural frequency is more significant. Optimization of the number of excursions is performed in the region of variation of the parameters ω_0 and γ which is determined from the conditions of the dampers' realizability.

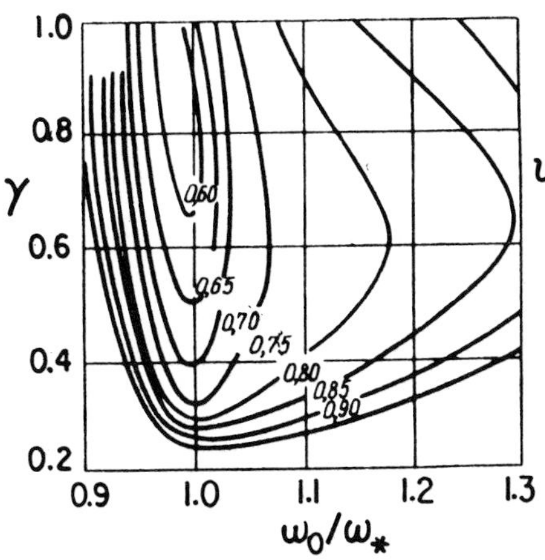

Figure 66

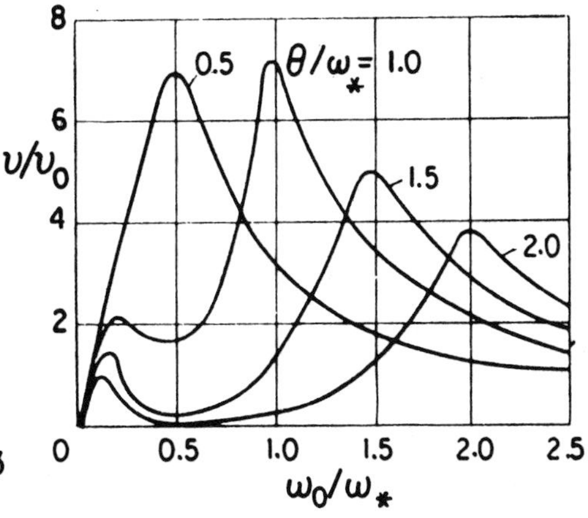

Figure 67

The data given above referred to the case when vibration of the base is a wideband process. Some results for a narrowband process at the input with the spectral density (1.20) are shown in Figure 67. Along the horizontal axis the ratio of the natural frequency ω_0 versus the typical frequency (7.99) is plotted. Along the vertical axis the ratio of the mathematical expectation of the number of excursions ν from the region (7.98) versus the mathematical expectation of the number of excursions ν_0 of the process $a_0(t)$ from the band $|a_0| < a_*$ is plotted. In this case,

$$\nu_0 = \frac{\Theta}{\pi}\left(1 + \frac{\alpha^2}{\Theta^2}\right)^{1/2} \exp\left(-\frac{a_*^2}{2\sigma_0^2}\right).$$

The numbers at the curves show values of the ratio Θ/ω_*. The remaining parameters are taken as $\langle a_0 \rangle = 0$, $a_* = 2\sigma_0$, $u_*\alpha^2 = 10^{-4}\sigma_0$, and $\gamma = 0.5$. The position of the extremum essentially depends on the value of the carry frequency Θ. The maxima here correspond to the resonance of the system with the carrying frequency and the minima, if there are any, are in the preresonance region. Further information can be found in the paper [85].

Vibration Protection of a System with Two Degrees of Freedom

Let us consider the system shown in Figure 68. Let the impacts of the masses m_1 and m_2 against each other be sufficiently rare events. In the absence of these impacts, the equations of the relative motion of this system can be taken in the form

$$\frac{d^2u_k}{dt^2} + 2\gamma_k\omega_k \frac{du_k}{dt} + \omega_k^2 u_k = -a_0(t) , \qquad (k = 1,2) ,$$

where the designations similar to (7.72) are used:

$$\omega_k = (c_k/m_k)^{1/2} , \qquad \gamma_k = b_k/(2\omega_k m_k) .$$

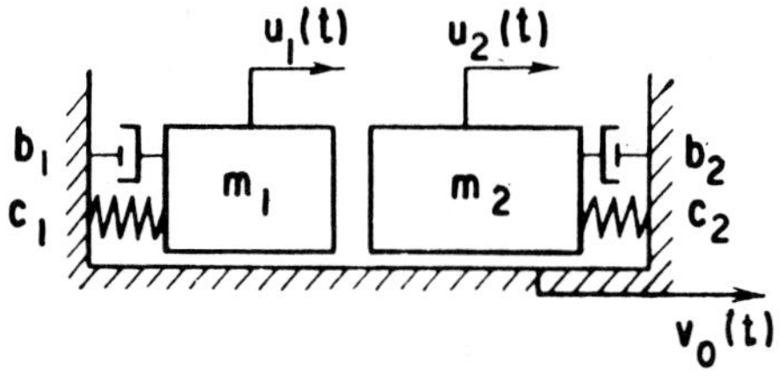

Figure 68

With the restrictions (7.90), the quality space $\underline{V}$ is three-dimensional, and, in this case, $v_1 = a_0+\ddot{u}_1$, $v_2 = a_0+\ddot{u}_2$, and $v_3 = u_1-u_2$. The admissible region Ω is a semi-infinite parallelepiped

$$|v_1| < a_{1*} , \qquad |v_2| < a_{2*} , \qquad v_3 < u_* . \tag{7.100}$$

Let the input action $a_0(t)$ be a stationary normal process with the spectral density (1.17). The mathematical expectation of the number of excursions from the region (7.100) can be found from a formula of the type (7.61), extended to the case of a three-dimensional quality space. Some results of calculations are shown in Figure 69. There it was assumed that $\langle a_0 \rangle = 0$, and $\omega_1 = 25s^{-1}$. The remaining parameters have the following values: $\gamma_1 = \gamma_2 = 0.25$, $\alpha = 1s^{-1}$, $\sigma_0 = 125m\cdot s^{-2}$, $a_{1*} = a_{2*} = 250m\cdot s^{-2}$ and $u_* = 10^{-2}m$.

On Nonlinear Vibration Protection Systems

The principle of maximum reliability is applicable to both linear and nonlinear systems. For an approximate solution of nonlinear problems, the method of statistical linearization described in Section 6.3, can, for example, be applied. In this case, we use the hypothesis that the output process has properties close to those of a normal process. Nonlinear stochastic equations are

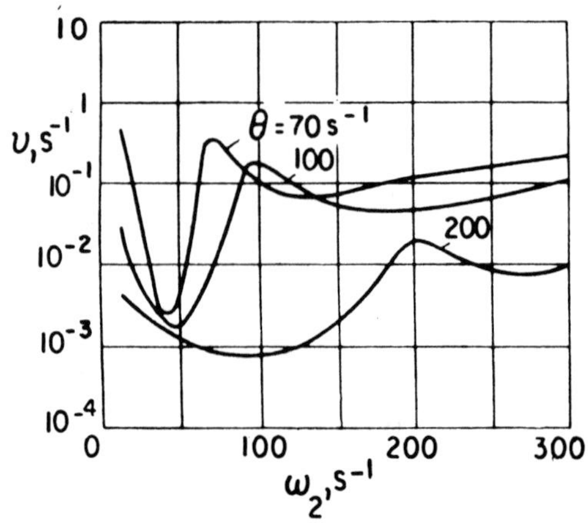

Figure 69

approximated by linear ones with coefficients depending on the mathematical expectations and the moments of second order of the analyzed processes. When the stochastic problem has been solved, and one-to-one correspondence between the parameters of the nonlinear problem and the equivalent linear problem have been found, the minimization of the number of excursions can be performed over the parameters of any of these problems. For example, in the case of a system with one degree of freedom whose equation of relative motion has the form

$$\frac{d^2u_1}{dt^2} + f\left(u_1, \frac{du_1}{dt}\right) = -a_0(t) \;, \tag{7.101}$$

the equivalent linearized equation is written in the form (7.97). Let $\langle a_0(t)\rangle \equiv 0$, and the function $f(u_1,\dot{u}_1)$ be represented as the sum $f(u_1,\dot{u}_1) = f_1(u_1)+f_2(\dot{u}_1)$, where $f_1(u_1)$ and $f_2(\dot{u}_1)$ are odd functions of their arguments. Then, we obtain from formulae (6.44) and (6.53)

$$\omega_0^2 = \frac{1}{m\sigma^2}\int_{-\infty}^{\infty} f_1(u_1)p(u_1)u_1du_1 \,, \qquad \gamma = \frac{1}{2m\omega_0 s^2}\int_{-\infty}^{\infty} f_2(\dot{u}_1)p(\dot{u}_1)\dot{u}_1d\dot{u}_1 \;. \tag{7.102}$$

Here, $p(u_1)$ and $p(\dot{u}_1)$ are probability densities of the process $u_1(t)$ and its first derivative, which are assumed to be normal, and σ^2 and s^2 are the corresponding variances.

We shall give the results of some calculations for the case of dry-friction dampers [24]. Let

$$f = h \operatorname{sign} \dot{u}_1 + \omega_0^2 u_1 ,$$

and $a_0(t)$ be a centred process with the spectral density (1.17). In this case, the equivalent frequency coincides with ω_0, and the equivalent damping factor, calculated from the second formula (7.102), is $\gamma = h[(2\pi)^{1/2}\omega_0 s]^{-1}$. Determining the spectral density $S_{u_1}(\omega)$ from equation (7.97), we shall find, with its help, the variance s^2, whence

$$\gamma = \frac{\eta^2}{1-\eta^2}\,\frac{1+\rho^2}{2\rho} , \qquad (7.103)$$

where the notations

$$\eta = (2/\pi)^{1/2} h/\sigma_0 , \qquad \rho = \alpha/\omega_0 ,$$

are used. As can be seen from formula (7.103), the latter has meaning if $\eta < 1$. This restriction can be considered as a necessary condition for the application of the statistical linearization method in the given problem. For large values of the dry-friction coefficient, the vibrations of the mass will essentially differ from a normal random process.

The graph of the dependence of the mathematical expectation of the number of excursions from the region (7.98) on the natural frequency ω_0 and on the non-dimensional dry-friction coefficient η is plotted in Figure 70. The curves in this graph can be considered as the mapping of the curves from Figure 66 by means of the transformation (7.103).

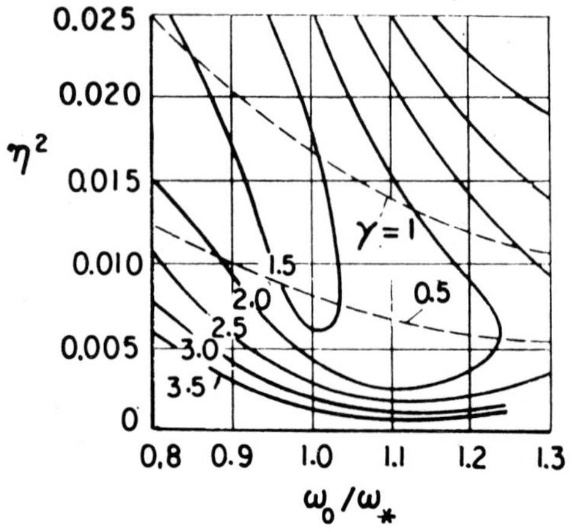

Figure 70

7.6 MODELS OF FATIGUE DAMAGE ACCUMULATION

Concepts of Phenomenological Models of Damage Accumulation

A considerable proportion of failures of members of structures and mechanisms undergoing random vibrations is caused by the accumulation of fatigue damage and breakdowns due to the development of major fatigue cracks. In this case, cyclic strength analyses are usually performed using phenomenological models of damage accumulation [10,53]. These models are based on the concept of a measure of damage, i.e., on a numerical characteristic of the degree of damage of a material, a machine member or a structural member. By its nature, this measure of damage is related to the physical picture of fracture (for example, to the density of dislocations or microcracks, to the intensity of plastic deformation, the depth of the major crack, to the degree of mechanical or corrosion wear, etc.). However, in phenomenological models, this measure is usually introduced axiomatically, not in connection with physics and mechanics of fracture. The simplest case is prescribing the measure of damage by means of a non-negative time function $v(t)$, which takes on values from the quantity v_0, corresponding to the initial state, up to the limiting quantity v_*, corresponding to fracture or, generally speaking, to failure. It is usually assumed that $v_0 = 0$ and $v_* = 1$, so that $0 \leq v(t) \leq 1$. If there are no processes of hardening or "healing" of damages, etc., damages received at different time intervals exhibit the property of additivity, so $v(t)$ is a nondecreasing time function. Furthermore, it is postulated that the measure of damage exhibits a Markov property: its increment depends only on the level of damage reached by the given time moment, and on the value of the load at the given time moment. Thus, we find that the measure of damage satisfies the equation

$$\frac{dv}{dt} = f(v, \underset{\sim}{s}) , \tag{7.104}$$

where the characteristic of the load $\underset{\sim}{s}$ is a vector-function $\underset{\sim}{s}(t)$ of time t. Equation (7.104) is integrated for the initial condition $v(0) = v_0$, and the time T to fracture (failure), which henceforth will be called longevity, is found from the condition

$$v(T) = v_* . \tag{7.105}$$

Along with equation (7.104), we shall consider its difference analog. Let the loading be performed by blocks or cycles. Then, the damage accumulated at the end of the k-th block satisfies the equation

$$v_k - v_{k-1} = w(v_{k-1}, \underset{\sim}{s}_k) , \qquad (k = 1,2,\ldots) . \tag{7.106}$$

The limiting number of blocks N is found from the condition $v_N = v_*$. In fatigue fracture, this number is sufficiently large, and the process of damage accumulation develops sufficiently slowly, so that for the description of such a process it is expedient to replace the difference equation (7.106) by its continuous analog, considering the number n as a continuous parameter:

$$\frac{dv}{dn} = w(v, \underset{\sim}{s}) \ . \tag{7.107}$$

Henceforth, as a rule, we shall not discriminate between the physical time t and the parameter n.

In the simple model described above, the unique parameter v was considered as a measure of damage. If this parameter is related to the volume element of the material, the damage of the body is described by the function $v(\underset{\sim}{x},t)$ of the coordinates $\underset{\sim}{x}$ and the time t. Then, the characteristic $\underset{\sim}{s}$ in equation (7.104) is the stress tensor for the given volume element, and equation (7.104) should be considered simultaneously with the equations of continuum mechanics. Since the parameters of the continuum depend on v, the result is a system of coupled equations of continuous fracture mechanics.*

In engineering design the measure of damage v is usually associated with a structural member or a machine member, i.e., the measure v characterizes the state of the body as a whole. The form of the functions $f(v,\underset{\sim}{s})$ and $w(v_{k-1},\underset{\sim}{s}_k)$ in the equations (7.104) and (7.106) is determined on the basis of tests on members or parts. As in the analyses of cyclic strength, the role of standard tests is played by fatigue tests with the following characteristics: the cycle is held constant, the cyclic loadings involve specified load blocks, etc. The function of phenomenological models of damage accumulation is to provide the engineer with a means for the analysis of reliability indices and those of longevity for variable (and random) loading regimes such that only a minimum of standard tests under the conditions of stationary loading need be

---

*Bolotin, V.V., "On Perspectives of the Phenomenological Approach to Problems of Fracture", *Coll. Works: Mekhanika Deformiruemikh Tel i Konstruktsii*, (Mechanics of Deformable Solids and Structures) Mashinostroenie, Moscow, 1975.

carried out. In order to verify phenomenological models, control tests for variable loading operations are used. In practice, one should take into consideration a significant scatter of strength and longevity in members of real structures and machines, the stochastic nature of loading, as well as the fact that one works with restricted samples in standard tests (even more so in control tests).

Rule of the Linear Summation of Damages

With certain restrictions on the right side of equation (7.104), the estimate of the lifetime (resource) T for loading according to an arbitrary program $\underset{\sim}{s} = \underset{\sim}{s}(t)$ can be reduced to the solution of the functional equation

$$\int_{o}^{T} \frac{dt}{T_s[\underset{\sim}{s}(t)]} = 1 \ . \qquad (7.108)$$

Here, $T_s(\underset{\sim}{s})$ is the lifetime (resource) for the standard test with the characteristic $\underset{\sim}{s}$ = const. Analogously, in the case of equation (7.106), we obtain the following condition for finding the limiting number of blocks N:

$$\Sigma \frac{n_k}{N_s(\underset{\sim}{s}_k)} = 1 \ . \qquad (7.109)$$

Here, n_k is the number of loading blocks, and $N_s(\underset{\sim}{s}_k)$ is the limiting number of blocks for standard tests with the characteristic $\underset{\sim}{s}_k$. The limiting number N is determined by the summation of all n_k.

Conditions (7.108) and (7.109) express the well-known "hypothesis of damage summation", which is widely used in engineering analyses. The hypothesis was first formulated by Palmgren as early as 1924 in connection with the processing of experimental data on the longevity of bearings. As applied to fatigue analyses, the hypothesis of damage summation is associated with the names of

Bakharev and Miner, and as applied to creep analyses and long-term strength analyses with the names of Robinson and Bailey.

Terms like "the hypothesis of damage summation", "the rule of linear summation", etc. are not quite suitable for application to the relations (7.108) and (7.109). The point is that the additivity of damages has actually been postulated in equations (7.104) and (7.106), whereas the relations (7.108) and (7.109) are derived from equations (7.104) and (7.106) by imposing certain restrictions on their right sides [10]. We shall proceed from the results of standard tests. Let the family of functions $v_s(t)$ for $\underset{\sim}{s}$ = const. be constructed, based on the results of these tests. We shall relate the function $v_s(t)$ to the nondimensional time $t/T_s(\underset{\sim}{s})$, where $T_s(\underset{\sim}{s})$ is the longevity for stationary loading. In general, $v_s = g[t/T_s(\underset{\sim}{s}),\underset{\sim}{s}]$. Assume that the measure of damage v_s is dependent only on $t/T_s(\underset{\sim}{s})$, but is not explicitly dependent on the loading characteristic $\underset{\sim}{s}$. We shall refer to relation

$$v_s = g[t/T_s(\underset{\sim}{s})] \tag{7.110}$$

as the condition of self-similarity of the damage accumulation process. Expressing the function $f(v,\underset{\sim}{s})$ in equation (7.104) in terms of the function (7.110), we find that

$$f(v,\underset{\sim}{s}) = g'[g^{-1}(v)]/T_s(\underset{\sim}{s}) \ , \tag{7.111}$$

where $g^{-1}(v)$ is the inverse of function $g(v)$. With the right side of (7.111), equation (7.104) is integrated by separation of variables, yielding

$$\int_0^T \frac{dt}{T_s[\underset{\sim}{s}(t)]} = \int_{v_0}^{v_*} \frac{dv}{g'[g^{-1}(v)]} = v_* - v_0 \ .$$

For $v_0 = 0$ and $v_* = 1$, we arrive at "the hypothesis of damage summation" (7.108). Note, that in this case damage accumulated per unit time is not equal to $T_s^{-1}(\underset{\sim}{s})$, but to the value obtained from the right side of formula (7.111), so that the rule (7.108), strictly speaking, does not express "damage summation". Henceforth, following the tradition, we shall nevertheless still use the term "rule of linear summation of damages".

Some Multi-Dimensional Generalizations

We introduce the damage vector $\underset{\sim}{v} = (v_1, v_2, \ldots, v_n)$. Depending on the physical meaning of a problem, values of the damage measure at various points or sections of a body as well as the damage measures relating to physically different mechanisms of damage can be components of this vector. In both cases, the basic equation has the form

$$\frac{d\underset{\sim}{v}}{dt} = f(\underset{\sim}{s}, \underset{\sim}{v}) \ . \qquad (7.112)$$

The critical time T is found as the minimum value of t at which the vector $\underset{\sim}{v}$ leaves for the first time some admissible region Ω. Thus interpreted, the problem of damage accumulation is very close, not only in form, but also in essence, to the basic problem of reliability theory. For example, suppose the limiting value of each component of the vector $\underset{\sim}{v}$ is equal to unity. Then, the region Ω is given by the relation $||\underset{\sim}{v}|| \leq 1$, where the norm of the vector $\underset{\sim}{v}$ is given by the relation (7.38). In most applications, the process of damage accumulation $\underset{\sim}{v}(t)$ has the properties of the quality vector in cumulative models of failure (Section 7.2)

The model of two stage fatigue damage [13] also belongs to the type (7.112). It is known that the process of fatigue fracture can be represented in two stages; in the first stage, a loosening of the material takes place and in the second, a major fatigue crack develops. Because of this, it is expedient to describe the process of fatigue damage accumulation by means of two functions. One of them, denoted by $v_1(t)$, will be interpreted as a measure for the loosening or a measure for the preparation of the material for the formation of a major fatigue crack. This measure equals zero for the initial state of the material and becomes equal to unity when the preparation stage is over. The second function, denoted by $v_2(t)$, is a measure for the development of a major crack. It is equal to zero for $v_1 \leq 1$, and it becomes unity when the length of the crack attains the critical value. Consider the corresponding generalization of equation (7.107). The two-stage process of damage accumulation is described by the system of equations

$$\frac{dv_1}{dn} = w_1(\underset{\sim}{s}, v_1) , \quad \frac{dv_2}{dn} = \begin{cases} 0 & (v_1 < 1) , \\ w_2(\underset{\sim}{s}, v_1, v_2) & (v_1 \geq 1) , \end{cases} \tag{7.113}$$

where w_1 and w_2 are functions which will be specified later.

By dividing the process of damage accumulation into two stages, we have more reasons to assume that for each stage separately the process is self-similar. We shall introduce the following two assumptions: (1) let us assume that the kinetics of the major crack development depends very little on the measure of loosening v_1, if $v_1 > 1$; (2) let us assume that for the stationary regime of loading, the measures v_1 and v_2 do not explicitly depend on the stress parameters $\underset{\sim}{s}$, but depend only on the ratios of the corresponding numbers of the cycles

$$v_{1s} = g_1\left[\frac{n}{N_{1s}(\underset{\sim}{s})}\right] , \qquad v_{2s} = g_2\left[\frac{n - N_{1s}(\underset{\sim}{s})}{N_s(\underset{\sim}{s}) - N_{1s}(\underset{\sim}{s})}\right] .$$

Here, $N_{1s}(\underset{\sim}{s})$ is the number of cycles necessary for nucleation of a fatigue crack, and $N_s(\underset{\sim}{s})$ is the limiting number of cycles. Unlike condition (7.110), here it is assumed that self-similarity only takes place for each stage separately. With due consideration for the hypothesis of self-similarity, let us integrate the first of the equations (7.113) from zero to N_1 (N_1 is the number of cycles necessary for crack nucleation), and the second equation from N_1 to N. As a result, we obtain two relations, each one analogous in structure to the rule of linear summation:

$$\int_{o}^{N_1} \frac{dn}{N_{1s}(s)} = 1 , \qquad \int_{N_1}^{N} \frac{dn}{N_s(\underset{\sim}{s})-N_{1s}(\underset{\sim}{s})} = 1 . \qquad (7.114)$$

From the first relation we find N_1, then from the second one the limiting number of cycles N.

On the Verification of the Rule of Linear Summation

Summation of fatigue damage occurs in the presence of significant statistical scatter of mechanical strength [91]. This scatter reflects the statistical character of the process of fracture; it does not disappear even when special precautions are taken (for example, all the samples are made of one piece and undergo additional rejection prior to tests). Since the rule of linear summation is intended for engineering analyses, it is necessary for its application and estimation to take into consideration the component of scatter which is associated with both nonhomogeneity and with the instability of technological regulations based on the existing level of standards of materials and acceptance tests. As a result, the longevity scatter (and it is this scatter that should be taken into account when applying the rule of summation) can be of the order one or more. The rules for damage calculations should be verified like any statistical hypothesis with all the corresponding requirements as to sample sizes, to methods of obtaining statistical estimates, and to the formulation of statistical conclusions. Unfortunately, from the point of view of mathematical statistics not one of the known experimental investigations of fatigue damage accumulation is quite adequate. In particular, sample sizes (number of tested samples) are completely insufficient.

There are two methods most frequently used. One of them is based on a two-stage loading program: first, the sample undergoes n_1 cycles at the level of loading s_1 = const., after which, at the level s_2 = const., it is brought up to fracture. In this case, if the rule of linear summation is correct, the experiments expect the numbers of cycles n_1 and n_2 of the first and second stages to be related by a linear equation following from (7.109). The other method allows any programs of loading. For these programs, we calculate the accumulated damage equal to the sum which is on the left side of the relation (7.109). The experimentors expect the accumulated damage to be equal to unity at the moment of fracture in the case when the rule of linear summation is fulfilled. However, due to irreproducibility of fracture phenomena of one and the same specimen, it is fundamentally impossible to directly verify the mentioned relations. For example, for two-stage program tests, the number of blocks (cycles) is specified, whereas the numbers n_2, $N_s(\underset{\sim}{s_1})$ and $N_s(\underset{\sim}{s_2})$ are actually stochastic values varying from one specimen to another.

Denote the random vector characterizing the mechanical properties of the specimen by $\underset{\sim}{r}$. To emphasize the fact that the relation (7.109) pertains to a single specimen, we shall write it, as applied to a two-stage loading, in the form

$$\frac{n_1}{N_s(\underset{\sim}{s_1}|\underset{\sim}{r})} + \frac{n_2}{N_s(\underset{\sim}{s_2}|\underset{\sim}{r})} = 1 \ . \tag{7.115}$$

Having fractured the specimen, we can find the realization of only one of three random values. In fact, experimentors seek only some regressive relations between the values n_1 and n_2, associating these values with some estimates for the mathematical expectations of limit numbers and expecting the analogous condition

$$\frac{n_1}{\langle N_s(\underset{\sim}{s_1})\rangle} + \frac{n_2}{\langle N_s(\underset{\sim}{s_2})\rangle} = 1 \tag{7.116}$$

to follow from (7.115). But, actually, condition (7.116) does not follow from (7.115). The correct relation following from (7.115) has the form

$$\frac{n_1}{\langle N_s(\underset{\sim}{s}_2)\rangle}\left\langle \frac{N_s(\underset{\sim}{s}_2)}{N_s(\underset{\sim}{s}_1)}\right\rangle + \frac{\langle n_2 | n_1\rangle}{\langle N_s(\underset{\sim}{s}_2)\rangle} = 1 \; . \tag{7.117}$$

All that has been stated above becomes more complicated due to the fact that averaging in (7.117) is performed not over all specimens, but only over those which satisfy certain conditions of the experiment: as a rule, experimentors do not take into account the specimens that do not reach the second stage of tests, and, therefore, are assumed to be "bad". Furthermore, all the specimens that had not been fractured by the time a certain standard number of blocks N_0 was reached, or the given limit duration of tests expired, are excluded from statistical processing. In experiments, the considered laws become less certain due to the scatter of experimental data when statistical sample sizes are insufficient. However, some problems can be investigated by numerical simulation. Assuming that the rule of linear summation for single specimens is fulfilled and that a family of stochastic fatigue curves is given, then using numerical simulation, it is possible to obtain arbitrarily large samples for the pairs of values n_1 and n_2, and with their help to construct regression lines, in order to estimate the width of confidence intervals, etc. Below we give an example of such an investigation.*

Example

For fatigue curves, the expression

$$N_s(s|r) = N_0\left(\frac{r}{s}\right)^m , \qquad (0 < s < \infty) \; , \tag{7.118}$$

*Bolotin, V.V., Yermolenko, A.F., *Summation of Fatigue Damages and Statistical Scatter of Strength,* Mashinovedenie, No. 1, 1979.

was taken, where the loading level is characterized by the numerical parameter s (for example, by the stress amplitude in the case of symmetric cycles). The positive constant N_0 will be identified with the standard number of cycles, and the parameter r will be considered as a random variable taking on positive values. This parameter, equal to the fracturing stress found from tests, can be interpreted as the individual fatigue limit of the specimen. For the fatigue limit r we shall take Weibull's distribution

$$F(r) = \begin{cases} 1 - \exp\left[- \left(\dfrac{r - r_0}{r_c} \right)^{\alpha} \right] & (r \geq r_0) , \\ 0 & (r < r_0) , \end{cases} \tag{7.119}$$

where $r_0 > 0$, $r_c > 0$, and $1 \leq \alpha < \infty$. The program, i.e., the transducer of pseudorandom numbers, processed sample values of the fatigue limit distributed according to (7.119). Then, the process of a two-stage loading was simulated. The realizations of the values $N_s(s_1|r)$ and $N_s(s_2|r)$ were found for the chosen specimen from the formula (7.118), after which the value n_2 corresponding to the given value n_1 was calculated from equation (7.115). All the results which did not satisfy the conditions $N_s(s_1|r) \geq n_1$ and $n_1 + n_2 \leq N_0$ were rejected.

Some results of the simulation are shown in Figures 71 and 72. The values of the parameters taken were $m = 8$, $\alpha = 4$ and $r_c/r_0 = 0.5$. The results of single tests are plotted in light circles. The values n_1 and n_2 are associated with the mathematical expectations of the limit numbers of the cycles $\langle N_s(s_1) \rangle$ and $\langle N_s(s_2) \rangle$, respectively. Averaged over 50 specimens, the values of n_2 are plotted in dark circles, and the dependence of the conditional mathematical expectation $\langle n_2 | n_1 \rangle$ on n_1 calculated from the equation (7.117), is plotted in a solid line. The dash-dot lines are the boundaries of the 90 percent confidence intervals for the averages of the results (over 50 specimens). Figure 71 corresponds to the case when $s_1/r_0 = 1.2$ and $s_2/r_0 = 1.6$, and it clearly shows the effect of the apparent hardening, which is explained mainly by the truncation of samples on the basis of the condition $N(s_1|r) \geq n_1$. Figure 72 is plotted for the inverse order of loading, i.e., for $s_1/r_0 = 1.6$ and $s_2/r_0 = 1.2$. The regression analysis shows a small apparent loss of strength if the first stage of loading is sufficiently short.

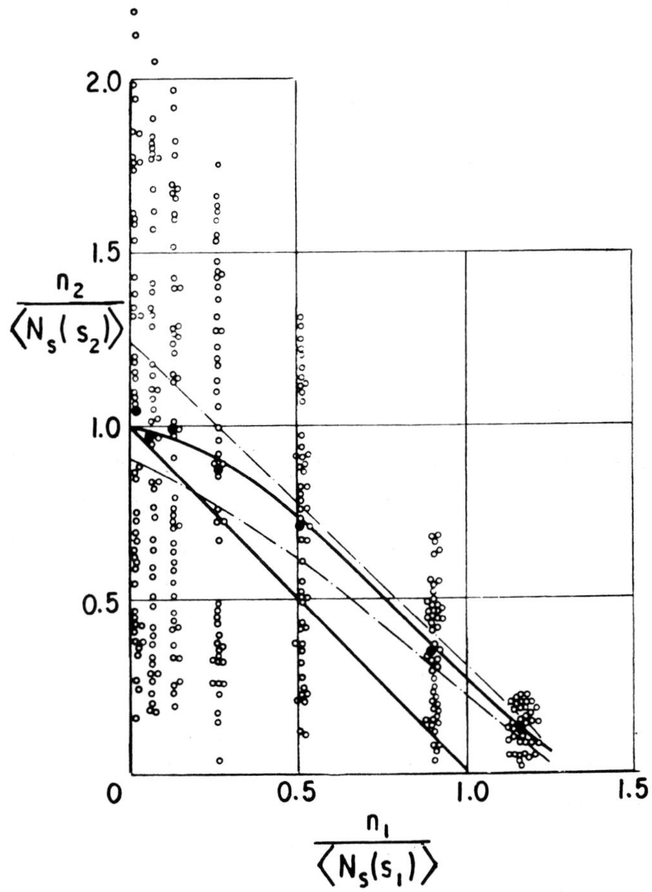

Figure 71

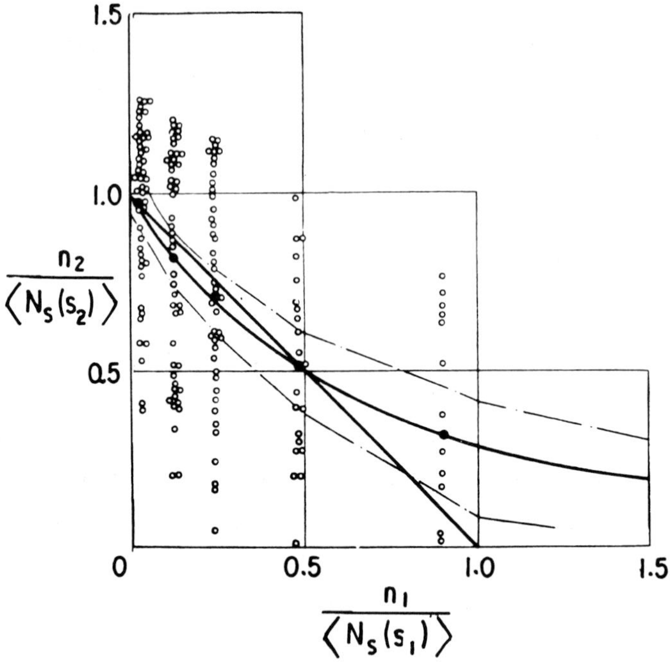

Figure 72

Stochastic simulation shows a significant scatter of the results for single tests. This scatter is already inherent in the distribution (7.119) for the fatigue limit. For the chosen values of numerical data, the coefficient of the fatigue limit variation is about 10 percent, which, with the index of the fatigue curve m = 8, gives for the coefficient of longevity variation a value of the order of unity. Such a significant scatter also takes place in the physical experiment unless special measures are taken to diminish it. Similar results are obtained if the tests are simulated in a two-stage program proceeding from nonlinear rules of summation, say, from the relation (7.114). In all the investigated cases, the deviation of the "true" law of damage accumulation from the linear law is not very significant. The number of tested specimens, the attitude of an experimentor to the "bad" specimens, the method of statistical processing and the showing of results of tests may prove to be no less essential for statistical conclusions than the influence of the loading history.

A similar situation is found when rules of linear summation are verified by means of a calculation of the accumulated damage. Actually, experimentors do not calculate the accumulated damage but the value

$$\check{v} = \Sigma \frac{n_k}{\overline{N_s(s_k)}}, \tag{7.120}$$

where $\overline{N_s(s)}$ is an estimate for the mathematical expectation of the limit number of cycles for s = const., with summation performed before the moment of fracture. It is easy to see that the value $\check{v}$ should not be equal to unity: it is a random value whose variation is of the same order as the variation of the limiting number of cycles $N_s(s)$. This conclusion is supported by numerous experimental data. A number of experimental results are given in the paper [118], where the realizations of the value $\check{v}$ vary from 0.11 to 4.88.

In summing up, we should note that it is difficult to find among the published experimental data results of such significance as to reject the rule of linear summation as a statistical hypothesis.

7.7 CHARACTERISTIC LONGEVITY IN THE CASE OF RANDOM VIBRATIONS

Concept of Characteristic Longevity

Assume that the loading process $\underset{\sim}{s}(t)$ is a random one. In this case, the damage accumulation process v(t) is also random, and the lifetime (resource) T is a random value. The distribution of this value F(T) is related to the reliability function R(t) by equation (7.12). Apart from the loading process, a stochastic element $\underset{\sim}{r}$, which characterizes the mechanical properties of the material and structure, enters into the analyses of reliability and longevity. In the majority of the analyses associated with random loadings, mechanical properties can be assumed not to vary with time, in which case $\underset{\sim}{r}$ is a numerical random vector with the distribution $p(\underset{\sim}{r})$ independent of the process $\underset{\sim}{s}(t)$. In this case, the solution of the problem is naturally divided into two stages. In the first stage, the characteristics of the conditional process $v(t|\underset{\sim}{r})$ are determined, as well as the conditional distribution of the lifetime $F(T|\underset{\sim}{r})$, the conditional reliability function $R(t|\underset{\sim}{r})$, etc. In the second stage, the scatter of the mechanical properties is taken into account by using the formulae of total probability of the type (7.5). Specifically, for probability density $p(T)$, we have the formula

$$p(T) = \int p(T|\underset{\sim}{r})p(\underset{\sim}{r})d\underset{\sim}{r} \ . \tag{7.121}$$

The simultaneous accounting of random properties of loads and cyclic strength presents considerable difficulties. Therefore, the contemplated scheme is not fully realized, not only in engineering analyses, but in applied research as well [53,91]. At best, the following simplified variant is used: the conditional distribution is represented as

$$p(T|\underset{\sim}{r}) \sim \delta[T-T_*(\underset{\sim}{r})] \ , \tag{7.122}$$

where $\delta(T)$ is the delta-function and $T_*(\underset{\sim}{r})$ is a certain typical (nonstochastic) value of the longevity index. After this, an approximate unconditional distribution of lifetime (resource) is calculated from formula (7.121). We shall denote the corresponding approximation by $p_0(T)$. Substituting the distribution (7.122) into formula (7.121) gives

$$p_0(T) = -\frac{d}{dT}\left[\int_{T<T_*(\underset{\sim}{r})} p(\underset{\sim}{r})d\underset{\sim}{r}\right] , \tag{7.123}$$

where integration is performed with respect to the set of values of the vector $\underset{\sim}{r}$ for which $T < T_*(\underset{\sim}{r})$.

We shall refer to the longevity index $T_*(\underset{\sim}{r})$ entered in the distribution (7.122) as characteristic longevity. This value can be found by using a model of fatigue damage accumulation. Let v be the damage measure taking on values from the interval $[v_0, v_*]$ where v_* is the critical value of the damage measure. Let us consider the process of damage accumulation $v(t|\underset{\sim}{r})$ in a structure with a given value of the vector $\underset{\sim}{r}$. The conditional longevity index $T(\underset{\sim}{r})$ is a random value determined from the relation

$$v(T|\underset{\sim}{r}) = v_* \; . \tag{7.124}$$

The characteristic conditional longevity $T_*(\underset{\sim}{r})$ can, for instance, be introduced as the root of the equation

$$\langle v(T_*|\underset{\sim}{r})\rangle = v_* \; , \tag{7.125}$$

where the angular brackets denote averaging over a set of realizations of the conditional process $v(t|\underset{\sim}{r})$.

Consider methods for estimating the characteristic conditional longevity, temporarily omitting the argument $\underset{\sim}{r}$ from the functions $v(t|\underset{\sim}{r})$, $T(\underset{\sim}{r})$, etc.

Proceeding from equation (7.104), we shall try to construct the equation for the mathematical expectation $\langle v(t)\rangle$. Averaging the equation (7.104) over a set of realizations of the process $\underset{\sim}{s}(t)$, we obtain

$$\frac{d\langle v\rangle}{dt} = \langle f(\underset{\sim}{s}, v)\rangle \; .$$

In order to close this equation for $\langle v\rangle$, we substitute on its right side $\langle f(\underset{\sim}{s}, v)\rangle \approx f(\underset{\sim}{s}, \langle v\rangle)$. The more compact the distribution of $p(v;t)$ is, the more exactly will this relation be satisfied. As a result, the equation will have the form

$$\frac{d\langle v\rangle}{dt} = f(\underset{\sim}{s},\langle v\rangle) \ . \qquad (7.126)$$

This equation is integrated with the initial condition being $\langle v(0)\rangle = v_0$, and the characteristic longevity T_* is then determined from the condition (7.125).

Thus, the stochastic problem of finding the distribution law p(T) of the random value T is replaced by the deterministic problem described by equation (7.126) and by the boundary conditions $\langle v(0)\rangle = v_0$, $\langle v(T_*)\rangle = v_*$. This is shown in Figure 73. The smaller the scatter of the longevity T is, i.e., the more compact the distribution p(T) is, the more reason we have to say that the stochastic problem and the deterministic one are close to each other. The characteristic longevity T_* generally does not coincide with the mathematical expectation of the longevity $\langle T\rangle$ (this will be considered further below).

The considerations given above serve as a basis for the engineering approach to the estimate of longevity under random vibrations. This approach is widely used in engineering practice [13,53], and the stochastic character of the process v(t) and of the longevity T is not emphasized. The mathematical expectation of the damage measure is identified with the damage v, and the characteristic longevity with the longevity T. In order not to contradict the universally accepted representation, we shall temporarily omit the angular brackets in $\langle v\rangle$ in equation (7.126) and in the boundary condition (7.125). Averaging the right side of equation (7.126) by means of the density $p(\underset{\sim}{s};t)$, we shall rewrite this equation in the form

$$\frac{dv}{dt} = \int f(\underset{\sim}{s},v)p(\underset{\sim}{s};t)d\underset{\sim}{s} \ , \qquad (7.127)$$

where integration is performed with respect to the entire region of the variation of the vector $\underset{\sim}{s}$.

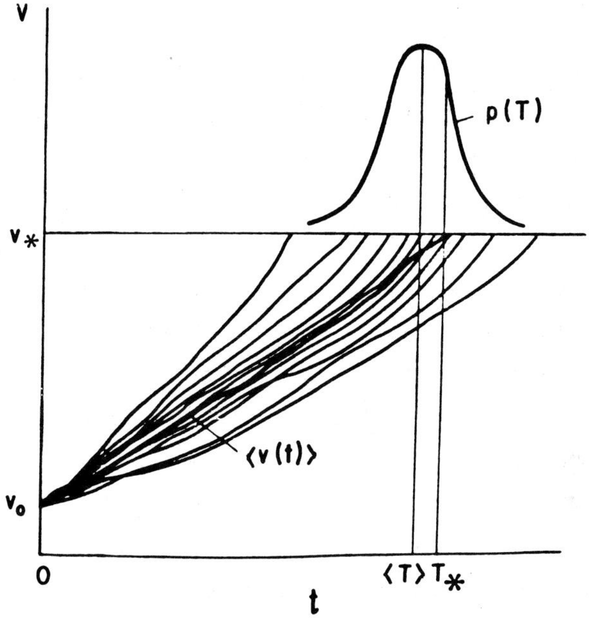

Figure 73

Formulae for the Characteristic Longevity under Stationary Random Vibrations

If $\underset{\sim}{s}(t)$ is a stationary random process, the variables v and t in equation (7.127) are separable. In order to find the characteristic longevity we have the formula [10]:

$$T_* = \int_{v_0}^{v_*} \frac{dv}{\int f(\underset{\sim}{s},v)p(\underset{\sim}{s})d\underset{\sim}{s}} \, . \tag{7.128}$$

For a self-similar process of damage accumulation, when relation (7.110) is satisfied, the formula (7.128) is transformed into the form ($v_0 = 0$, $v_* = 1$)

$$\frac{1}{T_*} = \int \frac{p(\underset{\sim}{s})d\underset{\sim}{s}}{T_s(\underset{\sim}{s})} \, . \tag{7.129}$$

Consider the application of formula (7.129) for an estimate of the fatigue longevity under stationary random loading [10,13]. Let $N_s(\tilde{s})$ be the number of cycles or blocks to fracture under programmed loading characterized by the vector $\tilde{s}$; and let $T_e(\tilde{s})$ be the duration of one cycle or block of loading, measured in units of time. Then, the time to fracture under programmed loading will be $T_s(\tilde{s}) = T_e(\tilde{s})N_s(\tilde{s})$, and formula (7.129) takes the form

$$\frac{1}{T_*} = \int \frac{p(\tilde{s})d\tilde{s}}{T_e(\tilde{s})N_s(\tilde{s})} , \tag{7.130}$$

where $p(\tilde{s})$ is the probability density for the vector $\tilde{s}$, and integration is performed with respect to all possible values of this vector. The vectorial character of $\tilde{s}$, which makes the formula (7.130) quite universal, should be emphasized. If s(t) is a one-dimensional narrowband stationary random process, we can take as a loading parameter the random amplitude s for almost symmetric cycles into which the realization of the narrowband process decomposes. The distribution p(s) coincides with the distribution of the envelope s(t) of the narrowband process.

Let us consider some formulae for the calculation of characteristic longevity. Let s(t) be a narrowband normal process with mathematical expectation equal to zero, with the variance σ_s^2, and with the carrying frequency ω_e (the effective period $T_e = 2\pi/\omega_e$). The amplitudes of this process are distributed according to Rayleigh's law

$$p(s) = \frac{s}{\sigma_s^2} \exp\left(-\frac{s^2}{2\sigma_s^2}\right) , \qquad (s \geq 0) . \tag{7.131}$$

For the fatigue curve $N_s(s)$, we take the approximation

$$N_s(s) = \begin{cases} N_0(r/s)^m , & (s \geq r) , \\ \infty & (s < r) , \end{cases} \tag{7.132}$$

where N_0 is a standard number of cycles and r is the fatigue limit for a symmetric cycle. In this context, r is a nonstochastic variable (the problem of the influence of interspecimen scatter on the longevity distribution under random loading will be considered in Section 7.8). Upon substituting the expressions (7.131) and (7.132) into the formula (7.130), we obtain

$$T_* = T_0 u_0^m / I_m(u_0) \,, \tag{7.133}$$

where the notations

$$T_0 = N_0 T_e \,, \qquad u_0 = \frac{r}{\sigma_s} \,, \qquad I_m(u_0) = \int_{u_0}^{\infty} u^{m+1} e^{-u^2/2} du \tag{7.134}$$

are used.

The integral $I_m(u_0)$ is expressed in terms of the incomplete gamma-function $\Gamma(\alpha,x)$. As a result, formula (7.133) takes the form

$$T_* = \frac{T_0 u_0^m}{2^{m/2}\Gamma(1+ \frac{1}{2}\, m,\, \frac{1}{2}\, u_0^2)} \,. \tag{7.135}$$

For calculation with this formula, tables of the incomplete gamma-function are required. If these tables are not available, it is possible to use the Pearson χ^2-distribution tables. These can be found in all reference books and manuals on mathematical statistics. Indeed, $\Gamma(1 + \frac{1}{2}\, m,\, \frac{1}{2}\, u_0^2) = \Gamma(1 + \frac{1}{2}\, m) P_{m+2}(u_0^2)$, where $\Gamma(x)$ is a complete gamma-function and $P_n(\chi^2)$ is the Pearson χ^2-distribution function with n degrees of freedom, or more exactly, the complement of this distribution function with respect to one. The formula for an estimate of the characteristic longevity with a symmetric cycle was given in [10] in the form

$$T_* = \frac{T_0 u_0^m}{2^{m/2}\Gamma(1 + \frac{1}{2}\, m) P_{m+2}(u_0^2)} \,. \tag{7.136}$$

The graph for the ratio T_*/T_0 as a function of m and $u_0 = r/\sigma_s$ is given in Figure 74.

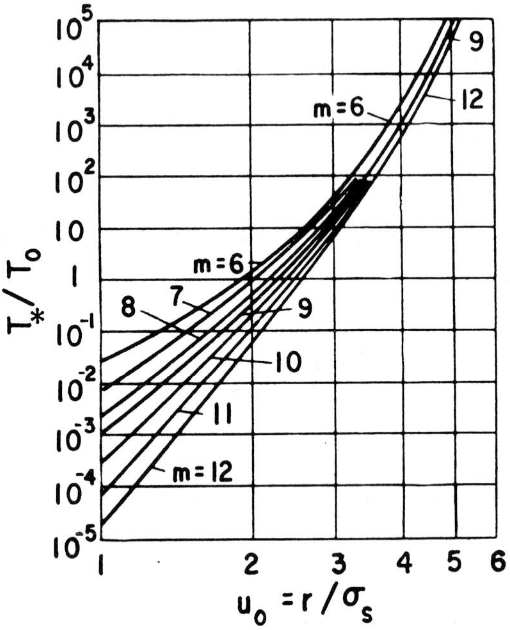

Figure 74

If the stress level is sufficiently high, i.e., $\sigma_s >> r$, it is possible to assume that $P_{m+2}(u_0^2) \approx 1$. Formula (7.136) is thus simplified to read

$$T_* = \frac{T_0}{2^{m/2}\Gamma(1 + \frac{1}{2}m)} \left(\frac{r}{\sigma_s}\right)^m . \qquad (7.137)$$

This formula coincides with the well-known formula of Miles [154] for the case when the fatigue curve is specified in the form (7.118), i.e., when there is no horizontal part in the fatigue curve.

Let $\sigma_s << r$, which corresponds to a very low stress level. By replacing $u^2/2 = z$, $u_0^2/2 = z_0$ and $m/2 = \mu$, we shall write the integral in (7.134) as

$$I_m(z_0) = 2^{\mu} \int_{z_0}^{\infty} z^{\mu} e^{-z} dx \ .$$

Integrating it by parts, we obtain an expansion in terms of decreasing powers of z_0, i.e.,

$$\int_{z_0}^{\infty} z^{\mu} e^{-z} dz = e^{-z_0} [z_0^{\mu} + \mu z_0^{\mu-1} + \mu(\mu-1) z^{\mu-2} + \cdots] \ .$$

Hence, returning to the original variables, we arrive at the asymptotic expansion

$$I_m(u_0) = e^{-1/2 u_0^2} [u_0^m + m u_0^{m-2} + m(m-2) u_0^{m-4} + \cdots] \ .$$

The stronger the inequality $u_0^2 >> m$ is, the faster this expansion will converge. For the characteristic longevity, we obtain the asymptotic formula

$$T_* \approx \frac{T_0 e^{1/2 u_0^2}}{1 + m u_0^{-2} + m(m-2) u_0^{-4} + \cdots} \ .$$

If $\sigma_s << r m^{-1/2}$, then,

$$T_* \approx T_0 \ \exp(r^2/(2\sigma_s^2)) \ . \qquad (7.138)$$

The fatigue curve index m does not enter into this formula. This is explained by the fact that at very low stress levels damage accumulates mainly due to those rare cycles that slightly exceed the fatigue limit r. Some generalizations of the formulae (7.130), (7.135), etc., with an allowance for complications, can be found in [10,13].

Characteristic Longevity for Nonsymmetric and Complex Cycles

So far, we have considered loading in the form of a narrowband random process with a mathematical expectation equal to zero.

Realizations of such a process consist of sequences of symmetric cycles. All the conclusions also hold true for narrowband loading about some constant mean value s_0. Here, s should be treated as the load amplitude (stress amplitude) s_a, and the mean load (mean stress) as a parameter. The fatigue curves $N_s(s_a, s_0)$ are determined in this case from program tests with constant mean values and constant load amplitudes.

Let the loading be a sequence of uncorrelated nonsymmetric cycles with maximum values s_{max} and minimum values s_{min}. Each cycle is characterized by the two-dimensional vector $\underset{\sim}{s} = (s_{max}, s_{min})$. For a calculation by means of the formula (7.130), it is necessary to know the joint probability density $p(s_{max}, s_{min})$ for consecutive extrema, and the family of fatigue curves $N = N_s(s_{max}, s_{min})$. The duration of one cycle T_e may also depend on the cycle's parameters. In this case, formula (7.130) takes the form

$$\frac{1}{T_*} = \iint \frac{p(s_{max}, s_{min}) ds_{max} ds_{min}}{T_e(s_{max}, s_{min}) N_s(s_{max}, s_{min})} . \tag{7.139}$$

It is often more convenient to describe the properties of the loading process s(t) by specifying the mean value s_0 and the amplitude value s_a. As an example, we can take the process formed by superimposing the rapidly varying narrowband process $s_1(t)$ on the slow process $s_0(t)$. Instead of (7.139), we obtain the analogous formula

$$\frac{1}{T_*} = \iint \frac{p(s_a, s_0) ds_a ds_0}{T_e(s_a, s_0) N_s(s_a, s_0)} ,$$

whose notations have meanings that are clear from the context. For the estimate of the longevity at nonsymmetric regimes of loading, it is necessary to have information about the fatigue surfaces $N = N_s(s_{max}, s_{min})$ or $N = N_s(s_a, s_0)$. So far, we have not obtained sufficient experimental data for the probabilistic description of these surfaces. The analyses have to be based on the data for the fatigue limit at nonsymmetric loading cycles, assuming that sections of fatigue surfaces with the planes $\sigma_{max}/\sigma_{min}$ = const. or σ_a/σ_0 = const. are a family of curves with equations of the type (7.118) or (7.132) for equal indices m.

Under the action of random vibrations, many natural modes can be excited simultaneously. If we consider stresses at some fixed point of a structure, they represent a wideband random process in which the so-called complex cycles play an essential part [13].

Consider the quasistationary wideband random process $s(t,\tau)$ where one can distinguish "rapid" time t and "slow" time τ. When differentiating with respect to t, and also when integrating and averaging with respect to time within intervals of the order $\Delta\tau$, the time τ can be considered as a parameter. We shall separate the slowly varying part $s_0(\tau)$ from the rapidly varying part $s_1(t,\tau)$ for the process $s(t)$. As we will see later, it is convenient to choose $s_0(\tau)$ so that the average number of zeros of the process $s_1(t,\tau)$ is as close as possible to the average number of extrema of this process. Here, we shall restrict ourselves to the cases when within each interval $\Delta\tau$ the function $s(t)$, and consequently $s_1(t,\tau)$, can be treated as a fully representative interval of realization of an ergodic stationary process. In this case, it is convenient to take for $s_0(\tau)$ the result of averaging the function $s(t)$ with respect to the "rapid" time.

The suggested representation is natural from the point of view of practical problems. Random stresses appearing in a structure under vibrations often consist of a constant or slowly varying component $s_0(\tau)$ and a rapidly varying component $s_1(t,\tau)$. The component $s_0(\tau)$ can be deterministic or stochastic. The component $s_1(t,\tau)$ is usually stochastic, with its probabilistic characteristics varying sufficiently slowly. Realizations of the process $s_1(t,\tau)$ with probabilistic characteristics varying as slowly as necessary for processing the results can be obtained for special tests.

In order to be able to apply the rule of damage summation to wideband quasistationary random processes, it is necessary to introduce the principle of division into intervals each of which could be considered with sufficient exactness as a closed cylce of stresses. It is obvious that such a division can be performed by various methods. Assume that damages caused by varying stresses $s_0(\tau)$ can be ignored. Then, the following method of division of the process $s(t)$ into cycles seems to be the most expedient: let us draw the line $s = s_0(\tau)$, and mark the points of intersection of the realization $s(t)$ with this line. The interval of realization $s(t)$ bounded by two adjacent intersections of the level $s = s_0(\tau)$ with a positive derivative will be referred to as a stress cycle. In this case, the average number of cycles per unit time concides with half of the average number of zeros of the process $s_1(t,\tau)$.

The most essential characteristics of the cycle, from the fatigue damage viewpoint, are the extremal values of stress. We shall classify cycles in accordance with the number of extrema. If a cycle contains one maximum and one minimum, it will be referred to as a simple cycle. If a cycle contains more than two extrema, it will be a complex cycle. If a cycle has four extrema, it will be referred to as a double cycle (since it contains an "inner" cycle), a cycle with six extrema - a triple cycle, etc.

A simple cycle is characterized by two extremal values of stresses ($s_{max}^{(1)}$ and $s_{min}^{(1)}$), a double cycle by four values ($s_{max}^{(1)}$, $s_{min}^{(1)}$, $s_{max}^{(2)}$ and $s_{min}^{(2)}$), etc. Another characteristic of a cycle is the stress s_0 at which the cycle "closes". Note that complex cycles, in their turn, can be divided into groups depending on the location of the minima with respect to the stress s_0. Thus, there are two types of double cycles (for $s_{min}^{(1)} > s_0$, and for $s_{min}^{(1)} < s_0$), three types of triple cycles, etc.

The relative contents of complex cycles depends primarily on spectral properties of the process $s_1(t,\tau)$ (it will be recalled that time τ is treated as a parameter in considering the high-frequency component). The narrower the process spectrum is, the smaller the part which the complex cycles play in it. In the limiting case of the process $s_1(t,\tau)$ with deterministic frequency (for which the spectral density has the form of a delta-function), all the cycles are simple ones.

To estimate longevity under vibrations with complex cycles of stresses, the relations (7.128) and (7.130) can be used. In this case, $\underset{\sim}{s}$ stands for the vector whose components are equal to the characteristic values of the loading process $\underset{\sim}{s} = (s_0, s_{max}^{(1)}, s_{min}^{(1)}, s_{max}^{(2)}, s_{min}^{(2)}, \ldots)$. If the cycle is a simple one, all the $s_{max}^{(k)}$ and $s_{min}^{(k)}$ starting from $k = 2$ should be assumed as equal to s_0. If the cycle has j inner cycles, all the components starting from $s_{max}^{(j+2)}$, $s_{min}^{(j+2)}$ should be assumed as equal to s_0. The probability density $p(\underset{\sim}{s})$ can be expressed in terms of the corresponding densities $p_I(s_0, s_{max}^{(1)}, s_{min}^{(1)})$ for simple cycles, $p_{II}(s_0, s_{max}^{(1)}, s_{min}^{(1)}, s_{max}^{(2)}, s_{min}^{(2)})$ for double cycles, etc., as well as in terms of the probability of a cycle taken at random belonging to a certain type. Analogously, the task of prescribing the function $N_s(\underset{\sim}{s})$ is reduced to prescribing the fatigue surfaces $N_I(s_{max}^{(1)}, s_{min}^{(1)})$, $N_{II}(s_{max}^{(1)}, s_{min}^{(1)}, s_{max}^{(2)}, s_{min}^{(2)})$, etc. for simple, double, etc. cycles. Details can be found in [13].

Engineering longevity analysis in the presence of complex cycles is based on replacing the given loading process by some process equivalent in "damageability", but consisting of only simple cycles. Strict equivalence is certainly not to be expected. The method of maxima, in which the number of simple cycles is assumed to be equal to the mean number of maxima of the process s(t), can serve as an example. The process s(t) is replaced by a symmetric one with respect to the level $s = s_0$. The amplitudes of the process are assumed to be equal to $s_a = |s_{max} - s_0|$, if $s_0 \geq 0$, and $s_a = |s_0 - s_{min}|$, if $s_0 < 0$. In the simplified variant of this method, the process s(t) is replaced by a sequence of symmetric cycles with the amplitude $s_a = |s_{max}|$ if $s_0 \geq 0$, and $s_a = |s_{min}|$ if $s_0 < 0$.

The maxima method usually gives a lower bound estimate for longevity, the second variant giving a too crude estimate. The method of excursions gives somewhat better results [10]. According to this method, the given process is replaced by a sequence of simple cycles, the number of which is equal to an average number of intersections of the level $s = s_0$ by the process $s(t)$. The amplitudes of an equivalent process are chosen in the same way as in the method of maxima. The process thus chosen may prove to be less damaging than the actually given process. For narrow-band processes, the method of maxima and the excursion method give close results. Some formulae for analyzing and comparing results obtained by the method of maxima and the method of excursions can be found in [13,53].

If a loading process is specified by its realizations, the method of amplitudes and the method of full cycles can be applied for longevity analysis [53]. According to the amplitude method, the amplitudes of an equivalent process are chosen to be equal to half the difference between the adjacent minimum and maximum of the process. According to the method of full cycles, inner cycles are successively eliminated from realizations of the process $s(t)$. It is obvious that the method of full cycles is close to the excursion method.

A systematic comparison of the methods listed above within the scope of a certain probabilistic model (a stationary normal loading process and the power law of fatigue) can be found in [37]. Figure 75 shows the dependence of the characteristic longevity T_* on the parameter of bandwidth β, determined from the formula

$$\beta = \frac{\left[\int_0^\infty S(\omega)\omega^4 d\omega \int_0^\infty S(\omega) d\omega\right]^{1/2}}{\int_0^\infty S(\omega)\omega^2 d\omega} .$$

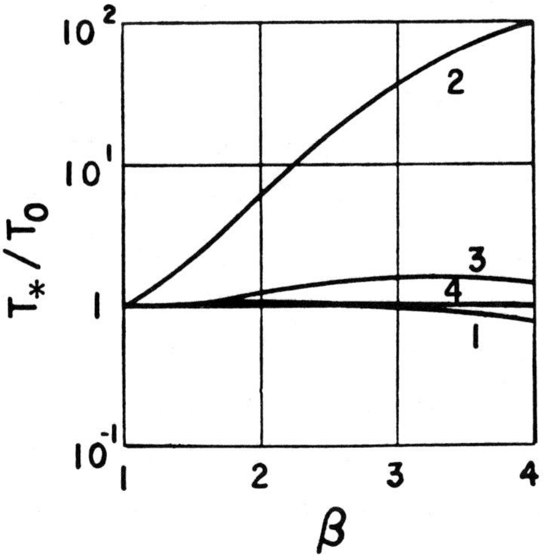

Figure 75

In this case, T_0 is the longevity calculated by the excursion method. Curve 1 is found by the method of maxima, curve 2 by the amplitude method, and curve 3 by the method of full cycles. The discrepancy between the results increases rapidly with the increase of the parameter of bandwidth β. This appears natural, because as β increases, the contents of complex cycles increases too. In this example, the excursion method and the method of full cycles give intermediate and, evidently, most realistic estimates for longevity.

The model considered in [37] does not involve the possibility of making allowance for damage caused by the high-frequency small-amplitude component. In fact, according to the experimental data, this component can cause essential additional damage. In general, the typical error of engineering methods of analysis is caused not only by a sketchy representation of the loading process, but by a sketchy representation of the fatigue curves as well. Also, the strength of a material under nonsymmetric and complex cycles is of great significance in this case.

7.8 DISTRIBUTION OF RELIABILITY AND LONGEVITY IN THE CASE OF RANDOM VIBRATIONS

Estimate of the Reliability Function by Characteristic Longevity

If we ignore that part of longevity scatter which is caused by the variability of loading realizations, we can express the distribution of longevity indices in terms of the conditional characteristic longevity $T_*(\tilde{r})$ and the distribution $p(\tilde{r})$ of the cyclic strength parameters. The approximate expression $p_0(T)$ for the probability density of the lifetime (resource) is calculated in this case by means of formula (7.123). The formula for the reliability function $R_0(r)$ takes on the form

$$R_0(t) = \int_{T_*(\tilde{r})>t} p(\tilde{r})d\tilde{r} , \qquad (7.140)$$

where the integration is performed with respect to the set of values of the vector $\tilde{r}$ for which $T_*(\tilde{r}) > t$.

If r is a one-dimensional random value with the distribution function $F(r)$, then, instead of (7.123) and (7.140), we obtain:

$$p_0(T) = \frac{d}{dt} F[r_*(T)] , \qquad R_0(t) = 1-F[r_*(t)] . \qquad (7.141)$$

Here, $r_*(t)$ is the root of the equation $T_*(r) = t$. For example, in the case of characteristic longevity determined from the approximate formula (7.137), we have

$$r_*(t) = \sigma_s (t/T_0)^{1/m} 2^{1/2} [\Gamma(1 + \frac{1}{2} m)]^{1/m} .$$

Let the parameter r obey Weibull's distribution (7.119) for $r_0 = 0$. Then, we easily find that the reliability function $R_0(t)$, determined from the second formula (7.141), obeys Weibull's law $R_0(t) = \exp[-(t/t_c)^{\beta}]$ with the parameters $t_c = T_*(r_c)$ and $\beta = \alpha/m$.

Application of the Central Limit Theorem

In the simplified scheme described above, random characteristics of the loading process are partially taken into consideration. However, the scatter of longevity $T(\underset{\sim}{r})$ is assumed to be negligibly small, so that the conditional distribution $p(T|\underset{\sim}{r})$ is approximated by the delta-function (7.122). If this scatter is actually small as compared to the contribution that is made to it by the variability of mechanical properties, this approximate approach is justified. However, it can be expected that in many applications scatter of conditional longevity $T(\underset{\sim}{r})$ cannot be considered as small. Unfortunately, in such a general formulation, the question of simultaneous influence of the variability of stresses in a vibrating structure and the variability of mechanical properties cannot have any other answer. For a more detailed analysis it is necessary to consider specific probability models. Further below, some considerations concerning cumulative models (Section 7.2) will be used.

Let the conditional process of damage accumulation $v(t|\underset{\sim}{r})$ satisfy the equation

$$\frac{dv}{dt} = f[\underset{\sim}{s}(t), \underset{\sim}{r}] \ , \qquad (7.142)$$

whose right side $f[\underset{\sim}{s}(t),\underset{\sim}{r}]$ depends on the random process of the loading $\underset{\sim}{s}(t)$. We take the initial and the critical values of the damage measure as $v_0 = 0$, $v_* = 1$, respectively. If the damage accumulation is a sufficiently slow process as compared to $\underset{\sim}{s}(t)$, the correlation time τ_c of the random vibrations being small as compared to the characteristic longevity $T_*(\underset{\sim}{r})$, we can assume that the values of the process $v(t|\underset{\sim}{r})$ are asymptotically normally distributed. Then, we obtain the approximate expression for the conditional distribution of the damage measure

$$p(v;t|\underset{\sim}{r}) \sim \frac{1}{\sqrt{2\pi}\sigma_v(t|\underset{\sim}{r})} \exp\left\{-\frac{[v-\langle v(t|\underset{\sim}{r})\rangle]^2}{2\sigma_v^2(t|\underset{\sim}{r})}\right\} . \qquad (7.143)$$

Here, $\langle v(t|\underset{\sim}{r})\rangle$ is the mathematical expectation and $\sigma_v^2(t|\underset{\sim}{r})$ is the variance of the process $v(t|\underset{\sim}{r})$. The function of conditional longevity distribution $T(\underset{\sim}{r})$ takes on the form

$$F(T|\underset{\sim}{r}) \sim 1 - \Phi\left[\frac{1-\langle v(T|\underset{\sim}{r})\rangle}{\sigma_v(T|\underset{\sim}{r})}\right] . \qquad (7.144)$$

If $\underset{\sim}{s}(t)$ is a stationary random process, then

$$\langle v(t|\underset{\sim}{r})\rangle = \mu(\underset{\sim}{r})t , \qquad \sigma_v^2(t|\underset{\sim}{r}) \approx \nu(\underset{\sim}{r})t , \qquad (7.145)$$

where $\mu(\underset{\sim}{r})$ and $\nu(\underset{\sim}{r})$ are some constants. The stronger the inequality $\tau_c << t$, the more exactly is the second (approximate) equality fulfilled. The constants $\mu(\underset{\sim}{r})$ and $\nu(\underset{\sim}{r})$ are given by the formulae

$$\mu(\underset{\sim}{r}) = \frac{1}{T_*(\underset{\sim}{r})} , \qquad \nu(\underset{\sim}{r}) = 2\int_0^\infty K_f(\tau|\underset{\sim}{r})d\tau . \qquad (7.146)$$

In this case, $K_f(\tau|\underset{\sim}{r})$ is the correlation function of the random process $f(t)$ entered in the right side of equation (7.142). The probability density for the conditional longevity takes the form

$$p(T|\underset{\sim}{r}) = \frac{1+\mu(\underset{\sim}{r})T}{2T\sqrt{2\pi\nu(\underset{\sim}{r})T}} \exp\left\{-\frac{[1-\mu(\underset{\sim}{r})T]^2}{2\nu(\underset{\sim}{r})T}\right\} . \qquad (7.147)$$

For a further analysis, the expressions for conditional moments

$$T_k(\underset{\sim}{r}) \equiv \langle T^k(\underset{\sim}{r})\rangle = \int_0^\infty T^k p(T|\underset{\sim}{r})dT \qquad (7.148)$$

will be needed. Substituting the distribution (7.147) into the right side of formula (7.148), and calculating the integrals, we find

$$T_1 = \frac{1}{\mu}\left(1 + \frac{\nu}{2\mu}\right),$$
$$T_2 = \frac{1}{\mu^2}\left(1 + 2\frac{\nu}{\mu} + \frac{3}{2}\frac{\nu^2}{\mu^2}\right). \tag{7.149}$$

The formulae (7.149) coincide with the formulae for the regeneration times in regeneration theory. The analogy between damage accumulation in accordance with the linear model and the regeneration process was considered in [106].

Influence of Load Variation on the Distribution of Longevity Indices

For the characteristic of longevity scatter caused by stress variability in a structure, we shall take the variable coefficient of conditional lifetime (conditional resource) as being

$$w_T = (T_2 - T_1^2)^{1/2}/T_1 . \tag{7.150}$$

Substituting here the expressions (7.149), we find that

$$w_T = \left(\frac{\nu}{\mu}\right)^{1/2} \frac{(1+5\nu/4\mu)^{1/2}}{1+\nu/2\mu} \approx \left(\frac{\nu}{\mu}\right)^{1/2} , \tag{7.151}$$

where the approximate equality corresponds to the case of small variability.

Further analysis requires specialization of the probabilistic model. Consider the results of calculations for the case when the stress $s(t)$ is a narrowband stationary normal process with the mathematical expectation equal to zero, and the right side of equation (7.142) having the form

$$f[s(t),r] = \frac{1}{T_0 N_s[s(t)\,|\,r]} , \tag{7.152}$$

where T_0 is the duration of the cycle and $N = N_s(s|r)$ is the equation of the fatigue curve for a symmetric cycle, which will be taken in the form (7.132). The strength characteristic r is, in general, a stochastic value, and in this context it is the characteristic of a single specimen. The amplitudes of the cycles s, entered in the formula (7.132) and (7.152), obey Rayleigh's distribution (7.131).

Direct calculation with regard to formulae (7.135) and (7.146) gives

$$\mu(r) = \frac{1}{T_0}\left(\frac{\sigma_s}{r}\right)^m 2^{m/2}\Gamma\left(1 + \frac{1}{2}m\,,\ \frac{r^2}{2\sigma_s^2}\right), \tag{7.153}$$

whereas for the calculation of the parameter $\nu(r)$ from the second formula in (7.146) it is assumed that the two-point distribution $p(s_1,s_2;\tau)$ for the Rayleigh process is known. This distribution can be represented in the form of an expansion into a series of Laguerre polynomials $L_n(x)$ of the power n and of zero order:

$$p(s_1,s_2;\tau) = \frac{s_1 s_2}{\sigma_s^2}\exp\left(-\frac{s_1^2+s_2^2}{2\sigma_s^2}\right)\sum_{n=o}^{\infty}\rho^{2n}(\tau)L_n\left(\frac{s_1^2}{2\sigma_s^2}\right)L_n\left(\frac{s_2^2}{2\sigma_s^2}\right). \tag{7.154}$$

Here, $\rho(\tau) = K_s(\tau)/\sigma_s^2$, i.e., the correlation function of the process s(t) normalized to unity. Taking into consideration the formulae (7.132), (7.152) and (7.154) we find that [43]

$$K_f(\tau) = \frac{1}{T_0^2}\left(\frac{\sigma_s}{r}\right)^{2m}\sum_{n=1}^{\infty} C_n^2\rho^{2n}(\tau)\ . \tag{7.155}$$

The coefficients C_n are represented as integrals which can be expressed in terms of the incomplete gamma-function:

$$C_n = 2^{m/2}\sum_{k=o}^{n}\frac{n!\,(-1)^{n+k}}{(k!)^2(n-k)!}\,\Gamma\left(1+k + \frac{1}{2}m\,,\ \frac{r^2}{2\sigma_s^2}\right). \tag{7.156}$$

Hence, from the second formula in (7.146), we find that

$$\nu = \frac{1}{T_0^2}\left(\frac{\sigma_s}{r}\right)^{2m} \sum_{n=1}^{\infty} C_n^2 \tau_n , \qquad \tau_n = 2 \int_0^{\infty} \rho^{2n}(\tau)\, d\tau . \qquad (7.157)$$

Calculations based on formulae (7.156) and (7.157) are rather cumbersome, and the results are complicated. For an estimate of the order of the value ν, we should note that $\nu \sim \sigma_f^2 \tau_c$, where σ_f^2 is the variance of the function (7.152) and τ_c is the correlation time of the process $s(t)$. Calculating the variance σ_f^2 approximately, we obtain

$$\nu \sim \frac{\tau_c}{T_0^2}\left(\frac{\sigma_s}{r}\right)^{2m} 2^m [\Gamma(1+m) - \Gamma^2(1 + \tfrac{1}{2} m)] . \qquad (7.158)$$

The estimate (7.158) agrees with formulae (7.156) and (7.157), provided we take into consideration that the constants τ_n in formulae (7.157) are of the order τ_c for small n. An estimate for μ of comparable accuracy will be obtained from the exact formula (7.153) if the incomplete gamma-function is replaced by the complete one, i.e., $\Gamma(1 + \frac{1}{2} m)$. Hence, from formula (7.151), we obtain an approximate estimate for the coefficient of conditional longevity variability:

$$w_T \sim \left(\frac{\tau_c}{T_0}\right)^{1/2} \left(\frac{\sigma_s}{r}\right)^{m/2} 2^{m/4} \left[\frac{\Gamma(1+m) - \Gamma^2(1 + \frac{1}{2} m)}{\Gamma(1 + \frac{1}{2} m)}\right] . \qquad (7.159)$$

Formula (7.159) characterizes (in the scope of the considered probabilistic model) that part of longevity scatter which is determined by the stochastic nature of the vibrations. The smaller the correlation time τ_c, i.e., the sooner the "mixing" of the process occurs, the lower the level of the stresses σ_s/r, and the smaller the fatigue curve index m is, the smaller the scatter will be.

Simultaneous Account of the Stochastic Properties of the Load and the Cyclic Strength

Below some results are given which were obtained by the method of numerical statistical simulation (Monte-Carlo). Loading by a stationary narrowband normal process s(t) with the correlation function (1.20) was considered. The fatigue curve $N = N_s(s|r)$ was taken in the form (7.132), and for the fatigue limit r, the Weibull distribution (7.119) was taken. Simulation was performed in the following sequence. First, random sequences of the amplitudes of the narrowband process $s_j = s(jT_e)$ and realizations of the stochastic value r were generated. Then, realizations of the random sequence $v_j = v(jT_e)$ were calculated, where

$$v_j = \begin{cases} v_{j-1} + \dfrac{1}{N_0}\left(\dfrac{s_j}{r}\right)^m & (s_j \geq r) , \\ v_{j-1} & (s_j < r) . \end{cases} \tag{7.160}$$

Realizations of the lifetime (resource) T were found by intersecting the critical value $v_* = 1$ with the sequence $\{v_j\}$. According to the ensemble of realizations T determined this way, the histogram $\tilde{p}(T)$, corresponding to the probability density p(T), was constructed.

In order to construct the histogram $\tilde{p}(T|r)$ corresponding to the conditional distribution $p(T|r)$, it is necessary to introduce only one change into the algorithm described above. The fatigue limit r in formula (7.160) should be considered as a given non-stochastic value. The first algorithm simulated real random load tests of specimens or products, with each realization of the load corresponding to each specimen. A similar scheme holds under random vibration conditions. The second algorithm reproduces an imaginary scheme of tests, in which one and the same specimen can be repeatedly brought to failure for various realizations of the loading process.

The following numerical data for the process s(t) were used: $T_e = 10^{-2}$s and $\tau_0 = 0.5$s. For the distribution (7.119), the values taken were: $r_0/r_c = 0.5$ and $\alpha = 0.4$. The stress level varied within the limits from $\sigma_s/r_c = 1.5$ to $\sigma_s/r_c = 2.5$. Parameters of the fatigue curve were taken as $N_0 = 10^5$ and m = 8. The longevity histograms for various ratios σ_s/r are given in Figures 76 - 78. The histogram of the conditional distribution $\tilde{p}(T|r)$ is constructed for the value r equal to the mathematical expectation of the fatigue limit $\langle r \rangle = r_0 + r_c\Gamma(1+1/\alpha)$. In addition, the curves of the theoretical distribution $p_0(T)$, which are constructed according to the first formula in (7.141), are plotted in these graphs.

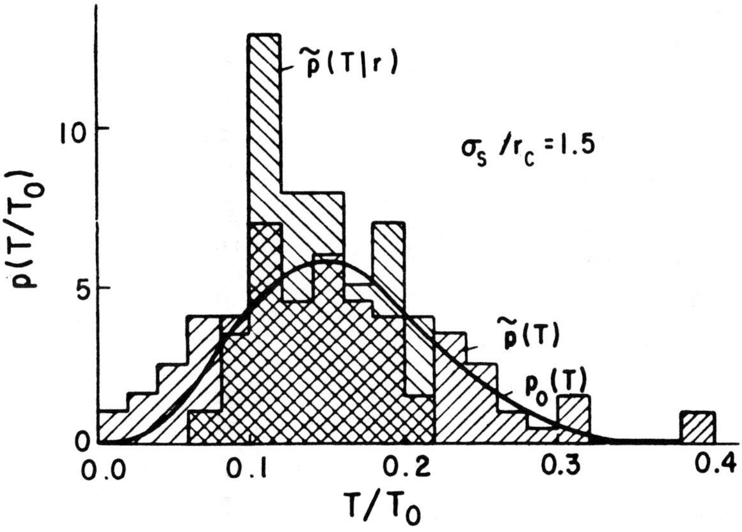

Figure 76

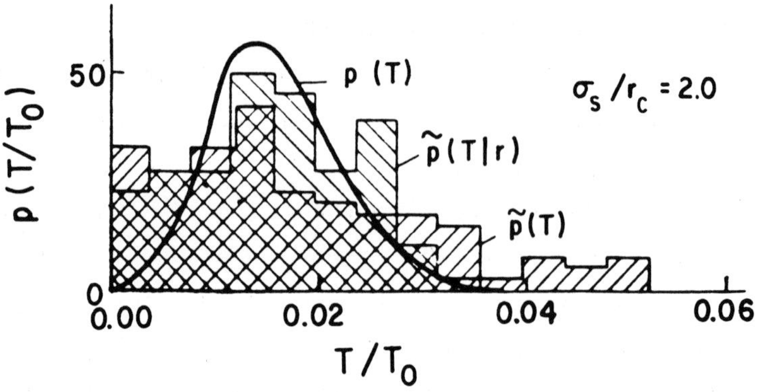

Figure 77

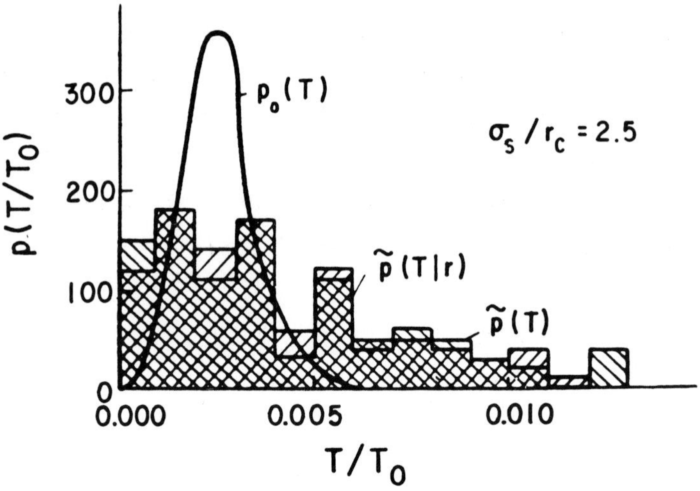

Figure 78

The analysis of the numerical results allows one to draw the following qualitative conclusions. If the stress level is sufficiently low, the influence of the variations in stresses under vibrations will be relatively small as compared to the influence of the scatter of the mechanical properties. In this case, the scatter of the unconditional longevity indices will be determined mainly by the scatter of the mechanical properties, so that the theoretical distribution (7.141) will correctly describe the longevity distribution in the first approximation (Figure 76).

But if, other things being equal, the stress level is sufficiently high, the influence of the longevity scatter (which is related to the scatter of the amplitudes of the vibration process) will increase. Then, the variability of stresses will make a basic contribution to the longevity distribution, and formula (7.141) will not give satisfactory results (Figure 78). The intermediate case is shown in Figure 77. Here, the contributions of stress scatter and cyclic strength are approximately identical.

The methods of reliability analysis and longevity analysis under conditions of random vibrations need further development. The main hindrance of effective application of these methods is the lack of statistical information on loads, materials and structures, given in a form suitable for analysis. The first requirement for a correct planning of experiments and their statistical processing is a clear idea of the contribution made by the various random factors into the longevity distribution.

Chapter 8
The Planning of Vibration Measurements in Structures under Random Vibrations

8.1 RECONSTRUCTION OF RANDOM FIELDS BY MEASUREMENTS TAKEN AT A FINITE NUMBER OF POINTS

Preliminary Remarks

Over many years, the techniques of vibration experiments have been developed mainly as applied to periodic (as a rule, harmonic) vibration processes. For the experimental study of random vibrations, it is necessary to have, apart from special equipment, sound methods of planning vibration measurements and of processing experimental data.

Random forces acting on a structure and random displacements, accelerations and stresses in structural members, as a rule, form space-time fields. However, measuring instruments, i.e., displacement sensors, acceleration sensors, strain gauge sensors, etc., only give information regarding the parameters' change at a finite number of points. The objective is to estimate the probabilistic characteristics of space-time random fields based on the selected data pertaining to this finite number of field points. Then, of course, there arises the problems of determining the number of sensors necessary to obtain sufficient information about the vibration field and of finding rational sites for them on a structure.

Methods of statistical analysis as applied to random processes (random functions of one independent variable) have been developed in great detail. A review of these methods can be found, for example, in [54,81]. There are no fundamental difficulties in extending these methods to functions of two or more variables. However, a purely statistical approach to investigations of space-time random fields requires the placing of sensors very close to each other, as well as a great number of channels for transmitting information and the processing of large arrays of numbers on computers, etc. In considering each specific structure, one nearly always has some additional a priori information about the properties of the field of loads and the vibration field. It is expedient to use this information for planning the vibration measurements so that the number of sensors can be decreased, as well as the amount of time and means for the statistical processing of the results.

This chapter deals with the method of setting up vibration measurements on structures under random vibrations [17,21]. Expansions of fields to be investigated in terms of certain coordinate functions will be used. The latter are chosen in regard to the properties of the structure and the random loads acting on it. Optimization principles for arranging the sensors on the structure are introduced; and the application of these principles is illustrated by examples. Making allowances for distortions of the vibration field due to the introduction of sensors is also considered. The problems listed so far are only a part of those associated with the planning of vibration measurements and the processing of their results. Specifically, problems of statistical processing of vibrometric data, methods of identification of vibration sources and methods of noise filtering in vibration measuring devices are not considered here.

Vibration Measurements in Structures under Random Vibrations

General Relations

Consider a space-time scalar field specified in some region $V \subset \underline{R}^m$. For definiteness, let it be the scalar field of the load $q(\underset{\sim}{x},t)$. In the region V, we choose n points with the coordinates $\underset{\sim}{x}_1, \underset{\sim}{x}_2, \ldots, \underset{\sim}{x}_n$. Measuring the realizations of the field $q(\underset{\sim}{x},t)$ at these points, and statistically processing the results, we find estimates of the mathematical expectations and joint moments for these points of the field:

$$\langle q(\underset{\sim}{x}_j,t)\rangle , \quad \langle q(\underset{\sim}{x}_j,t)q(\underset{\sim}{x}_k,t')\rangle , \quad \langle q(\underset{\sim}{x}_j,t)q(\underset{\sim}{x}_k,t')q(\underset{\sim}{x}_\ell,t'')\rangle , \cdots$$
$$(j,k,\ell,\ldots = 1,2,\ldots,n) .$$

The objective is to determine the number of sensors n, and their arrangement in the region V, necessary for the estimate of mathematical expections and joint moments of the field $q(\underset{\sim}{x},t)$ at all the points $\underset{\sim}{x} \in V$. An exact reconstruction of the field by means of measurements taken at separate points is a mathematically ill-posed problem. However, it is easy to make it well-posed, describing the field in an approximate way by means of a finite number of parameters. It is this method that is developed further below.

To solve this problem, we use the expansion of the field in terms of some system of deterministic coordinate functions $\phi_\alpha(\underset{\sim}{x})$. We choose this system so that almost any realization of the field $q(\underset{\sim}{x},t)$ could be approximated by means of the series

$$q(\underset{\sim}{x},t) = \sum_\alpha Q_\alpha(t)\phi_\alpha(\underset{\sim}{x}) . \tag{8.1}$$

Here, $Q_\alpha(t)$ are random time functions. If the $\phi_\alpha(\underset{\sim}{x})$ are coordinate functions for the expansion of the displacement function $u(\underset{\sim}{x},t)$, then the $Q_\alpha(t)$ have the meaning of generalized forces. The expansion (8.1) is, generally speaking, not stochastically orthogonal. For the mathematical expectation and the correlation function of the field q(x,t), we obtain the expressions

$$\langle q(\underset{\sim}{x},t)\rangle = \sum_{\alpha} \langle Q_\alpha(t)\rangle \phi_\alpha(\underset{\sim}{x}) , \qquad (8.2)$$

$$\langle \tilde{q}(\underset{\sim}{x},t)\tilde{q}(\underset{\sim}{x}',t')\rangle = \sum_{\alpha}\sum_{\beta} \langle \tilde{Q}_\alpha(t)\tilde{Q}_\beta(t')\rangle \phi_\alpha(\underset{\sim}{x})\phi_\beta(\underset{\sim}{x}') . \qquad (8.3)$$

Consider the relation (8.2). Assuming that $\underset{\sim}{x} = \underset{\sim}{x}_j$, $(j = 1,2,\ldots,n)$, and retaining n terms of the series, we obtain for the mathematical expectations $\langle Q_\alpha(t)\rangle$ the system of equations

$$\sum_{\alpha=1}^{n} a_{j\alpha}\langle Q_\alpha(t)\rangle = \langle q(\underset{\sim}{x}_j,t)\rangle , \qquad (j = 1,2,\ldots,n) . \quad (8.4)$$

The coefficients of this system $a_{j\alpha} = \phi_\alpha(\underset{\sim}{x}_j)$ form a square matrix A of dimensionality $n \times n$. This matrix is analogous to Vandermonde's matrix in interpolation theory

$$A = \begin{pmatrix} \phi_1(\underset{\sim}{x}_1) & \phi_2(\underset{\sim}{x}_1) & \cdots & \phi_n(\underset{\sim}{x}_1) \\ \phi_1(\underset{\sim}{x}_2) & \phi_2(\underset{\sim}{x}_2) & \cdots & \phi_n(\underset{\sim}{x}_2) \\ \cdots & \cdots & & \cdots \\ \phi_1(\underset{\sim}{x}_n) & \phi_2(\underset{\sim}{x}_n) & \cdots & \phi_n(\underset{\sim}{x}_n) \end{pmatrix} . \qquad (8.5)$$

If the determinant of matrix A is different from zero, we find the mathematical expectations of the generalized forces from equation (8.4). The time t is treated as a parameter. Substitution of the obtained $\langle Q_\alpha(t)\rangle$ into formula (8.2) makes it possible to calculate the mathematical expectation of the field $q(\underset{\sim}{x},t)$ at all the points $\underset{\sim}{x} \in V$. Thus, the required number of sensors is equal to the number of terms of the series (8.1) that is necessary for a satisfactory approximation of the field $q(\underset{\sim}{x},t)$.

We shall obtain similar equations for the reconstruction of the correlation function of the field $q(\underset{\sim}{x},t)$. Writing the relation (8.3) for the points $\underset{\sim}{x} = \underset{\sim}{x}_j$ and $\underset{\sim}{x}' = \underset{\sim}{x}_k$, and retaining n terms in the series (8.1), we obtain n^2 equations for the correlation functions of the generalized forces

$$\sum_{\alpha=1}^{n} \sum_{\beta=1}^{n} b_{jk\alpha\beta} \langle \tilde{Q}_\alpha(t)\tilde{Q}_\beta(t')\rangle = \langle \tilde{q}(\underset{\sim}{x}_j,t)\tilde{q}(\underset{\sim}{x}_k,t')\rangle \,, \qquad (j,k = 1,2,\ldots,n) \,. \tag{8.6}$$

The coefficients of these equations

$$b_{jk\alpha\beta} = \phi_\alpha(\underset{\sim}{x}_j)\phi_\beta(\underset{\sim}{x}_k) \,, \tag{8.7}$$

after the appropriate ordering of pairs of indices j,k and α,β, form the square matrix B of dimensionality $n^2 \times n^2$. If the determinant of this matrix is different from zero, we can find the correlation functions $\langle \tilde{Q}_\alpha(t)\tilde{Q}_\beta(t')\rangle$ from equation (8.6). The arguments t and t' are then considered as parameters. Furthermore, from formula (8.3), we reconstruct the correlation function in the remaining field points.

If it is known a priori that all

$$\langle \tilde{Q}_\alpha(t)\tilde{Q}_\beta(t')\rangle \equiv 0 \,, \qquad (\alpha \neq \beta),$$

i.e., that the expansion (8.1) is stochastically orthogonal, the number of unknowns in equations (8.6) will be equal to n. To determine these unknowns, it is sufficient to have the number of sensors equal to $\mathrm{Int}(\sqrt{n}+1)$. Henceforth, we assume that there is no a priori information on the stochastic independence of the generalized forces.

Comparing the formulae (8.5) and (8.7), we see that

$$b_{jk\alpha\beta} = a_{j\alpha}a_{k\beta} \,, \tag{8.8}$$

where the a_{jk} are the elements of Vandermonde's generalized matrix (8.5). Thus, the matrix B is a direct (Kronecker) square of the matrix A, i.e.,*

---

*Bellman, R., *Introduction to Matrix Theory*, Nauka, Moscow, 1969.

$$B = A \otimes A \equiv A^{[2]} \ . \tag{8.9}$$

The problem of reconstructing a moment function of higher order can be solved analogously. For example, the relationship connecting the moment function of order N with the correlation function is given by

$$\sum_{\alpha=1}^{n} \sum_{\beta=1}^{n} \sum_{\gamma=1}^{n} \cdots b_{jk\ell\ldots\alpha\beta\gamma\ldots} \langle Q_\alpha(t) Q_\beta(t') Q_\gamma(t'') \cdots \rangle$$

$$= \langle q(\underset{\sim}{x}_j, t) q(\underset{\sim}{x}_k, t') q(\underset{\sim}{x}_\ell, t'') \cdots \rangle \ ,$$

$$(j, k, \ell, \ldots = 1, 2, \ldots, n) \ ,$$

with the coefficients $b_{jk\ell\ldots\alpha\beta\gamma\ldots} = \phi_\alpha(\underset{\sim}{x}_j)\phi_\beta(\underset{\sim}{x}_k)\phi_\gamma(\underset{\sim}{x}_\ell)\ldots$. These coefficients make up a matrix of dimensionality $n^N \times n^N$, which is the N-th (Kronecker) power of the matrix A:

$$B_N = \underbrace{A \otimes A \otimes \cdots \otimes A}_{N} \equiv A^{[N]} \ .$$

The Field as a Stationary Random Time Function

If the field $q(\underset{\sim}{x}, t)$ is a stationary random time function, it is expedient to use spectral methods of analysis for processing the results of measurements at separate points of the field. As before, we shall represent the field $q(\underset{\sim}{x}, t)$ in the form of the expansion (8.1), and, for the estimate of the mathematical expectation of the field, we shall use the equations (8.2) and (8.4). Separating its mathematical expectation from the field $q(\underset{\sim}{x}, t)$, and representing the generalized forces $Q_\alpha(t)$ in the form of Fourier stochastic integrals, we obtain

$$q(\underset{\sim}{x}, t) = \langle q(\underset{\sim}{x}, t) \rangle + \sum_\alpha \phi_\alpha(\underset{\sim}{x}) \int_{-\infty}^{\infty} Q_\alpha(\omega) e^{i\omega t} d\omega \ . \tag{8.10}$$

Here, to be more economical with notations, the spectra $Q_\alpha(\omega)$ of the generalized forces $Q_\alpha(t)$ are denoted by the same letter. The functions $\phi_\alpha(\underset{\sim}{x})$ are real. For the time spectral density, we have the formula (1.26), from whence

$$S_q(\underset{\sim}{x},\underset{\sim}{x}';\omega) = \frac{1}{2\pi} \int_{-\infty}^{\infty} \langle\tilde{q}(\underset{\sim}{x},t)\tilde{q}(\underset{\sim}{x}'t+\tau)\rangle e^{-i\omega\tau} d\tau . \tag{8.11}$$

Let us also note that

$$S_q(\underset{\sim}{x},\underset{\sim}{x}';\omega) = \sum_\alpha \sum_\beta S_{Q_\alpha Q_\beta}(\omega)\phi_\alpha(\underset{\sim}{x})\phi_\beta(\underset{\sim}{x}') , \tag{8.12}$$

where $S_{Q_\alpha Q_\beta}(\omega)$ are joint spectral densities of the generalized forces.

Suppose the estimate for the time spectral density $S_q(\underset{\sim}{x},\underset{\sim}{x}';\omega)$ at the points $\underset{\sim}{x}_1,\underset{\sim}{x}_2,\ldots,\underset{\sim}{x}_n$ is obtained by realizations of the field $q(\underset{\sim}{x},t)$ at these points. Retaining n terms in the series (8.10), and assuming in the relation (8.12) that $\underset{\sim}{x} = \underset{\sim}{x}_j$, $\underset{\sim}{x}' = \underset{\sim}{x}_k$, $(j,k = 1,2,\ldots,n)$, we shall obtain for the joint spectral densities $S_{Q_\alpha Q_\beta}(\omega)$ the system of equations

$$\sum_{\alpha=1}^{n} \sum_{\beta=1}^{n} b_{jk\alpha\beta} S_{Q_\alpha Q_\beta}(\omega) = S_q(\underset{\sim}{x}_j,\underset{\sim}{x}_k;\omega) , \qquad (j,k = 1,2,\ldots,n). \tag{8.13}$$

The coefficients of this system are determined from the formula (8.7), i.e., they form the matrix (8.9). When solving the system, the frequency ω is treated as a parameter.

Writing the Equations in Real Form

Generally speaking, $S_q(\underset{\sim}{x},\underset{\sim}{x}';\omega)$ and $S_{Q_\alpha Q_\beta}(\omega)$ are complex-valued functions. Functions $\phi_\alpha(\underset{\sim}{x})$ are by definition real. Taking into account that

$$S_q(\underset{\sim}{x},\underset{\sim}{x}';\omega) = S_q^{(R)}(\underset{\sim}{x},\underset{\sim}{x}';\omega) + iS_q^{(I)}(\underset{\sim}{x},\underset{\sim}{x}';\omega) , \tag{8.14}$$

$$S_{Q_\alpha Q_\beta}(\omega) = S^{(R)}_{Q_\alpha Q_\beta}(\omega) + iS^{(I)}_{Q_\alpha Q_\beta}(\omega)\ ,$$
$$S_{Q_\beta Q_\alpha}(\omega) = S^{(R)}_{Q_\alpha Q_\beta}(\omega) - iS^{(I)}_{Q_\alpha Q_\beta}(\omega)\ , \tag{8.15}$$

we separate real and imaginary parts in equations (8.13). For the functions $S^{(R)}_{Q_\alpha Q_\beta}(\omega)$, we obtain the system of equations

$$\sum_{\alpha=1}^{n}\sum_{\beta\geq\alpha}^{n} c^{(R)}_{jk\alpha\beta} S^{(R)}_{Q_\alpha Q_\beta}(\omega) = S^{(R)}_{q}(\underset{\sim}{x}_j, \underset{\sim}{x}_k; \omega)\ , \quad (j,k = 1,2,\ldots,n;\ k \geq j)\ , \tag{8.16}$$

with coefficients

$$c^{(R)}_{jk\alpha\alpha} = b_{jk\alpha\alpha}\ , \quad c^{(R)}_{jk\alpha\beta} = b_{jk\alpha\beta} + b_{jk\beta\alpha}, \quad (\alpha \neq \beta)\ . \tag{8.17}$$

For $\beta > \alpha$, these equations involve n diagonal (principal) spectral densities $S_{Q_\alpha Q_\alpha}(\omega)$ and $\frac{1}{2}\, n(n-1)$ real parts of the joint spectral densities $S^{(R)}_{Q_\alpha Q_\beta}(\omega)$. The total number of unknowns, equal to the number of equations, is $\frac{1}{2}\, n(n+1)$.

Similarly, we obtain for the imaginary parts of the joint spectral densities the equations

$$\sum_{\alpha=1}^{n}\sum_{\beta>\alpha}^{n} c^{(I)}_{jk\alpha\beta} S^{(I)}_{Q_\alpha Q_\beta}(\omega) = S^{(I)}_{q}(\underset{\sim}{x}_j, \underset{\sim}{x}_k; \omega)\ , \quad (j,k = 1,2,\ldots,n;\ k > j), \tag{8.18}$$

where the notation

$$c^{(I)}_{jk\alpha\beta} = b_{jk\alpha\beta} - b_{jk\beta\alpha} \tag{8.19}$$

is used. The number of unknowns in equation (8.18) is equal to $\frac{1}{2}\, n(n-1)$. The total number of unknowns in the systems (8.16) and (8.18) is equal to n^2. The matrix of the unified system C is quasi-diagonal and of overall dimensionality $n^2 \times n^2$.

Reconstruction of the Random Vibration Field

So far, for definiteness, it was assumed that $q(\underset{\sim}{x},t)$ is a load field, and $Q_\alpha(t)$ are generalized forces. All the relations are valid for any vibration field. It can, for example, be a field of displacements, accelerations or stresses in a structure. Let us consider the displacement field $u(\underset{\sim}{x},t)$. Instead of (8.1), the expansion is assumed to be in the form

$$u(\underset{\sim}{x},t) = \sum_\alpha U_\alpha(t)\phi_\alpha(\underset{\sim}{x}) \ ,$$

where $U_\alpha(t)$ are the generalized displacements. Instead of the expansion (8.10), we have, correspondingly,

$$u(\underset{\sim}{x},t) = \langle u(\underset{\sim}{x},t)\rangle + \sum_\alpha \phi_\alpha(\underset{\sim}{x}) \int_{-\infty}^{\infty} U_\alpha(\omega)e^{i\omega t}d\omega \ .$$

The necessary formulae are obtained from those given above by changing the notations.

8.2 FUNDAMENTALS OF PLANNING VIBRATION MEASUREMENTS

Principles for the Arrangement of Sensors

Some of the principles were formulated earlier, when the systems of equations (8.4), (8.6), etc. were discussed. Now let us summarize and refine what was stated above.

(1) The minimal number of required sensors is equal to the number of terms of the series which are essential for the approximation of the investigated field. Generally speaking, the smaller the scale of the variation of a field in coordinates, the larger the number of required sensors. As an example, consider the measurements for an interval of a straight line of length ℓ. Let the functions $\phi_\alpha(x)$ be orthogonal in the interval $[0,\ell]$. Then, the

smallest scale of nonuniformity that can be taken into consideration by means of a series involving the first n functions will be of the order ℓ/n. The number of sensors should satisfy the inequality

$$n \geq c\ell/\lambda \ , \tag{8.20}$$

where c is a constant greater than unity, and λ is the nonuniformity scale of the measured field. Note that this requirement pertains to the case when the field is stochastically nonuniform. If the field is uniform and ergodic in coordinates, it is sufficient to arrange the sensors on an interval equal to several scales of correlation. The number of terms for the series (8.1) is determined from the condition of a sufficiently good approximation of the field in this interval. These considerations hold true for the cases of two-dimensional and three-dimensional regions. For example, if $V \subset \underline{R}^2$, then, instead of the formula (8.20), we have

$$n \geq cV/(\lambda_1\lambda_2) \ ,$$

where V is the area of the region, and λ_1, λ_2 are scales of nonuniformity in the two coordinates.

(2) Determinants of the equations (8.4), (8.6), (8.13), etc. should be different from zero. The determinant of the N-th Kronecker power of matrix A of dimensionality $n \times n$ is expressed in terms of the determinant of this matrix in the following way:

$$\det A^{[N]} = (\det a)^{Nn^{N-1}} \ . \tag{8.21}$$

Therefore, it is sufficient to state the condition

$$\det A \neq 0 \ . \tag{8.22}$$

Condition (8.22) is a warning against gross errors in the planning of experiments. Consider two examples of such errors. Suppose the pressure pulsations on a circular cylindrical shell are measured. If the field of pulsations is axially symmetric, the readings of sensors positioned in one cross-section of the shell duplicate one another. This is related to the vanishing of the determinants of the corresponding systems due to the equality of the elements of two or more rows. Another example is the positioning of the sensors on a zero line corresponding to one of the coordinate functions. Let it be the function $\phi_\gamma(\underset{\sim}{x})$. Then $\phi_\gamma(\underset{\sim}{x}_j) \equiv 0$, $(j = 1,2,\ldots,n)$ and the matrix A will have zero columns.

(3) Matrices of the equations (8.4), (8.6), (8.13), etc. should be sufficiently well conditioned. Then, small errors in measuring statistical characteristics of a field at isolated points will not lead to gross errors in reconstructing the field as a whole. It will be recalled that a matrix is referred to as badly conditioned if by its properties it is close to a singular (degenerate) matrix. Examples of well conditioned matrices are diagonal and orthogonal matrices. The requirement that the determinant of the matrix A should be sufficiently different from zero leads to a criterion for the arrangement of sensors, say,

$$|\det A| \to \max_{\underset{\sim}{x}_1,\underset{\sim}{x}_2,\ldots,\underset{\sim}{x}_n} . \tag{8.23}$$

Here, as well as henceforth, the extremum of the efficiency function is sought under certain restrictions imposed on the coordinates $\underset{\sim}{x}_1,\underset{\sim}{x}_2,\ldots,\underset{\sim}{x}_n$. They all should belong to the region V; in addition, other restrictions can be imposed, for example, the arrangements of sensors are chosen from a certain class.

Criteria Using Condition Numbers

The criterion (8.23) has no exact meaning. The determinant of a system of linear algebraic equations can be made arbitrarily large by multiplying the equations termwise by numbers greater than one. However, in this case, the condition of the system will not change. In numerical mathematics* condition numbers, invariant with respect to transformation of the type mentioned above, are introduced.

The Türing's condition numbers $\eta(A)$ of the real square matrix A are introduced in terms of the norm of this matrix in the following way:

$$\eta(A) = n^{-1}||A|| \; ||A^{-1}|| \; . \tag{8.24}$$

Here, n is the order of the matrix. Using, respectively, the spherical and the cubic norms

$$||A||_I = \left(\sum_\mu \sum_\nu a_{\mu\nu}^2 \right)^{1/2} \; , \quad ||A||_{II} = n \max_{\mu,\nu} |a_{\mu\nu}| \; , \tag{8.25}$$

we obtain the first $\eta_I(A)$ and the second $\eta_{II}(A)$ of the Türing numbers.

The Todd condition number $\eta_{III}(A)$ is expressed in terms of the eigenvalues α_μ of the matrix A:

$$\eta_{III}(A) = \max_\mu |\alpha_\mu| / \min |\alpha_\mu| \; . \tag{8.26}$$

Generally, these eigenvalues are complex numbers. Sometimes, it is more convenient to use another condition number, which is expressed in terms of the singular numbers λ_μ of the matrix A, equal to the arithmetic values of the square root of the eigenvalues of

---

*Faddeev, D.K., Faddeeva, V.N., *Computational Methods of Linear Algebra,* Phyzmatgiz, Moscow, 1960; Collatz, L., *Functional Analysis and Numerical Mathematics,* Mowcow, Mir, 1969.

the matrix A^TA. Here, A^T is the transposed matrix. Hence, we arrive at Todd's second condition number

$$\eta_{IV}(A) = \max_\mu \lambda_\mu / \min_\mu \lambda_\mu . \qquad (8.27)$$

If A is a symmetric matrix, then $\eta_{III}(A) = \eta_{IV}(A)$. All condition numbers are not smaller than unity and are related to one another by the inequalities

$$\eta_I(A) \leq \eta_{II}(A) \leq n^2\eta_I(A) ,$$

$$\eta_I(A) \leq \eta_{IV}(A) \leq n\eta_I(A) , \qquad (8.28)$$

$$\eta_{III}(A) \leq \eta_{IV}(A) .$$

The better the matrix A is conditioned, the closer to unity the numbers $\eta(A)$ are. With the help of these numbers, some criteria for the arrangement of sensors can be formulated. These criteria have the form

$$\eta(A) \to \min_{\underset{\sim}{x}_1,\underset{\sim}{x}_2,\ldots,\underset{\sim}{x}_n} , \qquad (8.29)$$

where $\eta(A)$ is one of the condition numbers (8.24), (8.26) or (8.27). Since the inequalities (8.28) are relatively narrow, the use of any of the three condition numbers will lead to similar results.

So far, only the condition of matrix A was considered. It is easy to show that condition numbers of the Kronecker powers $A^{[N]}$ are equal to the corresponding powers of these numbers for matrix A. Consider, for example, the number $\eta_I(A^{[N]})$. According to the first formula (8.25)

$$||A^{[N]}||_I = \left(\sum_{\mu_1,\mu_2,\ldots} \sum_{\nu_1,\nu_2,\ldots} a^2_{\mu_1\nu_1} a^2_{\mu_2\nu_2} \ldots a^2_{\mu_N\nu_N} \right)^{1/2}$$

$$= \left(\sum_{\mu} \sum_{\nu} a^2_{\mu\nu} \right)^{N/2} = ||A||_I^N .$$

Substituting into the formula (8.24) yields

$$\eta_I(A^{[N]}) = \eta_I^N(A) ,$$

which was to be proved. Thus, it is sufficient to restrict ourselves to the requirement of condition of matrix A with the elements determined by relations (8.5).

Relation to the Theory of Interpolation Functions

The problem of estimating probabilistic characteristics of random fields by results of measurements at a finite number of points can be interpreted as a problem of interpolation of deterministic functions. For example, the expression

$$\langle q(\underset{\sim}{x},t)\rangle = \sum_{\alpha=1}^{n} \langle Q_\alpha(t)\rangle \phi_\alpha(\underset{\sim}{x})$$

can be considered as a generalized interpolation polynomial in the region V, formed from the functions $\phi_1(\underset{\sim}{x}),\phi_2(\underset{\sim}{x}),\ldots,\phi_n(\underset{\sim}{x})$, with the points $\underset{\sim}{x}_1,\underset{\sim}{x}_2,\ldots,\underset{\sim}{x}_n$ as nodes of interpolation. Analogously, the reconstruction of probabilistic characteristics of the second order (i.e., correlation functions and spectral densities) can be interpreted as the interpolation of the function in the region $V \otimes V$. The generalized interpolational polynomials are formed from the functions $\phi_\alpha(\underset{\sim}{x})\phi_\beta(\underset{\sim}{x}')$ for $\alpha,\beta = 1,2,\ldots,n$. The nodes of interpolation are the n^2 pairs of points $(\underset{\sim}{x}_1,\underset{\sim}{x}'_1),(\underset{\sim}{x}_1,\underset{\sim}{x}'_2),\ldots,(\underset{\sim}{x}_n,\underset{\sim}{x}'_n)$. The reconstruction of probabilistic characteristics of the third order is interpreted as interpolation in the region $V \otimes V \otimes V$, etc.

The requirements for the arrangement of sensors can be formulated in terms of the theory of interpolation of functions. Specifically, the required number of sensors is chosen from the condition that the generalized interpolational polynomials should approximate the field probabilistic characteristics sufficiently well at points that are noncoinciding with the measurement points.

Examples

Consider the arrangement of pressure transducers for measuring large-scale turbulence on a high tower (Figure 79). For the description of these pulsations, we shall use the model of a random field, stationary in time, but nonuniform in height. We shall take the modes of flexural vibrations of a homogeneous beam, clamped at one end and free at the other, as the functions $\phi_\alpha(x)$ used in the expansion (8.1) - (8.3), etc.

$$\phi_\alpha(x) = \cosh\frac{\mu_\alpha x}{\ell} - \cos\frac{\mu_\alpha x}{\ell} - \frac{\cosh\mu_\alpha + \cos\mu_\alpha}{\sinh\mu_\alpha + \sin\mu_\alpha}\left(\sinh\frac{\mu_\alpha x}{\ell} - \sin\frac{\mu_\alpha x}{\ell}\right) .$$

Here, μ_α are roots of the frequency equation $\cosh\mu \cos\mu = -1$.

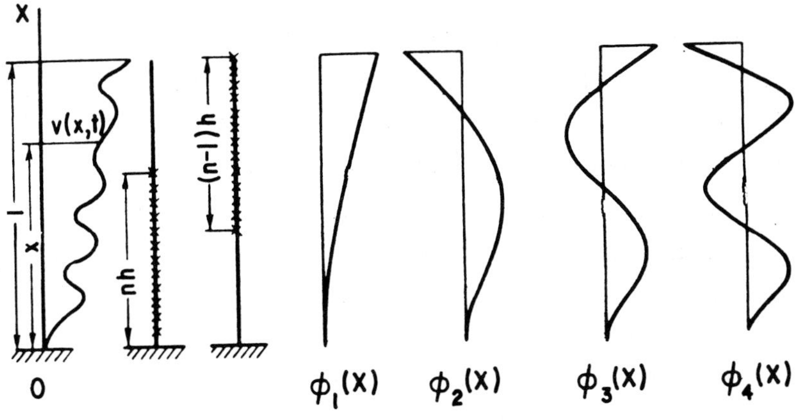

Figure 79

Since the intensity of the pressure pulsations is small at the tower base and increases with height, the form of the coordinate functions corresponds to the a priori picture of the non-uniformity of the field. If the estimate of a typical dimension of the large-scale turbulence is known, the required number of sensors is found from formula (8.20)

Figures 80 and 81 present the results of calculations for the case $n = 10$. The distance h between the adjacent sensors is taken as the only parameter characterizing the arrangement of the sensors (Figure 79). The graphs in Figure 80 correspond to the uniform arrangement of the sensors in the interval $[0,nh]$. Figure 81 is plotted on the assumption that the sensors are arranged uniformly in the interval $[\ell-(n-1)h,\ell]$. The determinant of the matrix A takes on the maximal value close to $h = \ell/n$. Close to this value, one can observe the minima of the condition numbers $\eta_I(A)$, $\eta_{II}(A)$ and $\eta_{IV}(A)$. This corresponds to the intuitive notion that a uniform arrangement of sensors along the entire length is close to optimal.

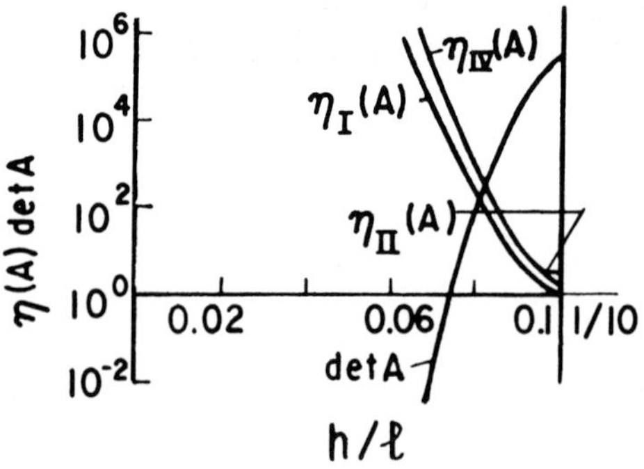

Figure 80

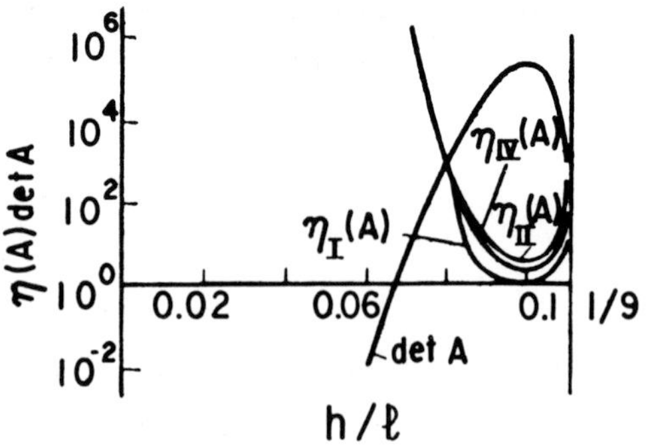

Figure 81

As another example, we shall consider a square plate with sides ℓ, taking the modes of vibrations of a simply supported plate as the functions for the expansions (8.1) and (8.10), namely,

$$\phi_{jk}(x,y) = \sin\frac{j\pi x}{\ell}\sin\frac{k\pi x}{\ell}, \qquad (j,k = 1,2,\ldots)\ . \quad (8.30)$$

For some classes of arrangement for sensors on a plate, it is easy to find "the best" and "the worst" arrangement. For example, let $n = 4$, with the functions $\phi_{jk}(x,y)$ being taken for $j = k = 1,2$. For sensors arranged as shown in Figure 82(a), the matrix A has the form

$$A = \begin{bmatrix} 3/4 & 3/4 & 3/4 & 3/4 \\ 3/4 & -3/4 & 3/4 & -3/4 \\ 3/4 & 3/4 & -3/4 & -3/4 \\ 3/4 & -3/4 & -3/4 & 3/4 \end{bmatrix}.$$

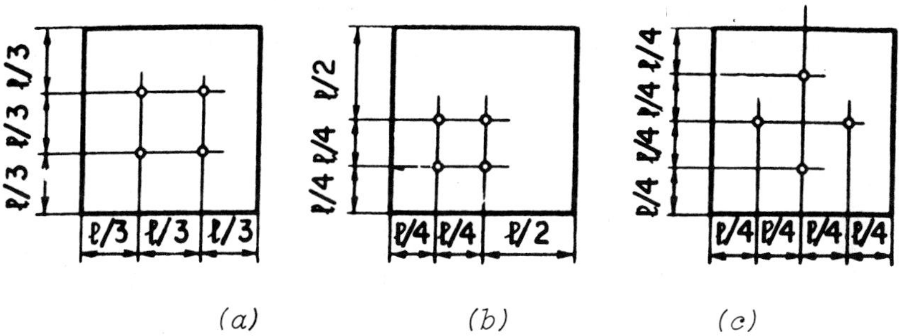

Figure 82

This matrix is orthogonal to an accuracy of a factor, and all of its condition numbers are equal to unity. If we arrange the sensors as in Figure 82(b), we obtain the triangular matrix

$$A = \begin{bmatrix} 1/2 & 1/\sqrt{2} & 1/\sqrt{2} & 1 \\ 1/\sqrt{2} & 0 & 1 & 0 \\ 1/\sqrt{2} & 1 & 0 & 0 \\ 1 & 0 & 0 & 0 \end{bmatrix},$$

in which all the eigenvalues are modulo equal to unity. For this matrix, $\eta_{III}(A) = 1$. The arrangement of sensors as shown in Figure 82(c) proves to be "the worst", i.e., the matrix A turns out to be degenerate.

8.3 RECONSTRUCTION OF A LOADING FIELD USING RESULTS OF MEASUREMENTS OF A VIBRATION FIELD

General Relations

In cases when the direct measurement of loads is impossible or difficult, there arises a problem of obtaining estimates of the probabilistic characteristics for the load field from the measurement results of a vibration field. For the solution of this problem, it is necessary to know the transfer characteristics of the structure. If the structure requires a dynamic analysis, these transfer characteristics are found in the process of analysis. This method is especially effective if external forces are represented in the form of expansions of the type (8.1) and (8.10) with a small number of series terms. Then, for the reconstruction of the load field, it is sufficient to take measurements of the vibration field at a small number of points.

A similar method can be used for the reconstruction of the vibration field from the results of measuring it at a small finite number of points. Let the external loads be specified accurately by a small number of random time functions or let them admit a close approximation with the help of a small number of terms of the series (8.1) and (8.10). For example, the force concentrated at a specified point is determined by the values of its components. The force concentrated at an unknown, generally "wandering" point is determined by the coordinates of this point and by the values of the components of the force, etc.

Suppose that the transfer characteristics of a system with respect to external loads of a given class are known. Then, it is expedient to estimate the probabilistic characteristics of the loads by means of the results of measurements of the vibration field. The probabilistic characteristics of the vibration field at the remaining points are then found by means of analysis. Let us formulate the method as applied to a linear stationary deterministic system subjected to forces which are stationary random time functions and arbitrary random functions of the coordinates. For simplicity, assume that the load field $q(\underset{\sim}{x},t)$ and the vibration field are scalar ones. For definiteness, we shall consider the displacement field $u(\underset{\sim}{x},t)$, though all the results extend to the case when velocities, accelerations and stresses in a structure etc. are measured. The relationship between the fields $q(\underset{\sim}{x},t)$ and $u(\underset{\sim}{x},t)$ is given by the operator equation

$$Lu = q \ . \tag{8.31}$$

Let the external forces be represented in the form (8.10). The solution of the stochastic equation (8.31) with the right side (8.10) is sought in the form

$$u(\underset{\sim}{x},t) = \langle u(\underset{\sim}{x},t)\rangle + \sum_{\alpha} \int_{-\infty}^{\infty} Q_{\alpha}(\omega)\psi_{\alpha}(\underset{\sim}{x},\omega)e^{i\omega t}d\omega \ . \tag{8.32}$$

Here, the $\psi_{\alpha}(\underset{\sim}{x},\omega)$ are the solutions of the auxiliary deterministic problem

$$L\psi_{\alpha}(\underset{\sim}{x},\omega)e^{i\omega t} = \phi_{\alpha}(\underset{\sim}{x})e^{i\omega t} \ . \tag{8.33}$$

Let us consider the problem of the reconstruction of the time spectral density $S_q(\underset{\sim}{x},\underset{\sim}{x}';\omega)$ by means of the results of measuring the time spectral density $S_u(\underset{\sim}{x},\underset{\sim}{x}';\omega)$ at a finite number of points. For this, we use the formula

$$S_u(\underset{\sim}{x},\underset{\sim}{x}';\omega) = \sum_{\alpha}\sum_{\beta} \psi^*_{\alpha}(\underset{\sim}{x},\omega)\psi_{\beta}(\underset{\sim}{x}',\omega)S_{Q_{\alpha}Q_{\beta}}(\omega) \ . \tag{8.34}$$

Retaining n terms in the series (8.10) and (8.32), and assuming that in the relations (8.34) $\underset{\sim}{x} = \underset{\sim}{x}_j$ and $\underset{\sim}{x}' = \underset{\sim}{x}_k$, $(j,k = 1,2,\ldots,n)$, we obtain a system of equations for the reciprocal spectral densities of the generalized forces:

$$\sum_{\alpha=1}^{n} \sum_{\beta=1}^{n} b_{jk\alpha\beta}(\omega) S_{Q_\alpha Q_\beta}(\omega) = S_u(\underset{\sim}{x}_j, \underset{\sim}{x}_k; \omega) , \qquad (j,k = 1,2,\ldots,n) . \tag{8.35}$$

The coefficients of this system, as distinguished from (8.13), are complex numbers. These coefficients are given by the relations

$$b_{jk\alpha\beta}(\omega) = \psi_\alpha^*(\underset{\sim}{x}_j, \omega)\psi_\beta(\underset{\sim}{x}_k, \omega) , \qquad (j,k,\alpha,\beta = 1,2,\ldots,n) . \tag{8.36}$$

Let the determinant of the system (8.35) be different from zero. We shall solve this system by treating the frequency ω as a parameter. The time spectral density of the load field and the time spectral density of the displacement field (8.34) can be represented by the spectral densities of the generalized forces obtained using (8.12).

Principles of Arrangement for Sensors

As can be seen from the above, the number of sensors should be equal to the number of terms when expanding the load field into the series (8.10). Under this condition, the number of unknowns in the relations (8.35) will be equal to the number of equations. Then, the matrix determinant consisting of the coefficients (8.36) is required to be different from zero for all the values of the frequency ω from the considered range $[\omega_*, \omega_{**}]$. We shall denote this matrix by $B(\omega)$. Thus, we arrive at the condition

$$\det B(\omega) \neq 0 , \qquad \omega \in [\omega_*, \omega_{**}] . \tag{8.37}$$

When choosing the arrangement of sensors, the maximum-modulus principle of the determinant of the matrix $B(\omega)$ can be taken, as in Section 8.2, as the criterion of optimality. But, unlike Section 8.2, the determinant is dependent on the frequency ω. Therefore, either the minimax criterion

$$\min_{\omega \in [\omega_*, \omega_{**}]} |\det B(\omega)| \to \max_{\underset{\sim}{x}_1, \underset{\sim}{x}_2, \ldots, \underset{\sim}{x}_n}, \tag{8.38}$$

or a criterion of the integral type is taken, for example,

$$\int_{\omega_*}^{\omega_{**}} |\det B(\omega)| d\omega \to \max_{\underset{\sim}{x}_1, \underset{\sim}{x}_2, \ldots, \underset{\sim}{x}_n}. \tag{8.39}$$

When measuring narrowband processes, it is sufficient to take the optimal arrangement of sensors that corresponds to the dominating frequency.

The criteria involving condition numbers are introduced in a similar way. The criterion of the minimax type takes the form

$$\max_{\omega \in [\omega_*, \omega_{**}]} \eta[B(\omega)] \to \min_{\underset{\sim}{x}_1, \underset{\sim}{x}_2, \ldots, \underset{\sim}{x}_n}, \tag{8.40}$$

and the integral criterion takes the form

$$\int_{\omega_*}^{\omega_{**}} \eta[B(\omega)] d\omega \to \min_{\underset{\sim}{x}_1, \underset{\sim}{x}_2, \ldots, \underset{\sim}{x}_n}, \tag{8.41}$$

where $\eta[B(\omega)]$ is one of condition numbers of the matrix $B(\omega)$. These numbers are determined by the formulae in Section 8.2, extended to matrices with complex elements. For example, the spherical norm of the matrix $B(\omega)$ is introduced as

$$||B(\omega)||_I = \left(\sum_\mu \sum_\nu |b_{\mu\nu}(\omega)|^2 \right)^{1/2} , \tag{8.42}$$

where $b_{\mu\nu}(\omega)$ are the elements of the matrix $B(\omega)$. The singular numbers of matrix $\lambda_\mu(\omega)$ are equal to the arithmetic values of the square root of the eigenvalues of matrix $B^*(\omega)B(\omega)$, where $B^*(\omega)$ is the Hermitian conjugate matrix. A detailed description can be found in [17].

Some Simplifications Resulting from the Properties of Matrix B(ω)

Introduce the matrix $A(\omega)$ with complex elements

$$a_{j\alpha}(\omega) = \psi_\alpha(\underset{\sim}{x}_j, \omega) , \qquad (j, \alpha = 1, 2, \ldots, n) . \tag{8.43}$$

Using the notion of the Kronecker product, and taking formula (8.36) into consideration, we find that

$$B(\omega) = A^*(\omega) \otimes A(\omega) . \tag{8.44}$$

This relation allows one to obtain a number of criteria of the types (8.37) - (8.41), in which, instead of the matrix $B(\omega)$ with dimensionality $n^2 \times n^2$, the matrix $A(\omega)$ of dimensionality $n \times n$ is used. For example, taking into consideration that

$$\det A^*(\omega) = [\det A(\omega)]^* ,$$

we shall reduce the criterion (8.37) to the form

$$\det A(\omega) \neq 0 , \qquad \omega \in [\omega_*, \omega_{**}] . \tag{8.45}$$

Instead of (8.38) we obtain

$$\min_{\omega \in [\omega_*, \omega_{**}]} |\det A(\omega)|^2 \to \max_{\underset{\sim}{x}_1, \underset{\sim}{x}_2, \ldots, \underset{\sim}{x}_n} , \tag{8.46}$$

etc. Then, it is easy to show that the condition numbers of the matrix $B(\omega)$ are related to the corresponding numbers of the matrix $A(\omega)$ by the expression

$$\eta[B(\omega)] = \eta^2[A(\omega)] \ . \tag{8.47}$$

For example, consider the number $\eta_{III}[B(\omega)]$. The eigenvalues of the Kronecker product are known to be equal to all possible products of the eigenvalues of the matrices' cofactors. We shall designate the eigenvalues of the matrix $A(\omega)$ by $\alpha_\mu(\omega)$. From a formula of the type (8.26)

$$\eta_{III}[B(\omega)] = \frac{\max\limits_{\mu,\nu}|\alpha_\mu^*(\omega)\alpha_\nu(\omega)|}{\min\limits_{\mu,\nu}|\alpha^*(\omega)\alpha_\nu(\omega)|} = \frac{\max\limits_{\mu}|\alpha_\mu(\omega)|^2}{\min\limits_{\mu}|\alpha_\mu(\omega)|^2} \ ,$$

we obtain the formula (8.47). Unlike equation (8.13), the coefficients $b_{jk\alpha\beta}(\omega)$ are in this case complex numbers. Therefore, after turning to real variables, the unknowns $S^{(R)}_{Q_\alpha Q_\beta}(\omega)$ and $S^{(I)}_{Q_\alpha Q_\beta}(\omega)$ do not turn up in separate equations. The formulae for the elements of the matrix $C(\omega)$, relating the unknowns mentioned above to the measured variables $S_u^{(R)}(\underset{\sim}{x}_j, \underset{\sim}{x}_k;\omega)$ and $S_u^{(I)}(x_j, x_k;\omega)$, can be found in [17].

Examples

We consider two examples illustrating the application of the suggested criteria. As a first example, we shall take a circular thin plate of a linear viscoelastic material clamped along its contour. Let the plate be subjected to the normal axisymmetric load $q(r,t)$, which is a stationary random time function and an arbitrary random function of the radius r $(0 \leq r \leq R)$. The displacements at n points of the plate are reported by n sensors. The objective is to determine the probabilistic characteristics of the load $q(r,t)$ by the results of these measurements.

We shall take the modes of the plate's natural vibrations $\phi_\alpha(r) = J_0(\mu_\alpha r) - J_0(\mu_\alpha R) I_0(u_\alpha r)/I_0(\mu_\alpha R)$ as the functions $\phi_\alpha(r)$ in the expansion (8.10). The numbers μ_α are obtained from the frequency equation $J_0(\kappa) I_0'(\kappa) - J_0'(\kappa) I_0(\kappa) = 0$, where $\kappa = \mu R$. The functions $\psi_\alpha(r,\omega)$ are calculated by means of equation (3.53) and the corresponding boundary conditions for $r = R$. After some simple calculations, we find that

$$\psi_\alpha(r,\omega) = \frac{\psi_\alpha(r)}{\mu_\alpha^4 D_r (1 + i\chi \text{ sign } \omega) - \rho h \omega^2}, \tag{8.48}$$

where D_r is the real part of the cylindric stiffness and $\chi(\omega)$ is the loss tangent. Considering formula (8.48), we find the elements of the matrix $C(\omega)$, and by them the values of the determinants and the condition numbers. Some results of calculations for the case $h = 0.04R$, $\chi = 0.05$, $\nu = 0.3$ and $n = 2$ are shown in Figures 83 and 84. Figure 83 shows the dependence of Türing's number $\eta_I(C)$ on the coordinates of the sensors r_1 and r_2. The graph is plotted for the frequency $\omega = 2.739\omega_1$, where ω_1 is the lowest natural frequency of the plate. There is only one arrangement of sensors which, for the specified frequency, reduces the number $\eta_I(C)$ to a minimum. The condition of the matrix C decreases with the sensors arranged close to the lines $r_1 = r_2$, $r_1 = R$ and $r_2 = R$. The natural solution (based on intuition) to place one of the sensors in the centre of the plate, does not correspond to the minimum of the condition number. Indeed, the smallest value of the number $\eta_I(C)$ for $r = 0$ is about three, whereas the absolute minimum does not exceed two. However, as far as the condition of matrix C is concerned the difference between these two values is not significant.

Figure 84 shows the dependence of the condition numbers $\eta_I[C(\omega)]$, $\eta_{II}[C(\omega)]$ and $\eta_{IV}[C(\omega)]$ on the frequency ω for $r_1 = 0.6R$, and $r_2 = 0.3R$. The maxima of the condition numbers are close to the natural frequencies ω_1 and ω_2. The nature of the variation of all three condition numbers is identical; in fact, the frequencies corresponding to their minima coincide.

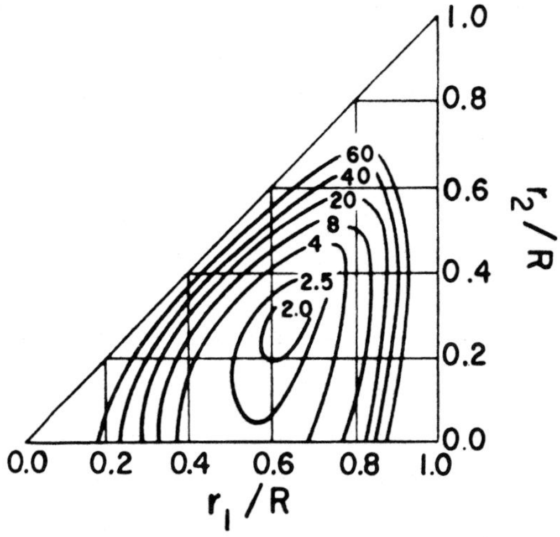

Figure 83

As a second example, we take a semi-infinite circular shell of a linear viscoelastic material. On the end-wall of the shell $x = 0$, the random normal displacement $w_0(t)$ and the rotation of the normal $\phi_0(t)$, both independent of the circumferencial coordinate, are specified (see Section 3.6). The objective is to find the probabilistic characteristics of the random functions $w_0(t)$ and $\phi_0(t)$ from the displacements $w(x,t)$, measured at the two points with coordinates x_1 and x_2.

Figures 85 and 86 show the numerical results for the following data: $h = 2\text{mm}$, $R = 500\text{mm}$, $E_r = 0.7 \cdot 10^{11} \text{N} \cdot \text{m}^{-2}$, $\nu = 0.25$, $\chi = 0.05$ and $\rho = 2700 \text{ kg} \cdot \text{m}^{-3}$. One of the displacement sensors is placed in the section $x = x_1 = 0$, while the coordinate x_2 of another sensor is varied. The dependence of the condition number $\eta_I[C(\omega)]$ on the dimensionless coordinate of the sensor x_2/λ_0 and on the dimensionless frequency ω/ω_0 is shown in Figure 85. Here, λ_0 and ω_0 are determined according to (3.140). With the exception of

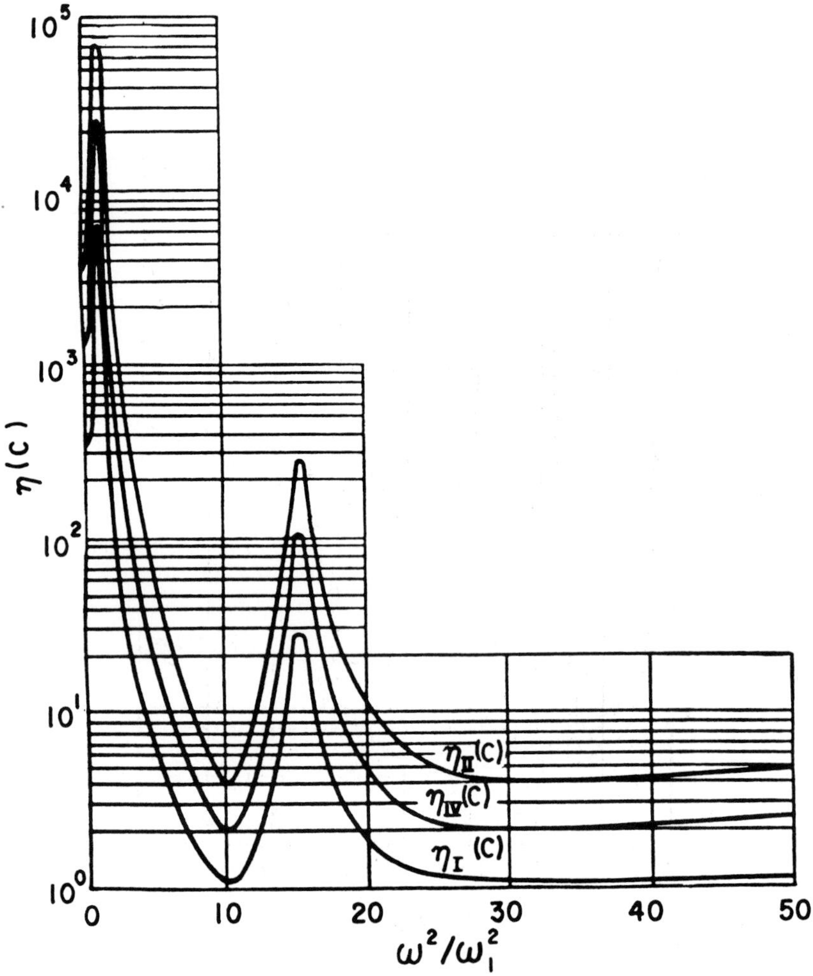

Figure 84

the vicinity of the frequency ω_0, the minimum of the condition number corresponds to having the second sensor located in the zone of the boundary effect. Removing the sensor from this zone causes the condition of the matrix (especially at low frequencies) to deteriorate. Dependence of the determinant of matrix $C(\omega)$ on the parameters x_2/λ_0 and ω/ω_0 is shown in Figure 86. Although the location of the minima of number $\eta_I[C(\omega)]$ and the location of the maxima of det $C(\omega)$ do not coincide, the two approaches lead to very similar results.

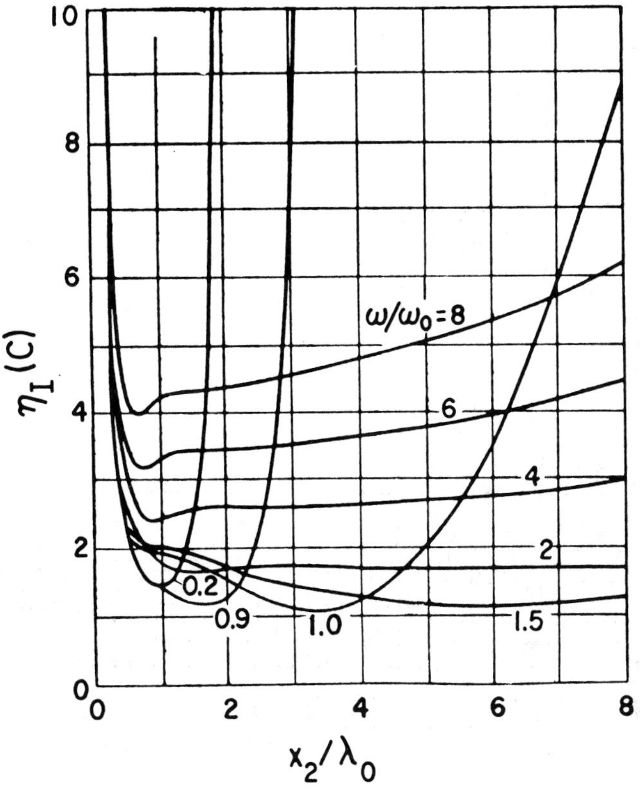

Figure 85

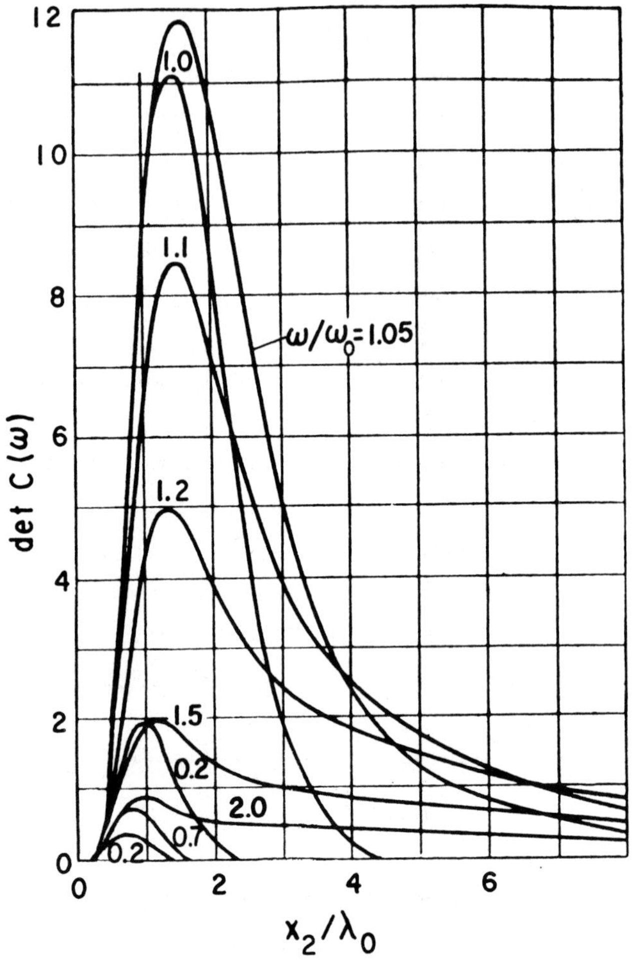

Figure 86

8.4 SOME PROBLEMS OF A POSTERIORI PLANNING OF VIBRATION MEASUREMENTS

The Notion of a Posteriori Planning

The approaches described above for the planning of measurements on vibrating structures can be called a priori in the sense that only some tentative, a priori information about the vibration field properties is used to carry them out. This information includes the data on the scale of the field variation, on frequency ranges, on the nature of external loads, etc. Based on these data, the suitable coordinate functions, that approximate fields or displacements and/or loads are chosen. The number of sensors is assumed to be equal to the number of series terms essential for the satisfactory approximation of the field of displacements, accelerations and/or loads. After this, the objective of planning is to find an arrangement of sensors that will either provide the best interpolation of the vibration field or will correspond to that scheme of measuring which is the least susceptible to errors in measurements. Using the notion of the best condition of matrices, it is possible to formalize and combine both approaches.

Along with the a priori planning of vibration measurements, the a posteriori planning, which makes use of results obtained during the process of measuring, is also of practical interest. The number of sensors and channels of information-measuring devices is often limited, which makes it necessary to seek ways for the most economical use of measuring devices. Therefore, if in the process of measuring there is some excess of information, the experimentor may wish to change the arrangement of sensors so as to get as much information as possible without increasing the number of sensors.

Let us illustrate the statement of the problem of a posteriori planning by the example of measuring the stationary field $u(\underset{\sim}{x},t)$ at two points $\underset{\sim}{x}_1$ and $\underset{\sim}{x}_2$. Suppose an estimate for the correlation matrix of the process $\{u(\underset{\sim}{x}_1,t),u(\underset{\sim}{x}_2,t)\}$ in the coinciding time moments, obtained as a result of measurements, is

$$K = \begin{pmatrix} K_{11} & K_{12} \\ K_{21} & K_{22} \end{pmatrix}. \tag{8.49}$$

It is natural to take the value of the correlation coefficient as the measure of interrelation between the values of the field at two points, i.e.,

$$\rho_{12} = K_{12}/(K_{11}K_{22})^{1/2} . \tag{8.50}$$

By taking an absence of correlation between the readings of transducers as an indication for their best arrangement, we arrive at the criterion

$$|\rho_{12}| \to \min_{\underset{\sim}{x}_1, \underset{\sim}{x}_2} . \tag{8.51}$$

Another approach is based on some concepts of information theory. The quantity of information on the variable u_1 obtained as a result of observing the variable u_2 is determined as

$$I_{u_2}[u_1] = \int_{-\infty}^{\infty} \int_{-\infty}^{\infty} p(u_1,u_2) \ \ell og \frac{p(u_1|u_2)}{p(u_1)} du_1 du_2 ,$$

where $p(u_1,u_2)$, $p(u_1|u_2)$, and $p(u_1)$ are the probability densities of the variables $u_1 \equiv u(\underset{\sim}{x}_1,t)$ and $u_2 \equiv u(\underset{\sim}{x}_2,t)$. Among all the arrangements of sensors pertaining to a given class, the best will be the one that gives to the quantity of information $I_{u_2}[u_1]$ a minimal value:

$$I_{u_2}[u_1] \to \min_{\underset{\sim}{x}_1, \underset{\sim}{x}_2} . \tag{8.52}$$

If the field $u(\underset{\sim}{x},t)$ is normal, the quantity of information is expressed in terms of the correlation matrix elements [31], i.e.,

$$I_{u_2}[u_1] = \ell og \sqrt{\frac{K_{11}K_{22}}{K_{11}K_{22}-K_{12}^2}} = \ell og \frac{1}{\sqrt{1-\rho_{12}^2}} .$$

The criterion of the minimum of conditional information (8.52) is equivalent in this case to the criterion of the minimum of the modulus of the correlation coefficient (8.51).

The considerations given above extend to the multidimensional case [31,50]. Here, instead of the correlation coefficient, certain correlation measures averaged over all sensors are used.

Information criteria are also related (at least for normal fields) to the properties of the correlation matrix. Further below, a sufficiently general criterion, in which, as in a priori criteria, the notion of the condition of the matrix is used, will be formulated. It will be shown that this criterion can be treated as a generalization of the criteria formulated above.

Criterion for the Best Condition of the Correlation Matrix

We shall begin with the case $n = 2$. Calculating Türing's condition numbers from the formulae (8.24) and (8.25), we find that the criteria (8.51) and (8.52) can be formulated in terms of condition numbers. This approach can naturally be extended to the multidimensional case. However, there arises in this case the following difficulty: the correlation matrix

$$K(t,t') = \begin{bmatrix} K_{11}(t,t') & K_{12}(t,t') & \cdots & K_{1n}(t,t') \\ K_{21}(t,t') & K_{22}(t,t') & \cdots & K_{2n}(t,t') \\ \cdots & \cdots & \cdots & \cdots \\ K_{n1}(t,t') & K_{n2}(t,t') & \cdots & K_{nn}(t,t') \end{bmatrix} \qquad (8.53)$$

can be well conditioned on the whole, but there may be a considerable correlation between some groups of sensors. For example, the condition of matrix (8.53) for $n > 2$ does not generally ensure a smallness of correlation between the measurements at the points $\tilde{x}_1$ and $\tilde{x}_2$. In order to consider this situation, it is necessary to extend the notion of condition, including in it the condition of the corresponding system of linear algebraic equations in relation to any group of variables. Henceforth, we shall not write out the arguments of the matrix $K(t,t')$, assuming that the optimization of the sensor arrangement is performed with some values of the arguments fixed (for example, for the stationary field at $t = t'$).

Let us form, from the elements of the matrix K, the principal minors, which are obtained by crossing out rows and columns with identical numbers. We shall designate the principal minors by $K_{jk\ell}...$, where the indices are equal to the numbers of crossed out rows and columns. The total number of principal minors of a matrix of the order n is equal to $2^n - n - 2$.

Let us introduce generalized condition numbers which are equal to the maximal values on a set of condition numbers of the given matrix and all the principal minors:

$$\zeta(K) = \max\{\eta(K), \eta(K_1), \ldots, \eta(K_{12\cdots(n-2)})\} . \tag{8.54}$$

As a criterion for the most proper arrangement of the sensors, we shall take the condition

$$\zeta(K) \to \min_{\underset{\sim}{x_1}, \underset{\sim}{x_2}, \ldots, \underset{\sim}{x_n}} . \tag{8.55}$$

For n = 2, this criterion coincides with the requirement for the best condition in the usual sense. For n > 2, the criterion (8.55) also involves the requirement of the minimum of pairwise correlation between the readings of all sensors, as well as some requirements reducing the multiple correlation to a minimum.

Criterion of the Best Condition of the Spectral Matrix

Let the field $u(\underset{\sim}{x}, t)$ be a stationary one, and let the estimate for the values of the spectral density $S_u(\underset{\sim}{x_j}, \underset{\sim}{x_k}; \omega)$ at n points be the result of measuring the field at these points. These values make up the spectral matrix

$$S(\omega) = \begin{bmatrix} S_{11}(\omega) & S_{12}(\omega) & \cdots & S_{1n}(\omega) \\ S_{21}(\omega) & S_{22}(\omega) & \cdots & S_{2n}(\omega) \\ \cdots & \cdots & \cdots & \cdots \\ S_{n1}(\omega) & S_{n2}(\omega) & \cdots & S_{nn}(\omega) \end{bmatrix} . \tag{8.56}$$

As with matrix K, we shall introduce the set of all principal minors $S_{jk\ell}\ldots$, and also the generalized condition number

$$\zeta(S) = \max\{\eta(S), \eta(S_1), \ldots, \eta(S_{12\cdots(n-2)})\} . \tag{8.57}$$

The criterion for the choice of the transducer arrangement can be taken in a form analogous to (8.40) or (8.41), for example,

$$\max_{\omega \in [\omega_*, \omega_{**}]} \zeta[S(\omega)] \to \min_{\underset{\sim}{x}_1, \underset{\sim}{x}_2, \ldots, \underset{\sim}{x}_n} . \tag{8.58}$$

It is easy to see the connection between a posteriori criteria and a priori ones introduced in Sections 8.2 and 8.3. Take, for example, the criterion of the best condition of the spectral matrix (8.56). The spectral density of the displacement field is related to the spectral density of the external load by the type of relations that were considered in Chapter 3. By writing these relations in matrix form, we find that the spectral matrix (8.56) is expressed as the product of several matrices, among which are the transfer matrix of the system and the Vandermode generalized matrix (8.5). For the condition numbers of the product of two matrices, we have the relation $\eta(AB) \leq n\eta(A)\eta(B)$, whence it follows that a good condition of the matrix-factors leads to a good condition of the product. Thus, optimizing the sensors' arrangement under the conditions of best interpolation of the field or under the conditions of best stability of the system's "generalized forces and generalized displacements" in relation to errors of measurements, we improve at the same time the condition of the spectral matrix (8.56) and, hence, decrease the excess of information in the sensors' readings.

Some Practical Conclusions

The analysis of the sensor arrangement based on the methods described above is very labourious if there are many sensors Performing such analysis for specific engineering problems is not expedient. At the same time, however, the fundamental considerations on which the analysis is based can be used in planning vibration measurements.

The first conclusion that can be drawn from all that was stated above is to point out the significance of the part the joint correlation functions and joint spectral densities play in the analysis of the results of vibration measurements. Often, experimentors do not determine joint correlative and spectral characteristics of the vibration fields, whereas these characteristics contain very valuable information. If we reject this information, we must increase the number of sensors. For example, if we try to estimate the joint spectral densities of the generalized forces by using the relations (8.34), without using the joint correlations of the vibration field, the numbers of sensors required will be $m = n^2$. Consideration of joint correlation reduces the number of required sensors to n. If the generalized forces are stochastically orthogonal, then, without consideration for joint correlations of the vibration field, the number of required sensors will be n. With consideration of joint correlations it is sufficient to have the number of sensors equal to $\mathrm{Int}(\sqrt{n}) + 1$.

8.5 CONSIDERATION OF EFFECTS ON THE VIBRATION FIELD DUE TO INSTALLATION OF SENSORS

Statement of the Problem

The installation of sensors on a structure can change its dynamic properties and, thus, cause distortions in the vibration field. For example, the mass of piezoelectric sensors is of an order of 10g. If the walls of the structure are sufficiently thin, this mass can prove to be comparable to the typical mass of the structure under vibration. At high frequencies, distortions can occur due to the deformations of a sensor or its suspension.

In connection with this, a number of problems in the theory of random vibrations appear. How are the random fields of displacements and accelerations in the elastic system affected when a concentrated mass is added? Under what conditions can these changes be ignored? How can one recalculate the measured values of spectral densities and correlation functions for the case when there are no concentrated masses? Under what conditions will the correction following an introduction of sensors depend only on the properties of the structure and the sensors and not on the probabilisitc characteristics of the external forces? These problems, as applied to plates and shells under stationary random vibrations, were considered by V. Chirkov [104,105]. Some results will be given further below.

Random Vibrations of an Infinite Plate with a Concentrated Mass

Let an infinite plate of constant thickness undergo flexural vibrations under the action of the normal load $q(\underset{\sim}{x},t)$. The concentrated mass M is rigidly fixed to the plate at a point $\underset{\sim}{x} = \underset{\sim}{0}$. The equation of the system's vibrations has the form

$$\underline{D}\Delta\Delta w + \rho h \frac{\partial^2 w}{\partial t^2} + M\delta(\underset{\sim}{x}) \frac{\partial^2 w}{\partial t^2} = q \, , \qquad (8.59)$$

where $\delta(\underset{\sim}{x})$ is a two-dimensional delta-function. Consider a load which is a centred stationary random time function and a homogeneous random function of the coordinates. Representing the load in terms of the integral canonical expansion

$$q(\underset{\sim}{x},t) = \int_{-\infty}^{\infty}\int_{-\infty}^{\infty} Q(\underset{\sim}{k},\omega)e^{i(\underset{\sim}{k}\underset{\sim}{x}+\omega t)}d\underset{\sim}{k}d\omega \;, \tag{8.60}$$

we seek a solution to the equation (8.59) in the form

$$w(\underset{\sim}{x},t) = \int_{-\infty}^{\infty}\int_{-\infty}^{\infty} W(\underset{\sim}{k},\omega)e^{i(\underset{\sim}{k}\underset{\sim}{x}+\omega t)}d\underset{\sim}{k}d\omega \;. \tag{8.61}$$

By substituting the expressions (8.60) and (8.61) into (8.59) and using the formal expansion of the delta-function into a Fourier integral

$$\delta(\underset{\sim}{x}) = \frac{1}{4\pi^2}\int_{-\infty}^{\infty} e^{i\underset{\sim}{k}\underset{\sim}{x}}d\underset{\sim}{k} \;,$$

we find that the generalized random functions $W(\underset{\sim}{k},\omega)$ satisfy the integral equation

$$W(\underset{\sim}{k},\omega) - \frac{M\omega^2}{4\pi^2 F(\underset{\sim}{k},\omega)}\int_{-\infty}^{\infty} W(\underset{\sim}{k}',\omega)d\underset{\sim}{k}' = \frac{Q(\underset{\sim}{k},\omega)}{F(\underset{\sim}{k},\omega)} \;. \tag{8.62}$$

Here, the notations

$$F(\underset{\sim}{k},\omega) = (k_1^2+k_2^2)^2 D(\omega) - \rho h\omega^2 \;,$$

$$D(\omega) = D_r(1+i\chi \text{ sign } \omega) \;, \tag{8.63}$$

are used. The solution of equation (8.62) has the form

$$W(\underset{\sim}{k},\omega) = \frac{1}{F(\underset{\sim}{k},\omega)}\left[Q(\underset{\sim}{k},\omega) + \frac{M\omega^2}{4\pi^2[1-B(\omega)]}\int_{-\infty}^{\infty}\frac{Q(\underset{\sim}{k}',\omega)d\underset{\sim}{k}'}{F(\underset{\sim}{k}',\omega)}\right] \;. \tag{8.64}$$

In this case

$$B(\omega) = \frac{M\omega^2}{4\pi^2} \int_{-\infty}^{\infty} \frac{d\underset{\sim}{k}}{F(\underset{\sim}{k},\omega)} . \tag{8.65}$$

By applying formula (8.64), we find the relation between the time spectral density of the displacements $S_w(\underset{\sim}{x},\underset{\sim}{x}';\omega)$ and the time spectral density of the load $S_q(\underset{\sim}{x},\underset{\sim}{x}';\omega)$. Without dwelling on details, we write down the final result as

$$\begin{aligned} S_w(\underset{\sim}{x},\underset{\sim}{x}';\omega) &= \int_{-\infty}^{\infty} \frac{S_q(\underset{\sim}{k},\omega)}{|F(\underset{\sim}{k},\omega)|^2} H^*(\underset{\sim}{x},\underset{\sim}{k};\omega)H(\underset{\sim}{x}',\underset{\sim}{k};\omega)\,d\underset{\sim}{k} , \\ H(\underset{\sim}{x},\underset{\sim}{k};\omega) &= e^{i\underset{\sim}{k}\underset{\sim}{x}} + \frac{M\omega^2}{4\pi^2[1-B(\omega)]} \int_{-\infty}^{\infty} \frac{e^{i\underset{\sim}{k}'\underset{\sim}{x}}\,d\underset{\sim}{k}'}{F(\underset{\sim}{k}',\omega)} . \end{aligned} \tag{8.66}$$

Correction for the Installation of a Single Sensor

Denote the time spectral density of displacements calculated in the absence of a concentrated mass by $S_w^0(\underset{\sim}{x},\underset{\sim}{x}';\omega)$. In the notations of formula (8.66), this spectral density may be expressed as

$$S_w^0(\underset{\sim}{x},\underset{\sim}{x}';\omega) = \int_{-\infty}^{\infty} \frac{S_q(\underset{\sim}{k},\omega)e^{i\underset{\sim}{k}(\underset{\sim}{x}'-\underset{\sim}{x})}\,d\underset{\sim}{k}}{|F(\underset{\sim}{k},\omega)|^2} . \tag{8.67}$$

The ratio of the spectral densities (8.66) and (8.67) is of some interest. It characterizes the degree of distortion of the vibration field due to the installation of a single sensor, as well as the magnitude of the correction needed to eliminate this distortion. Consider the ratio of the spectral densities at the point of installation of the sensor $\underset{\sim}{x} = \underset{\sim}{0}$. Applying the formulae (8.66) and (8.67) we find that

$$\gamma(\omega) = \frac{S_w(0,0;\omega)}{S_w^0(0,0;\omega)} = \frac{1}{|1-B(\omega)|^2} . \tag{8.68}$$

Thus, the coefficient $\gamma(\omega)$ is independent of the type of the spectral density of the load. In other words, the correction for the installation of a sensor is of a universal character.

Let us analyze formula (8.68). Taking into account the relations (8.63) and (8.65), we obtain

$$B(\omega) = \frac{M\omega^2}{4\pi^2} \int_{-\infty}^{\infty} \frac{d\tilde{k}}{(k_1^2+k_2^2)^2 D(\omega) - \rho h\omega^2} = -\frac{iM\omega}{8\sqrt{\rho h D(\omega)}} .$$

Substituting the result into formula (8.68) gives

$$\gamma(\omega) = [1+2\beta \sin \xi + \beta^2]^{-1}, \tag{8.69}$$

where the notations used are

$$\beta = \frac{M}{\rho h^3} \frac{\omega}{\omega_h} \frac{\sqrt{3(1-\nu^2)}}{4 \sqrt[4]{1+\chi^2}}, \quad \xi = \frac{1}{2} \tan^{-1} \chi(\omega), \quad \omega_h = \frac{1}{h}\left(\frac{E_r}{\rho}\right)^{1/2}. \tag{8.70}$$

The graph of the function (8.69) is plotted in Figure 87. The graph is plotted for $\nu = 0.3$ and $\chi = 0.05$. It follows from the graph and from the relations (8.69) and (8.70) that the error due to the sensor's mass increases with the increase of the mass, increase of the frequency and with the decrease of the thickness, of the modulus of elasticity, and of the density of the plate's material. For example, for an aluminum alloy plate of thickness $h = 2\text{mm}$, a sensor's mass $M = 10\text{g}$, and a frequency $\omega = 2500\text{s}^{-1}$, the error will be about 5 percent.

Random Vibrations of a Plate of Finite Dimensions

The results given above also hold true for a plate of finite dimensions. Consider a wider class of elastic systems whose vibrations are described by the operator equation

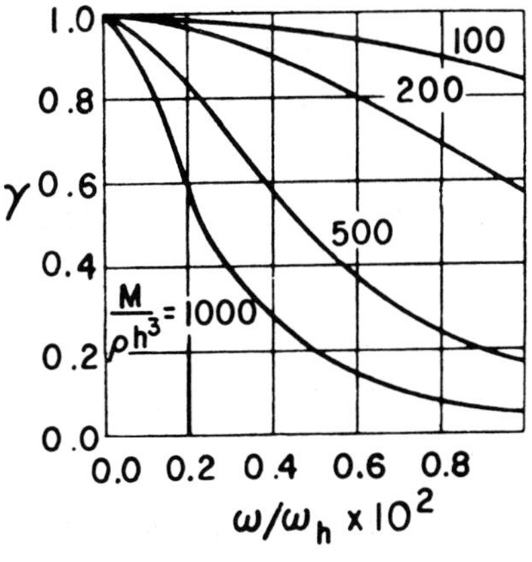

Figure 87

$$A \frac{\partial^2 w}{\partial t^2} + \underline{C}w + M\delta(\underset{\sim}{x}) \frac{\partial^2 w}{\partial t^2} = q(\underset{\sim}{x},t) \ . \tag{8.71}$$

Here, A is an inertial operator in the absence of the concentrated mass, $\underline{C}$ is a linear viscoelastic operator, $w(\underset{\sim}{x},t)$ and $q(\underset{\sim}{x},t)$ are centred scalar fields. Let these fields be stationary time functions, and let the viscoelastic operator $\underline{C}$ be such that the restrictions from Section 3.2 are satisfied. We represent the right side of equation (8.71) and its solution as a series in terms of the natural modes of the elastic system without the concentrated mass

$$\begin{aligned} q(\underset{\sim}{x},t) &= \sum_\alpha \phi_\alpha(\underset{\sim}{x}) \int_{-\infty}^{\infty} Q_\alpha(\omega) e^{i\omega t} d\omega, \\ w(\underset{\sim}{x},t) &= \sum_\alpha \phi_\alpha(\underset{\sim}{x}) \int_{-\infty}^{\infty} W_\alpha(\omega) e^{i\omega t} d\omega \ . \end{aligned} \tag{8.72}$$

Substituting these expressions into (8.71) and calculating, we again arrive at formula (8.68). As distinguished from (8.65), the function $B(\omega)$ is determined by the relation

$$B(\omega) = \sum_\alpha \frac{M\omega^2 \phi_\alpha^2(\underset{\sim}{0})}{(A\phi_\alpha, \phi_\alpha)\,[\Omega_\alpha^2(\omega) - \omega^2]} \ . \tag{8.73}$$

Here, $\Omega_\alpha^2(\omega)$ are the eigenvalues of the equation

$$[C(\omega) - \omega^2 A]\phi = 0 \ ,$$

where $C(\omega)$ is the Fourier transform of the viscoelastic operator $\underset{\sim}{C}$. If $C(\omega) = C_r[1+i\chi(\omega) \text{ sign } \omega]$, then $\Omega_\alpha^2 = \omega_\alpha^2[1+i\chi(\omega) \text{ sign } \omega]$, where ω_α are the frequencies of the natural vibrations of the elastic system with the elastic operator C_r.

The magnitude of the correction required by the introduction of the concentrated mass depends substantially on the relations between the frequency ω and the natural frequencies ω_α. For small dissipation, this correction may turn out to be quite large. To illustrate this, Figure 88 shows the coefficient $\gamma(\omega)$ for a rectangular slab. The graph is plotted for the following values of the parameters: the sides of the slab are a = 50 cm, b = 62.5 cm and the thickness is h = 2 mm. The mass is placed in the centre of the slab. The material parameters are: $E_r = 0.7 \cdot 10^{11} \text{N} \cdot \text{m}^{-2}$, $\chi = 0.05$, $\nu = 0.25$ and $\rho = 2700 \text{ kg} \cdot \text{m}^{-3}$.

Consideration of the Elasticity of the Suspension

If the frequencies ω are comparable to the partial natural frequencies of the sensor, it is essential to take into consideration the influence of the sensor's deformations on the vibration field in the structure. Let the mass M be attached to the structure by a viscoelastic member with a stiffness coefficient c and a viscosity coefficient b. Let the suspension be such that the mass M will have one degree of freedom, its displacement u(t) being colinear to the displacement $w(\underset{\sim}{x},t)$ at the point of suspension, $\underset{\sim}{x} = \underset{\sim}{0}$. Under these conditions, instead of the equation (8.71), we obtain a system of two equations, i.e.,

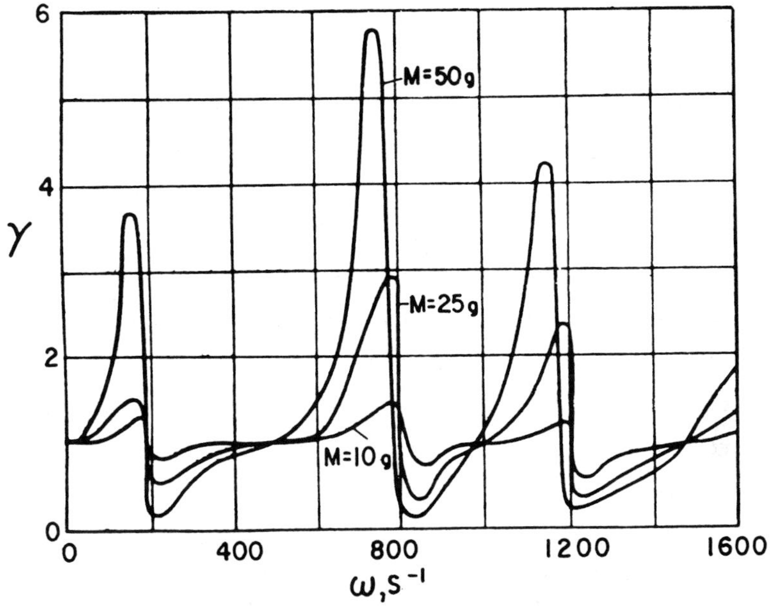

Figure 88

$$A \frac{\partial^2 w}{\partial t^2} + \underline{C}w + \delta(\underset{\sim}{x}) \left(c + b \frac{\partial}{\partial t} \right) (w-u) = q ,$$

$$M \frac{\partial^2 u}{\partial t^2} = \left(c + b \frac{\partial}{\partial t} \right) (w-u) , \qquad (\underset{\sim}{x} = \underset{\sim}{0}) .$$

If a system undergoes stationary random vibrations, the influence of the deformation of the suspension can be determined from the formulae (8.68) and (8.73), the difference being that the mass M is replaced by the effective mass

$$M_* = M \frac{c+i\omega b}{-M\omega^2+c+i\omega b}$$

Some numerical results can be found in [105].

References

[1] ALEKSEEV, V. and VALEEV, K., "A Study in Oscillations of Linear Systems with Random Coefficients", *Radiophysics*, Izvestia Vuzov, Vol. 14, 1971, No. 12.

[2] ANDRONOV, A., PONTRIAGIN, L. and VITT, A., "On Statistical Considerations of Dynamic Systems", *Collected Works*, edited by A. Andronov, Moscow, Akademia Nauk, 1956.

[3] ASLANIAN, A., KUZINA, Z., LIDSKY, V. and TULOVSKY, V., "Distribution of Natural Frequencies of a Thin Elastic Shell of Arbitrary Outline", *Prikladnaia Matematika i Mekhanika*, Vol. 37, 1973, No. 4.

[4] BABITSKY, V., *Theory of Vibration Shock Systems: Approximate Methods*, Moscow, Nauka, 1978.

[5] BARCHENKOV, A., *Dynamic Analysis of Highway Bridges*, Moscow, Transport, 1976.

[6] BELIAEV, Yu., "On Fluctuations of Random Fields", *Dokladi Akademii Nauk, U.S.S.R.*, Vol. 176, 1967, No. 3.

[7] BOLOTIN, V.V., "Statistical Theory of Seismic Stability of Structures", *Mekhanika i Mashinostroenie*, Izvestia Akademii Nauk, U.S.S.R., 1959, No. 4.

[8] BOLOTIN, V.V., "Boundary Effect in Vibrations of Elastic Shells", *Prikladnaia Matematika i Mekhanika*, Vol. 24, 1960, No. 5.

[9] BOLOTIN, V.V., "Asymptotic Method of Investigation of Problems of Eigenvalues for Rectangular Regions", *Problemi Mekhanika Sploshnoi Sredi, (Problems in Continuum Mechanics)*, Akademia Nauk, Moscow, 1961.

[10] BOLOTIN, V.V., "Strength and Damage Accumulation under Random Loads", *Collected Works: Rascheti na Prochnost, (Strength Analysis)*, Mashinostroenie, Moscow, 1961, No. 7.

[11] BOLOTIN, V.V., "On Elastic Vibrations Excited by Forces with Wide Spectrum", *Mashinostroenie*, Izvestia VUZAV, 1963, No. 4.

[12] BOLOTIN, V.V., "On Frequency Densities of Natural Vibrations of Thin Elastic Shells", *Prikladnaia Matematika i Mekhanika*, Vol. 27, 1963, No. 2.

[13] BOLOTIN, V.V., *Statistical Methods in Structural Mechanics*, Stroyizdat, Moscow, 1965.

[14] BOLOTIN, V.V. and MAKAROV, B., "On Approximate Solution of Some Problems of Statistical Dynamics", Izvestia Akad. Nauk, U.S.S.R., *Mekhanika*, 1965, No. 3.

[15] BOLOTIN, V.V., "Mechanics of Solids and Reliability Theory", *Trudi II Vsesoyuzny Syesd po Teoreticheskoi i Prikladnoi Mekhanike, (Proc. 2nd Soviet Congress Theor. and Appl. Mech.)*, M.T.T., Moscow, Nauka, 1966.

[16] BOLOTIN, V.V., "Pressure Fields Inside Shells under Random Vibrations", *Inzhenerny Zhurnal*, M.T.T., 1968, No. 1.

[17] BOLOTIN, V.V., "Planning Vibration Measurements for Structures under Random Vibrations", *Izvestia Akademii Nauk*, M.T.T., 1970, No. 1.

[18] BOLOTIN, V.V., "Theory of Optimal Vibration Protection for Random Actions", *Trudi Moskovski Energeticheski Inst.*, 1970, No. 74.

[19] BOLOTIN, V.V., *Application of Methods of Probability Theory and Reliability Theory to Analyses of Structures*, Moscow, Stroyizdat, 1971.

[20] BOLOTIN, V.V., "Theory of Distribution of Natural Frequencies of Elastic Bodies and Its Application to Random Vibration Problems", *Prikladnaia Mekhanika*, Vol. 8, 1972, No. 4.

[21] BOLOTIN, V.V., "Optimal Arrangement of Sensors for Measuring Random Fiekds", *Mekhanika Deformiruemikh Tel i Konstruktsi, (Mech. Def. Solids and Structures)*, Mashinostroenie, Moscow, 1975.

[22] BIKOV, V., *Digital Simulation in Statistical Radio Engineering*, Sovetskoie Radio, Moscow, 1971.

[23] VALEEV, K., "Oscillations of Cylindrical Panel under Random Loading", *Teoria Plastin i Obolochek, (Theory of Plates and Shells)*, Nauka, Moscow, 1971.

[24] VOLOKHOVSKY, V. and RADIN, V., "On Choosing Optimal Parameters of Nonlinear Vibration Protection Systems under Random Actions", *Izvestia Akademii Nauk, U.S.S.R.*, M.T.T., 1972, No. 2.

[25] VOLOKHOVSKY, V., "Investigation of Estimates for Reliability Function of Continuous Mechanical Systems by Method of Statistical Simulation", *Izvestia Akademii Nauk, U.S.S.R.*, M.T.T., 1973, No. 5.

[26] VOLMIR, A. and KILDIBEKOV, I., "Probabilistic Characteristics of Behaviour of Cylindrical Shell under Acoustic Loading", *Prikladnaia Mekhanika*, Vol. 1, 1965, No. 3.

[27] VOLMIR, A., *Stability of Deformable Systems*, Nauka, Moscow, 1967.

[28] VOLMIR, A., *Nonlinear Dynamics of Plates and Shells*, Nauka, Moscow, 1972.

[29] VOLMIR, A. and KULTERBAEV, KH., "Investigation of Cylindrical Panels under Action of Wind", *Prikladnaia Mekhanika*, Vol. 10, 1974, No. 3.

[30] GAVRILOV, Yu., "Determination of Frequencies of Natural Vibrations of Thin Elastic Cylindrical Shells", *Izvestia Akademii Nauk, U.S.S.R.*, Mekhanika i Mashinostroenie, 1961, No. 1.

[31] GANIEV, R. and KUZMA, V., "Optimal Arrangement of Vibration Sensors on Randomly Vibrating Structures According to Information Theory", *Prikladnaia Mekhanika*, Vol. 9, 1973, No. 12.

[32] GANIEV, R. and FROLOV, K., "On Problem of Vibration Damping of Devices and Machinery in Nonlinear Formulation", *Kolebania i Ustoichivost Priborov, Mashin i Elementov System Upravlenia, (Vib. and Stab. of Devices, Machines and Control System Components)*, Nauka, Moscow, 1968.

[33] GNEDENKO, B., BELIAEV, Yu. and SOLOVIEV, A., *Mathematical Methods in Reliability Theory*, Nauka, Moscow, 1965.

[34] GONCHARENKO, V., "Buckling of Panels under Random Forces", *Teoria Obolochek i Plastin, (Theory of Plates and Shells)*, Erevan, Akademii Nauk Armianskoi S.S.R., 1964.

[35] GONCHARENKO, V., "On Dynamic Problems of Statistical Stability Theory of Elastic Systems", *Problemi Ustoichivosti v Stroitelnoi Mekhanike, (Stability Problems in Struct. Mech.)*, Stroyizdat, Moscow, 1965.

[36] GONCHARENKO, V., "Vibrations of Plates of Nonlinear Elastic Material in Homogeneous Field of Random Pressures", *Problemi Nadezhnosti v Stroitelnoi Mekhanike, (Reliability Problems in Struct. Mech.)*, Vilnus, RINTIP, 1968.

[37] GUSEV, A., "On Distribution of Amplitudes in Wideband Random Pressures in their Schematization by Method of Complete Cycles", Izvestia Akademii Nauk, U.S.S.R., *Mashinovedenie*, 1974, No. 1.

[38] DIMENTBERG, M., "Forced Oscillations of Plates under Loading which is a Space-Time Random Process", *Inzhenerny Zhurnal*, Vol. 1, 1961, No. 2.

[39] DIMENTBERG, M., "On Lower Bound Estimate for Longevity for Stationary Random Loads", *Izvestia Akademii Nauk, U.S.S.R., Mekhanika i Mashinostroenie*, 1962, No. 3.

[40] DIMENTBERG, M., "Nonlinear Vibrations of Elastic Panels under Random Actions", *Izvestia Akademii Nauk, U.S.S.R., Mekhanika i Mashinostroenie*, 1962, No. 5.

[41] DIMENTBERG, M., "Some Problems of Stability of Shells Subjected to Random Perturbations", *Problemi Ustoichivosti v Stroitelnoi Mekhanike, (Stability Problems in Struct. Mech.)*, Stroyizdat, Moscow, 1965.

[42] YERMAKOV, S. and MIKHAILOV, G., *Course in Statistical Simulation*, Nauka, Moscow, 1976.

[43] YERMOLENKO, A., "Correlation Function and Variance of Measure and Damage for Stationary Random Action", *Trudi Mosk. Energ. Inst.*, 1974, No. 184.

[44] YEFIMTSOV, B. and MOSKALENKO, V., "Excitation of Multispan Plates in Random Acoustic Field", *Problemi Nadezhnosti v Stroit. Mekhanike, (Reliability Problems in Struct. Mech.)*, Vilnus, RINTIP, 1968.

[45] YEFIMTSOV, B., "Vibrations of Plates in a Field of Close-to-Wall Pressure Pulsations", *Problemi Nadezhnosti v Stroit. Mekhanike, (Reliability Problems in Struct. Mech.)*, No. 2, Vilnus, RINTIP, 1971.

[46] ZHINZHER, N. and KHROMATOV, V., "Application of Asymptotic Method to Investigation of Spectra of Orthotropical Circular Cylindrical Shell Oscillations", *Izvestia Akademii Nauk U.S.S.R., M.T.T.*, No. 6, 1971.

[47] ZHINZHER, N., "Dynamic Boundary Effects in Orthotropic Elastic Shells", *Prikladnaia Matematika i Mekhanika*, Vol. 39, 1975, No. 4.

[48] ZAREMBO, L. and KRASILNIKOV, V., *Introduction to Nonlinear Acoustics*, Nauka, Moscow, 1966.

[49] ILYINSKY, V., *Protection of Devices from Dynamic Actions*, Energia, Moscow, 1970.

[50] ILYICHEV, V., "Mathematical Planning, Analysis and Data Generalization in Parametric Investigations", *Trudi TSAGI*, (Central Aero-Hydrodynamic Institute), No. 995, 1966.

[51] KILDIBEKOV, I. and MITSUK, A., "Investigation in Carrying Capacity of Reinforced Panels in Acoustic Field", *Prikladnaia Matematika i Mekhanika*, Vol. 7, 1971, No. 12.

[52] KILDIBEKOV, I., "Probabilistic Characteristics of Circular Cylindrical Shell Behaviour under Acoustic Pressure and Combined Static Loading", *Izvestia Akademii Nauk, M.T.T.*, No. 4, 1974.

[53] KOGAEV, V., *Strength Analyses under Stresses Variable in Time*, Mashinostroenie, Moscow, 1977.

[54] COX, D. and LEWIS, P., *Statistical Analysis of Sequences of Events*, Mir, Moscow, 1969.

[55] KOLOVSKY, M., *Nonlinear Theory of Vibration Protection Systems*, Nauka, Moscow, 1963.

[56] KOLOVSKY, M., *Automatic Control of Vibration Protection Systems*, Nauka, Moscow, 1976.

[57] KOMAR, N. and OKOPNY, Yu., "Experimental Verification of Estimate for Reliability Function on Electronic Analogs", *Problemi Nadezhnosti v Stroit. Mekhanike, (Reliability Problems in Struct. Mech.)*, No. 2, Vilnus, RINTIP, 1971.

[58] KOMAR, N. and OKOPNY, Yu., "Investigation of Nonlinear Vibration Protection Systems on Electronic Analogs", *Izvestia Akademii Nauk, Mashinovedenie,* No. 1, 1972.

[59] KRILOV, Yu., STREKALOV, S. and TSIPLUKHIN, V., *Wind Waves and Their Action on Structures,* Leningrad, Hydrometeoizdat, 1976.

[60] KUZMA, V., "Dynamic Instability of Random Vibrations of a Rod", *Prikladnaia Mekhanika,* Vol. 2, 1966, No. 6.

[61] LARIN, V., "Selection of Damper's Free Travel under Random Vibrations", *Inzhenerny Zhurnal,* M.T.T., No. 1, 1968.

[62] LOMAKIN, V., *Statistical Problems of Mechanics of Solids,* Nauka, Moscow, 1970.

[63] MAKAREVSKY, A., et al, *Strength of Aircraft: Normalization Methods of Analysis of Aircraft's Strength,* Mashinostroenie, Moscow, 1975.

[64] MAKSIMOV, L., "On Analysis of Passive Vibration Isolation for Actions in the Form of Stationary Random Process", *Inzhenerny Zhurnal,* M.T.T., 1966, No. 3.

[65] MONIN, A., and YAGLOM, A., *Statistical Fluid Mechanics,* Nauka, Moscow, 1965, p. 1; and 1967, p. 2.

[66] MOSKALENKO, V., "On Application of More Precise Theories of Plate Bending to Problems of Natural Vibrations", *Inzhenerny Zhurnal,* 1961, No. 3.

[67] MOSKALENKO, V. and CHEN-DE-LYN, "On Natural Vibrations of Multispan Uncut Plates", *Prikladnaia Mekhanika,* Vol. 1, 1965, No. 3.

[68] MOSKALENKO, V., "Random Vibrations of Multispan Plates", *Izvestia Akademii Nauk, U.S.S.R., M.T.T.,* 1968, No. 4.

[69] MOSKALENKO, V., "On Vibrations of Multispan Plates", *Rascheti na Prochnost, (Stress Analysis),* Mashinostroenie, Moscow, 1969, No. 14.

[70] MOSKALENKO, V., "On Spectra of Natural Vibrations of Shells of Revolution", *Prikladnaia Matematika i Mekhanika,* Vol. 36, 1972, No. 2.

[71] MOSKVIN, V. and SMIRNOV, A., "On Stability of Linear Stochastic Systems", *Izvestia Akademii Nauk, M.T.T.,* 1975, No. 4.

[72] MOSKVIN, V. and SMIRNOV, A., "Oscillations of Mechanical Systems under Random Parametric Actions", *Trudi Moskovsk. Energ. Inst.,* 1978, No. 353.

[73] NIKOLAENKO, N., *Probabilistic Methods of Dynamic Analysis of Machine Building Structures,* Mashinostroenie, Moscow, 1967.

[74] PALMOV, V., "Thin Plates Subjected to Wideband Random Loading", *Trudi Leningradsk. Polytekhnichesk. Inst.,* 1965, No. 252.

[75] PALMOV, V., "Thin Shells Subjected to Wideband Random Loading", *Prikladnaia Matematika i Mekhanika*, Vol. 29, 1965, No. 4.

[76] PALMOV, V., "Propagation of Random Vibrations in Viscoelastic Rod", *Problemi Nadezhnosti v Stroitelnoi Mekhanike, (Reliability Problems in Struct. Mech.)*, Vilnus, RINTIP, 1968.

[77] PALMOV, V., *Vibrations of Elastic-Plastic Bodies*, Nauka, Moscow, 1976.

[78] PLAKHOV, D., "Correlation Relations in Acoustic Field of Infinite Plates Subjected to Random Pressure Fluctuations", *Akustichesky Zhurnal*, Vol. 14, 1968, No. 2.

[79] POPOV, E. and PALTOV, I., *Approximate Methods of Investigations of Nonlinear Automatic Systems*, Phyzmatgiz, Moscow, 1960.

[80] PROKHOROV, Yu., and ROSANOV, Yu., *Probability Theory: Fundamental Concepts, Limit Theorems, Random Processes*, Nauka, Moscow, 1973.

[81] PUGACHEV, V., *Theory of Random Functions and Its Application to Problems of Automatic Control*, Phyzmatgiz, Moscow, 1962.

[82] PUPIREV, V., "On Relations of Acoustic and Vibration Actions on Elastic Plate", *Izvestia Akademii Nauk, U.S.S.R., M.T.T.*, 1968, No. 1.

[83] RADIN, V. and CHIRKOV, V., "Transmission of Random Vibrations in Semi-Infinite Cylindrical Shell from Forces Acting on Cylinder Base", *Trudi Moskovsk. Energetich. Institute*, 1970, No. 74.

[84] RADIN, V. and CHIRKOV, V., "On Transmission of Random Vibrations through Thin-Walled Structures", *Problemi Nadezhnosti v Stroitelnoi Mekhanike, (Reliability Problems in Struct. Mech.)*, No. 2, Vilnus, RINTIP, 1971.

[85] RADIN, V., "On Optimization of Linear Vibration Protection Systems with Respect to Reliability", *Prikladnaia Mekhanika*, Vol. 8, 1972, No. 9.

[86] ROSANOV, Yu., *Random Processes (Brief Course)*, Nauka, Moscow, 1971.

[87] ROMANOV, Yu., "On Possibility of Representing Seismic Action as Stationary Random Process", *Stroitelnaia Mekhanika i Raschet Sooruzhenii*, No. 5, 1963.

[88] SAVITSKY, G., *Analysis of Antenna Structures: Physics*, Sviaz, Moscow, 1978.

[89] SVETLITSKY, V., *Random Vibrations of Mechanical Systems*, Mashinostroenie, Moscow, 1976.

[90] SVESHNIKOV, A., *Applied Methods of Theory of Random Functions*, Nauka, Moscow, 1968.

[91] SERENSEN, S. and KOGAEV, V., "Probabilistic Methods of Analysis for Variable Loads", *Mekhanicheskaia Ustalost' v Statisticheskom Aspekte, (Statistical Aspects of Mech. Fatigue)*, Nauka, Moscow, 1969.

[92] SILAEV, A., *Spectral Theory of Suspension of Road Vehicles*, Mashinostroenie, Moscow, 1972.

[93] STRATONOVICH, R., *Some Questions of Theory of Fluctuations in Radio Engineering*, Sovetskoe Radio, Moscow, 1961.

[94] STRATONOVICH, R., "Stochastic Integrals and Equations Written in a New Form", *Vestnik Moskovsk. Gosuniversiteta, Seria Matematika i Mekhanika*, No. 1, 1964.

[95] TIKHONOV, V., *Fluctuations of Random Processes*, Nauka, Moscow, 1970.

[96] TIKHONOV, V. and MIRONOV, M., *Markov Processes*, Sovetskoe Radio, Moscow, 1977.

[97] TOVSTIK, P., "On Density of Vibration Frequencies of Thin Shells of Revolution", *Prikladnaia Matematika i Mekhanika*, Vol. 36, 1972, No. 2.

[98] TROITSKY, V., "On Synthesis of Optimal Dampers", *Prikladnaia Matematika i Mekhanika*, Vol. 31, 1967, No. 4.

[99] FEDOROV, Yu., "Vibrations of Closed Circular Cylindrical Shell in Field of Random Acoustic Pressures", *Inzhenerny Zhurnal*, Vol. 3, 1963, No. 3.

[100] FEDOROV, Yu., "On Nonlinear Vibrations of Rectangular Plate Subjected to Random Forces", *Inzhenerny Zhurnal*, Vol. 4, 1964, No. 3.

[101] KHASMINSKY, R., *Stability of Systems of Differential Equations in Random Perturbations of their Parameters*, Nauka, Moscow, 1969.

[102] KHROMATOV, V., "Properties of Spectra of Thin Circular Cylindrical Shells, Vibrating Close to Membrane Stress State", *Izvestia Akademii Nauk, U.S.S.R., M.T.T.*, No. 2, 1972.

[103] KHROMATOV, V., "Wideband Random Vibrations of Thin Shallow Shells and Their Connection with Problems of Distribution of Natural Frequencies", *Trudi 10 Vsesoyuznoi Konferentsii po Teorii Obolochek i Plastin, (Proc. 10th Sov. Conf. Theory of Shells and Plates)*, Kutaisi, 1975, Tbilisi, Metsniereba, 1975.

[104] CHIRKOV, V., "On Distortions of Random Vibration Field Caused by Installation of Vibration Transducers", *Trudi Moskovsk. Energet. Institute*, No. 74, 1970.

[105] CHIRKOV, V., "Random Vibrations of Thin-Walled Structures Carrying Concentrated Masses", *Izvestia Akademii Nauk, U.S.S.R., M.T.T.*, No. 3, 1975.

[106] SHUKAILO, V., "Some Questions of Theory of Recovery and Fatigue Reliability of Mechanical Members", *O Nadezhnosti Slozhnikh Tekhnicheskikh Sistem, (On Reliability of Complex Engineering Systems)*, Sovetskoe Radio, Moscow, 1966.

[107] ARIARATNAM, S.T., "Dynamic Stability of a Column under Random Loading", *Dynamic Stability of Structures*, Oxford a.o., Pergamon Press, 1967.

[108] ARIARATNAM, S.T. and TAM, D.S.F., "Moment Stability of Couples Linear Systems under Combined Harmonic and Stochastic Excitation", *Proc. IUTAM Symposium on Stochastic Problems in Dynamics*, University of Southampton, 1976.

[109] BOGDANOFF, J.L., COTE, L.J. and KOZIN, F., "Introduction to the Statistical Theory of Land Locomotion", *J. of Terramechanics*, Vol. 2, 1965, No. 3.

[110] BOGDANOFF, J.L. and SCHIFF, A., "Earthquake Effects in the Safety and Reliability Analysis of Engineering Structures", *Int. Conf. on Structural Safety and Reliability*, Oxford a.o., Pergamon Press, 1972.

[111] BOLOTIN, V.V., "Statistical Theory of Aseismic Design of Structures", *Proc. of the 2nd World Conf. on Earthquake Engineering*, Vol. 2, Tokyo, 1961.

[112] BOLOTIN, V.V., "An Asymptotic Method for the Study of the Eigenvalues Problem for Rectangular Regions", *Problems of Continuum Mechanics*, SIAM, Philadelphia, 1961.

[113] BOLOTIN, V.V., "On the Broadband Random Vibration of Elastic Systems", *Proc. of the 11th Int. Congress of Appl. Mech.*, Springer-Verlag, Berlin, 1964.

[114] BOLOTIN, V.V., "The Density of Eigenvalues in Vibration Problems of Elastic Plates and Shells", *Proc. Vibr. Problems*, Vol. 6, 1965, No. 4.

[115] BOLOTIN, V.V., "Broadband Random Vibration of Elastic Systems", *Int. J. Solids and Structures*, Vol. 2, 1966, No. 1.

[116] BOLOTIN, V.V., "Reliability Theory and Stochastic Stability", *SM Study No. 6 on Stability*, Solid Mechanics Division, University of Waterloo Press, 1971.

[117] BOURGINE, A., "Méthodes de calcul de la response d'une structure soumise a un environnement aléatoire", *Rech. Aérospat.*, 1970, No. 6.

[118] BROOKS, R.D., "Structural Fatigue Research and Its Relation to Design", *Fatigue in Aircraft Structures*, Academic Press, New York, 1956.

[119] CAUGHEY, T.K. and GRAY, A.H., Jr., "On the Almost Sure Stability of Linear Dynamic Systems with Stochastic Coefficients", *J. of Applied Mechanics*, Vol. 32, 1965, No. 2.

[120] CAUGHEY, T.K., "Nonlinear Theory of Random Vibrations", *Advances in Appl. Mech.*, 1971, No. 11.

[121] CORNELL, C.A. and RASCON, O.A., *Strong Motion Earthquake Simulation,* MIT Press, Cambridge, Massachusetts, 1968.

[122] CRANDALL, S.H., *Random Vibration,* MIT Press, Cambridge, Massachusetts, Vol. 1, 1958; Vol. 2, 1963.

[123] CRANDALL, S.H. and MARK, W.D., *Random Vibration in Mechanical Systems,* Academic Press, New York, 1963.

[124] CRANDALL, S.H., "The Role of Damping in Vibration Theory", *J. of Sound and Vibration,* Vol. 11, 1970, No. 1.

[125] CRANDALL, S.H., "Structural Response Patterns Due to Wide-Band Random Excitation", *Proc. IUTAM Symposium on Stochastic Problems in Dynamics,* University of Southampton, 1976.

[126] DICKINSON, S.M. and WARBURTON, G.B., "Natural Frequencies of Plate Systems Using the Edge Effect Method", *J. Mech. Eng. Sci.,* Vol. 9, 1967, No. 4.

[127] DICKINSON, S.M., "Bolotin's Method Applied to the Buckling and Lateral Vibration of Stressed Plates", *AIAA Journal,* Vol. 13, 1975, No. 1.

[128] DICKINSON, S.M., "Modified Bolotin's Method Applied to Buckling and Vibration of Stressed Plates", *AIAA Journal,* Vol. 13, 1975, No. 12.

[129] DOWELL, E.H., "Transmission of Noise from a Turbulent Layer through a Flexible Plate into a Closed Cavity", *J. Acoustical Society of America,* Vol. 46, 1969, No. 1.

[130] DOWELL, E.H., "Noise or Flutter or Both?", *J. Sound and Vibration,* Vol. 11, 1970, No. 2.

[131] FAHY, F.I., "Response of a Cylinder to Random Sound in the Contained Fluid", *J. Sound and Vibration,* Vol. 13, 1970, No. 2.

[132] FREUDENTHAL, A.M. and SHINOZUKA, M., "Probability of Structural Failure under Earthquake Acceleration", *Trans. Japan Soc. Civil Engineers,* 1965, No. 118.

[133] FRÝBA, L., *Vibration of Solids and Structures under Moving Loads,* Academia, Prague, 1972.

[134] FRÝBA, L., "Non-Stationary Response of a Beam to a Moving Random Force", *J. of Sound and Vibration,* Vol. 46, 1976, No. 3.

[135] GRAEFE, P.M., "Stability of a Linear Second Order System under Random Excitation", *Ing. Arch.,* Vol. 35, 1966, No. 3.

[136] GRAY, A.H., Jr., "Frequency Dependent Almost Sure Stability Conditions for a Parametrically Excited Random Vibrational System", *J. of Appl. Mech.,* Vol. 34, 1967, No. 4.

[137] HEIMANN, B., "Strenge Abschirmbedingung für mechanische Schwingungssysteme unter zufälliger Einwirkungen", *Maschinenbautechnik,* Vol. 20, 1971, No. 6.

[138] HENNIG, K., "Zur Berechnung von stationären Zufällschwingungen flacher Schalen", *ZAMM,* Vol. 52, 1972, No. 10.

[139] IJENGAR, R.N. and IJENGAR, K.T.S.R., "A Nonstationary Process Model for Earthquake Accelerograms", *Bull. Seism. Soc. America,* Vol. 59, 1969, No. 3.

[140] INFANTE, E.F., "On the Stability of Some Linear Nonautonomous Random Systems", *J. of Appl. Mech.,* Vol. 35, 1968, No. 1.

[141] KAUL, M.K. and PENZIEN, J., "Stochastic Seismic Analysis of Yielding Offshore Towers", *J. of Eng. Mech. Division, ASCE,* Vol. 100, 1974, No. 5.

[142] KING, W.W. and LIN, C.C., "Application of Bolotin's Method to Vibrations of Plates", *AIAA J.,* Vol. 12, 1974, No. 3.

[143] KISTNER, A., "Über die Differenzialgleichungen für die Momente linearer Systeme", *ZAMM,* Vol. 56, 1976, No. 3.

[144] KISTNER, A., "On the Moments of Linear Systems Excited by a Coloured Noise Process", *Proc. Sym. on Stochastic Problems in Dynamics,* University of Southampton, 1976.

[145] KONISHI, I., "Safety and Reliability of Suspension Bridges", *Proc. Int. Conf. on Structural Safety and Reliability,* Oxford a.o., Pergamon Press, 1972.

[146] KOZIN, F., "On Almost Sure Stability of Linear Systems with Random Coefficients", *J. Math. Phys.,* Vol. 2, 1963, No. 1.

[147] KUSHNER, H.J., *Stochastic Stability and Control,* Academic Press, London, New York, 1967.

[148] LIGHTHILL, M.J., "Sound Generated Aerodynamically", *Proc. Royal Society, (London),* Vol. A267, 1962.

[149] LIN, C.C. and KING, W.W., "Free Transverse Vibrations of Rectangular Unsymmetrically Laminated Plates", *J. of Sound and Vibration,* Vol. 36, 1974, No. 1.

[150] LIN, Y.K.M., *Probabilistic Theory of Structural Dynamics,* McGraw-Hill, New York, 1967.

[151] LYON, R.H., "Spatial Response Concentrations in Extended Structures", *J. Engineering Industry,* Vol. 89, 1967, No. 4.

[152] MAESTRELLO, L. and LINDEN, T.L.J., "Response of an Acoustically Loaded Panel Excited by Supersonically Convected Turbulence", *J. of Sound and Vibration,* Vol. 16, 1971, No. 3.

[153] MENKES, E.G. and HOUBOLT, J.C., "Evaluation of Aerothermoelasticity Problems for Unmanned Mars-Entry Vehicles", *J. of Spacecraft,* Vol. 6, 1969, No. 2.

[154] MILES, J.W., "On the Structural Fatigue Under Random Loading", *J. of Aeron. Sci.*, Vol. 21, 1954, No. 11.

[155] MILLER, D.K. and HART, F.D., "Modal Density of Thin Circular Cylinders", *NASA Report No. CR-897*, 1970.

[156] NEMAT-NASSER, S., "On the Response of Shallow Thin Shells to Random Excitations", *AIAA Journal*, Vol. 6, 1968, No. 7.

[157] NEWMARK, N.M. and ROSENBLUETH, E., *Fundamentals of Earthquake Engineering*, Prentice-Hall Inc., Englewood Cliffs, New Jersey, 1971.

[158] SAGIROW, P., "Zur Abschliessung der Momentengleichungen linearer Systeme mit stochastischer Parametererregung, *ZAMM*, Vol. 56, 1976, No. 3.

[159] SHINOZUKA, M., "Probability of Structural Failure under Random Loading", *J. Eng. Mech. Div., ASCE*, Vol. 90, 1964, No. 5.

[160] SHINOZUKA, M. and HENRY, L., "Random Vibration of a Beam Column", *J. Eng. Mech. Div., ASCE*, Vol. 91, 1965, No. 5.

[161] SHINOZUKA, M., "Simulation of Multivariate and Multidimensional Random Processes", *J. Acoust. Soc. America*, Vol. 49, 1971, No. 1.

[162] STEARN, S.M., "Spatial Variation of Stress, Strain and Acceleration in Structures Subjected to Broad Frequency Band Excitation", *J. of Sound and Vibration*, Vol. 12, 1970, No. 1.

[163] TRIFUNAC, M.D., "Response Envelope Spectrum and Interpretation of Strong Earthquake Ground Motion", *Bull. Seism. Soc. America*, Vol. 61, 1971, No. 2.

[164] VAICAITIS, R., YAN, C.M. and SHINOZUKA, M., "Nonlinear Panel Response from a Turbulent Boundary Layer", *AIAA J.*, Vol. 10, 1972, No. 7.

[165] VAICAITIS, R., "Nonlinear Panel Response to Nonstationary Wind Forces", *J. Eng. Mech. Div., ASCE*, Vol. 101, 1975, No. 4.

[166] VIJAYAKUMAR, K. and RAMAIAH, G.K., "Analysis of Vibration of Clamped Square Plates by the Rayleigh-Ritz Method with Asymptotic Solution from a Modified Bolotin Method", *J. of Sound and Vibration*, Vol. 56, 1978, No. 1.

[167] WAN, F.Y.M., "A Direct Method for Linear Dynamic Problems in Continuum Mechanics with Random Loads", *J. Math. Phys.*, Vol. 52, 1973.

[168] WAN, F.Y.M. and LAKSHMIKANTHAM, C., "Spatial Correlation Method and a Time-Varying Flexible Structure", *AIAA J.*, Vol. 12, 1974, No. 5.

[169] WEDIG, W., "Stabilitätsbedingungen für parametererregte Schwingungssysteme mit schmalbandigen Zufallserregungen", *ZAMM*, Vol. 52, 1972, No. 3.

[170] WEIDENHAMMER, F., "Stabilitätsbedingungen für Schwinger mit zufälligen Parameterregungen", *Ing. Arch.*, Vol. 33, 1964, No. 6.

[171] WEIDENHAMMER, F., "Auswanderungserscheinungen unter Zufallserregungne", *ZAMM*, Vol. 48, 1968, No. 5.

[172] WILKINSON, I.P.D., "Modal Densities of Certain Shallow Structural Elements", *J. Acoust. Soc. of America*, Vol. 43, 1968, No. 2.

[173] WIRSCHING, P.H. and HAUGEN, E.B., "General Statistic Model for Random Fatigue", *J. Eng. Materials and Tech.*, Vol. 96, 1974, No. 1.

Index